Explaining Photosynthesis

History, Philosophy and Theory of the Life Sciences

Volume 8

For further volumes:
http://www.springer.com/series/8916

Kärin Nickelsen

Explaining Photosynthesis

Models of Biochemical Mechanisms, 1840-1960

Springer

Kärin Nickelsen
History of Science
LMU Munich
Munich
Germany

ISSN 2211-1948 ISSN 2211-1956 (electronic)
History, Philosophy and Theory of the Life Sciences
ISBN 978-94-017-9581-4 ISBN 978-94-017-9582-1 (eBook)
DOI 10.1007/978-94-017-9582-1

Library of Congress Control Number: 2015934603

Springer Dordrecht Heidelberg New York London

Printed on acid-free paper

Springer is part of Springer Science+Business Media (www.springer.com)

Preface

This book is a study of 'light and darkness'—as was the title of the habilitation thesis on which it is based: metaphorically speaking, in terms of dark failures and bright successes in the middle of conceptual fog; and in a very literal sense, referring to the chemical effects of light and darkness on the green parts of plants. It deals with the elucidation of the photosynthetic mechanism, which has not been the subject of a book-length study before. This means that a lot of primary material had to be analysed, processed and brought into a sensible arrangement.

At the same time, the book aims to contribute to the longstanding philosophical question of how knowledge is generated in science. It has repeatedly been argued, by historians of science, by philosophers of science and in particular by those who consider themselves affiliated to the 'History and Philosophy of Science' (HPS) that this question can only be answered by scrutinising actual cases and trying to understand from these some characteristic features of scientific research.[1] This is what I am attempting to do in this study. The result is richer in historical detail than many philosophers of science might wish to engage with; at the same time, its core questions are more philosophically oriented than many historians of science might expect. The main issue is of a methodological nature: How and for what reasons do scientists do what they do, while they are investigating a problem, such as the mechanism of photosynthesis? How do they set their priorities when preferring one option to another? According to which conventions, habits and expectations do they organise and (re-)direct their work? It is, to sum it up, a study in the heuristics of scientific research.

Before I explain these thoughts in more detail, I shall spend the rest of this preface on some highly appropriate words of acknowledgement and gratitude. First and foremost, I would like to thank Gerd Graßhoff and all the members and students of the History and Philosophy of Science Division of the University of Bern's Institute of Philosophy. The particular research profile of this division—which, unfortunately,

[1] Cf., e.g., Schickore and Steinle (2006); arguments along these lines can also be found in Mauskopf and Schmaltz (2011), which is the outcome of one of the recently instituted conferences on &HPS, that is, 'Integrated History and Philosophy of Science'.

has been abolished since—strongly influenced my work. The project not only benefited from stimulating discussions with Gerd Graßhoff and others but also from the sabbatical leave that I was generously permitted to take in the autumn of 2008. I spent this leave at the Max Planck Institute for the History of Science in Berlin, and I am particularly grateful to Lorraine Daston for her tremendous hospitality during my stay there. This study also profited enormously from my acquaintance with Govindjee (no first name) of the University of Illinois at Urbana–Champaign, who so generously shared his immense knowledge of, and his contagious enthusiasm for, the history of photosynthesis research with me, and who meticulously went through the details of this book's manuscript. My extended research stay at Urbana–Champaign proved exceedingly fruitful and pleasurable, thanks largely to the hospitality and practical help given to me by Govindjee and his wife, Rajni Govindjee. I am also grateful to have participated in the DFG-funded network 'Philosophy of the Life Sciences', organised by Maria Kronfeldner, which has provided a valuable source of inspiration over the past years.

Furthermore, I am grateful to a large number of friends and colleagues for discussions at various occasions, useful hints and valuable material; (in alphabetical order) Christina Brandt, Angela Creager, Petra Gentz-Werner, Mathias Grote, Ekkehard Höxtermann, Jeremiah James, Christian Joas, Fabian Krämer, Gianna Pomata, Tilman Sauer, Raphael Scholl, Phillip Sloan, Richard Staley, Friedrich Steinle, Gerhard Wagenitz, Marcel Weber, Volker Wissemann and Adrian Wüthrich. The same holds true for the enormously helpful staff at the various archives that I consulted, namely the archive departments of: the ETH Zurich; the Max Planck Society; the Berlin-Brandenburg Academy of Sciences and Humanities; the Berlin State Library; the University of Illinois at Urbana–Champaign; the University of Chicago; the University of Cambridge, UK; and the Braunschweig/Berlin-based Physikalisch-Technische Bundesanstalt. Besides the University of Bern's Institute of Philosophy, I would also like to thank the following institutions for their generous financial support: The Young Academy at the Berlin-Brandenburg Academy of Sciences and Humanities and the German Academy of Sciences Leopoldina; the Hochschulstiftung of the Burgergemeinde Bern; and the Mittelbaufonds of the University of Bern. I also would like to thank Margareta Simons, who carefully edited the book; Basil Marti, who managed the archival data bases, straightened out the chemistry and gave the habilitation a final read; Josephine Musil-Gutsch and Claus Spenninger, who were of enormous practical help; and, in particular, Caterina Schürch, who greatly improved the graphs, went through the bibliography and proofread the whole manuscript to its enormous benefit.

Finally, I am indebted to the History Department of the Ludwig Maximilians University Munich for granting me a sabbatical leave very soon after I started my tenured position there in 2011. This gave me the indispensable freedom to focus on finalising this book for publication. And, of course, there are the usual suspects among friends and family who know very well, I hope, how much their support has been cherished.

Contents

1 Introduction 1
- 1.1 Models of Mechanisms 2
- 1.2 Heuristics and Strategies in Photosynthesis Research 6
- 1.3 The Individual and the Community 8
- 1.4 Survey of the Book 12

2 In Pursuit of a Pathway (1843–1918) 15
- 2.1 The Nineteenth-Century Conception of Photosynthesis 16
- 2.2 Finding the Chemical Mechanism 18
  - 2.2.1 Justus Liebig and the Organic Acid Model 19
  - 2.2.2 Adolf von Baeyer and the Formaldehyde Model 22
  - 2.2.3 Testing and Modification 25
  - 2.2.4 The Formaldehyde Problem 28
- 2.3 Reconciling the First Approaches 30
  - 2.3.1 The Formic Acid Model 31
  - 2.3.2 The Organic Acid–Formaldehyde Hypothesis 32
  - 2.3.3 The Water Cleavage Model 32
  - 2.3.4 The Chlorophyll Complex Model 34
- 2.4 Features Common to the Models 36
  - 2.4.1 Simplicity and Simplification 36
  - 2.4.2 Incomplete Empirical Support 39
  - 2.4.3 Selected Focus and Modules 40
- 2.5 Model-Building Heuristics 41
  - 2.5.1 Extending a Prototype 41
  - 2.5.2 The Transfer of Causal Knowledge 42
  - 2.5.3 The Building Block Strategy 44
  - 2.5.4 The Principle of Plurality 46
- 2.6 Collective Versus Individual Goals 48
  - 2.6.1 Photosynthesis as a Side Issue 48
  - 2.6.2 Constructive Research Opportunism 51

3 Otto Warburg and the Turn to Manometry (1912–1925) 55
3.1 New Materials and Methods 56
3.1.1 Otto Warburg (1883–1970) 56
3.1.2 Manometry 58
3.1.3 Chlorella 60
3.1.4 Buffer Solution 63
3.1.5 Bolometry 63
3.1.6 Rotating Sectors: The Flashing Light Technique 64
3.1.7 Inhibitors 64
3.2 Warburg's Early Photosynthesis Model 65
3.2.1 Experimental Findings and the Interpretations thereof.... 66
3.2.2 Photosynthesis Framed as Photolysis 73
3.3 The Efficiency of the Process 75
3.4 Father, Son and Photosynthesis 78
3.4.1 Emil Warburg and Photochemistry 78
3.4.2 Influences on Otto Warburg 80
3.5 Studies in Cell Respiration 83
3.6 Comparison with the Chlorophyll-Complex Model 87
3.7 Otto Warburg's Building Blocks 89
3.8 Warburg's Impact on Photosynthesis Research 90

4 Struggling with the Standard Model (1930–1941) 93
4.1 Fluorescence and the Standard Model......................... 94
4.1.1 The Kautsky Effect.................................... 94
4.1.2 James Franck and Photosynthesis 96
4.1.3 The New Standard Model 99
4.2 The Crucial Experiments of 1932 108
4.2.1 Robert Emerson: Harvard, Berlin, Caltech and Stanford .. 108
4.2.2 Emerson, Arnold and 2500 Molecules of Chlorophyll 110
4.3 The Generalised Equation for Photosynthesis 113
4.3.1 Cornelis B. Van Niel and General Microbiology......... 113
4.3.2 Bacterial Photosynthesis and the Consequences 116
4.4 The Photosynthetic Unit 120
4.4.1 Context and Scope of the 1936 Paper.................. 122
4.4.2 Critique of the Standard Model........................ 124
4.4.3 The Unit as Explanatory Alternative................... 125
4.5 Franck's Conservative Alternatives 126
4.5.1 Franck and Herzfeld's Proposal 126
4.5.2 Franck's Further Attempts 129
4.6 Isolated Chloroplasts and Water Splitting................... 132
4.6.1 Robert (Robin) Hill and the Chloroplast Reaction 132
4.6.2 Implications of the Findings 137
4.7 On the Verge of New Perspectives 138
4.7.1 Biological Studies Generalised 138
4.7.2 The Main Lines of Thought............................ 140
4.7.3 A New Conception of Photosynthesis 146

5 The Maximum Quantum Yield Controversy (1937–1955) 149
5.1 Photochemical Quantum Yield Measurements 150
5.2 First Opponents of the Warburg–Negelein Value 151
5.3 Emerson and Lewis's Challenge 153
5.3.1 The Carbon Dioxide Burst (1939–1941) 153
5.3.2 The Red Drop (1943) 156
5.4 Wartime 157
5.5 Attempts to Find a Solution 161
5.5.1 The Photosynthesis Project at Urbana 161
5.5.2 Warburg Comes to the United States 161
5.5.3 The Time Spent at Urbana 164
5.5.4 The Rediscovery of the Maximum Quantum Yield 167
5.5.5 Franck's Attempt to Find a Compromise 171
5.5.6 Further Exchanges of Blows 172
5.5.7 Controversial Themes Around 1950 174
5.6 A Hardening of the Fronts 178
5.6.1 The One-Quantum Mechanism 178
5.6.2 The Gatlinburg Conference on Photosynthesis, 1952 181
5.6.3 Emerson Strikes Back (1955) 187
5.7 The Aftermath 190
5.7.1 The Enhancement Effect 190
5.7.2 Emerson's Death and Beyond 193
5.8 How Controversies End 194

6 The Path of Carbon in Photosynthesis (1937–1954) 201
6.1 Early Radiotracer Experiments 202
6.1.1 Kamen Meets the Cyclotron 202
6.1.2 Kamen Meets Ruben 204
6.1.3 The First Metabolism Experiments 207
6.1.4 Carbon-14 and the End of a Collaboration 212
6.2 The Bio-Organic Chemistry Group 214
6.3 The First Cyclic Model 220
6.4 New Methods and a New Model 226
6.4.1 Paper Chromatography 226
6.5 The Second Cyclic Model (1950) 229
6.6 In Need of Re-Orientation 232
6.6.1 Two Unexpected Compounds (1952) 232
6.6.2 The Degradation Studies 234
6.7 The Saturation Experiments 236
6.8 The Path XXI Paper 240
6.8.1 The Fraction 1 Protein 243
6.9 The Heuristics of the Berkeley Group 246

7 Elucidating the Light Reactions (1950–1961) 251
7.1 The Synthesis of Reducing Power 253
7.2 Photophosphorylation 257
7.2.1 Energy-Rich Phosphate Bonds and Photosynthesis 257
7.2.2 The "Compartment Model" Challenged 262
7.3 Cytochromes and Ferredoxin 268
7.3.1 Background 268
7.3.2 Photosynthetic Cytochromes 270
7.3.3 Ferredoxin 273
7.4 Photosynthetic Electron Transport Chains 277
7.4.1 Arnon's First Model 277
7.5 Energy Migration, Chlorophyll *a* and P700 282
7.5.1 Early Observations 282
7.5.2 Fluorescence Resonance Transfer 285
7.5.3 Special Pigments and the Antagonistic Light Effect 290
7.6 A Solution That Was in the Air 292
7.6.1 Far-Sighted Ideas at Urbana 293
7.6.2 The Thermodynamic Approach 296
7.6.3 Spectroscopy and Two Photosystems 299
7.6.4 Two Light Reactions, One Pigment 302
7.6.5 New Flashing Light Experiments 303
7.6.6 The Z-Scheme 305
7.7 Convergent Research Pathways 307

Epilogue 311

References 317

Index 347

Chapter 1
Introduction

"The importance of work on carbon assimilation depends not merely on its value in plant physiology and on its application in agriculture, but also (is of interest) for the utilisation of radiant energy"[1]—this is how research into the photosynthetic mechanism was hailed in 1916: with reference to the variety of interests it would serve. On a similar note, the study of how this mechanism was searched for and found is both of historical interest (since this episode has been one of the blind spots in the history of science) and of philosophical interest, as the investigation contributes to crucial methodological issues.

The present book reconstructs the history of photosynthesis research with a special focus on the actors' heuristics. It starts with the first chemical speculations in the mid-nineteenth century, but concentrates on the work that was done between 1919 and 1960. This is a long stretch of time; hence, rather than striving for completeness, the chapters of this book will highlight crucial episodes and milestones, while many issues inevitably remain unexplored. No in depth study of this field of research has been undertaken so far—which is quite extraordinary, given the importance of photosynthesis as a natural process and the enormous amount of research that has been spent on it. Shorter treatments can be found in otherwise science-oriented books;[2] photosynthesis research is touched upon in general works on the history of biochemistry;[3] some surveys are available,[4] as well as scattered articles,[5] and a number of autobiographical pieces by participants and reviews written in retrospect of former achievements.[6] But so far no attempt has been made to put these approaches together into a more comprehensive whole.

[1] Jörgensen and Stiles (1916b, p. 92).

[2] See, e.g., the pertinent sections in Rabinowitch (1945), Loomis (1960) and Walker (1992a).

[3] See, e.g., Fruton (1999).

[4] See, e.g., Myers (1974), Höxtermann (1992), Huzisige and Ke (1993), Gest and Blankenship (2004), Govindjee and Krogmann (2004) and Nickelsen (2008b).

[5] See for fine examples, e.g., Zallen (1993b) and Sloan (2009).

[6] The journal *Photosynthesis Research* regularly publishes tributes, obituaries and personal recollections. Many of these contributions have more recently been collated into a volume entitled "Discoveries in Photosynthesis"; see Govindjee et al. (2005).

K. Nickelsen, *Explaining Photosynthesis*, History, Philosophy and Theory of the Life Sciences 8, DOI 10.1007/978-94-017-9582-1_1

To some extent, this lack of treatment confirms the observation that the history of experimental biology and biochemistry has been overshadowed by work on the history of genetics and the history of evolutionary theory. This pertains particularly to anything that has to do with plants. However, another reason may be the fact that photosynthesis research was carried out by actors from very different backgrounds, who were using techniques that included photochemical analyses, manometrical measurements, the culturing of freshwater algae, spectrophotometry, quantum physical calculations and radiotracer studies. Photosynthesis research does not conform to the matrix of traditional scientific disciplines; it is one of those fields of study that perpetually stayed in-between, without ever acquiring disciplinary identity itself.[7] This might have contributed to the fact that photosynthesis research has failed to spark the interest of most historians of biology, chemistry or physics. At the same time, it is exactly the entanglement of photosynthesis research with numerous other fields that makes its investigation so rewarding, as, hopefully, is demonstrated in this book.

1.1 Models of Mechanisms

Today, photosynthesis is known as the process by which solar energy is converted into energy that can be used in biochemical reactions. It is fundamental to life on earth, and the way organisms accomplish this task has intrigued scientists for quite some time. The first tentative models of the photosynthetic mechanism, in the sense of a chemical pathway, were developed by organic chemists around 1840, but it was only from the 1920s onwards that noticeable progress was achieved in this respect. In particular, great strides were made after 1945, and by 1960 an elaborate model of photosynthesis at a molecular level had been established. This model included a set of light reactions, with two different photochemical systems, which was linked to a light-independent sequence of dark reactions that formed a cyclic pathway. Almost immediately the model became widely accepted, and in the decades that followed work on extending and refining it dominated the field.

Not all aspects of photosynthesis research can possibly be covered in one single book. This study focuses on the attempts to elucidate the underlying biochemical and biophysical mechanism. The latter term was frequently used by the actors themselves, while it has also received increasing attention as an analytical category within the philosophy of science. Already in 1972, William Wimsatt famously claimed that "[a]t least in biology, most scientists see their work as explaining types of phenomena by discovering mechanisms".[8] This line of thought was picked up 20 years later

[7] On the different institutional and conceptual paths of cross-disciplinary fields of this type, see, e.g., Bechtel (1986b).

[8] Wimsatt (1972, p. 67).

by William Bechtel and Robert Richardson, and the literature has proliferated enormously since.[9] Different definitions of "mechanisms" are available, but one common denominator, which is also adopted in this study, is that mechanisms consist of specific ensembles of entities in interaction; and in line with Wimsatt it is assumed that it is often mechanisms rather than theories that scientists refer to in their attempts to *explain* certain phenomena. "Mechanisms are sought to explain how a phenomenon comes about or how some significant process works", is one of the claims in the often-cited paper by Peter Machamer, Lindley Darden and Carl Craver.[10] Stuart Glennan elaborated on this and defined a mechanism for a certain behaviour (or phenomenon) as "a complex system that produces that behaviour by the interaction of a number of parts", while it is the causal interactions of the system's parts that receive particular attention.[11] If one wants to drop the requirement of a system (as it presupposes a level of stability that in some cases might not exist), mechanisms may also simply be characterised as "entities and activities organised in such a way that they are responsible for the phenomenon".[12]

It is, however, not the photosynthetic mechanism itself that forms the subject of this book. It is a book about the *models of this mechanism* that scientists developed over the years. Glennan introduced the notion of a "mechanical model"—which is a model of a mechanism—that "consists of (i) a description of the mechanism's behaviour (the behavioural description); and (ii) a description of the mechanism that accounts for that behaviour (the mechanical description)".[13] This is similar to what Machamer, Darden and Craver called a "mechanism scheme". However, one of the reasons why Glennan preferred the term "model", which I also adopted in this study, is that scientists rather use this term than talk of schemes.[14] The disadvantage, of course, is that the term "model" is notoriously ambiguous;[15] while at least in this study, whenever the term is used as an analytical category, it refers to the model of a mechanism as described above. This type of model has two characteristic elements.

[9] Classic references include, e.g., Bechtel and Richardson (1993), Glennan (1996), Machamer et al. (2000), Glennan (2002), Bechtel and Abrahamsen (2005), Glennan (2005), Craver (2006), Bechtel (2006), Bogen (2008) and Darden (2008). For a forthcoming "field guide" on the more recent literature, see Andersen (2014).

[10] Machamer et al. (2000 p. 2).

[11] See Tabery (2004) for the argument that Glennan's definition is not contradictory but complementary to the proposal by Machamer, Darden and Craver.

[12] Illari and Williamson (2012, p. 119).

[13] Glennan (2005, pp. 445, 446). A slightly different proposal of accounts of mechanisms is given in Woodward (2002, p. 375).

[14] Glennan (2005, p. 449).

[15] The literature on models is immense. Morrison (2006) summarises parts of the debate and discusses the value of theories in comparison with models. Bailer-Jones (2009) gives a survey of how models were treated in the history of philosophy of science. Toon (2012) recently offered an account of models as "make-believe". A more comprehensive and highly useful overview of the field is provided by Frigg and Hartmann (2012), with an extended bibliography.

On the one hand, it captures the properties and changes of a phenomenon and, perhaps, strives to represent its regularities (which is what many other models do as well, such as the model of a pendulum). On the other hand, models of mechanisms additionally require a description of *how* these changes were brought about, through an analysis of the concrete interactions of the parts of an organised system. Consequently, an evaluation of these models will have to consider both: its "behavioural" adequacy (does the model accurately predict the behaviour of the mechanism and the phenomena it produces?) and its "mechanical" adequacy (does the model present a convincing account of how the parts of the mechanism interact in order to produce the phenomena?).[16]

I will argue that a mechanical model along these lines is exactly what photosynthesis researchers were heading for, in the period under study here: they were searching for a continuous sequence of compounds, in causal interaction with each other and their cellular environment, that eventually gave rise to the end products. The first attempts started with a behavioural description, that is, a diagnosis of the changes that the process of photosynthesis brought about: carbon dioxide and water were converted into carbohydrates and oxygen. In this respect, most of the models were in close agreement. However, what the scientists tried to achieve was an actual understanding of the mechanism behind these changes, that is, they wanted to know *how* carbohydrates and oxygen were produced, and in this respect the models were widely divergent. Only if the scientists had succeeded in spelling out the sequence of compounds and the kind of interactions, then, it was agreed, they would have "explained" photosynthesis.[17] The gist of these explanations is not always easy to grasp. In order to clarify their content and make it easier to compare them with each other, many of the models of the photosynthetic mechanism were reformulated in this book as graphs.[18] The graph notation became limited, however, as soon as the models got too complicated and began to include, starting in the 1950s, the role of structural components of the cell in addition to the sequence of chemical compounds. For these cases the (often graphical) mode of representation chosen by the actors themselves was retained.

While there is a prolific philosophical discussion on the question what mechanisms are and how the mechanistic perspective relates to classical issues such as laws, regularities and causation, surprisingly little work has been done on the question of how models of mechanisms are constructed. This is the systematic focus of this book.

[16] Glennan (2005, p. 457).

[17] They were certainly convinced of the reality of the reconstructed mechanism; that is, their epistemic explanation (the development of which is followed in this book) was parasitic upon the fact that the mechanism physically produced the phenomena in question. See on this point Illari and Williamson (2011).

[18] This notation was developed to represent "causal graphs"; see, May (1999), Graßhoff and May (2001) and Baumgartner and Graßhoff (2004). The extension of the underlying approach to the analysis of experiments was provided in Graßhoff et al. (2000).

Bechtel and Richardson, who were pioneers in this respect, identified the "decomposition" and "localisation" of the mechanism's parts as typical strategies in dealing with the overwhelming complexity of living systems. Both of these strategies, they argued, work towards "a hierarchical analysis [of the mechanism under study] into functional components", and their efficiency was traced in a number of highly interesting cases.[19] Lindley Darden became interested in similar questions. She defined a "strategy" very broadly as "a method, a procedure, a practice, a principle"—anything that scientists use to produce new ideas, assess them and improve them.[20] Darden preferred to speak of strategies rather than "heuristics" because the actors in science tend to use this term, while she had to admit that drawing a line between these two was difficult. In this book, both terms are used, adopting Darden's broad vision of what the terms refer to; heuristics is assumed to be the overarching term, within which several specific strategies are discernible.

Three strategies, Darden argued, are particularly prominent in the "discovery of mechanisms", as she put it: schema instantiation, modular subassembly and forward and backward chaining.[21] Schema instantiation means that a mechanism scheme (in the terminology here: a model) is imported from the work on related phenomena; modular subassembly refers to the cobbling together of known or suspected components of the mechanism under study so that a "plausible mechanism candidate" is formed; and the technique of chaining, either forward or backward, implies that the scientists know about some entities or activities that are part of the mechanism and try to link them up in a continuous chain. The latter usually relies strongly on the body of generally accepted knowledge within the field. Together with Carl Craver, Darden explored and elaborated this approach, most prominently at the example of research in protein synthesis.[22] This study, to some extent, continues Darden's and Craver's project in view of a different field of research, while it goes much deeper into historical detail and context than their work. It also widens the perspective to the communities of scientists and looks at collective research heuristics at different points in time and within different disciplinary contexts.[23]

---

19 Bechtel and Richardson (1993, p. 7); see also Bechtel and Abrahamsen (2005). A similar strategy was identified by Bailer-Jones (1999) in the discipline of astrophysics.

20 Darden (1991, pp. 20–21).

21 Darden (2002).

22 Darden and Craver (2002). The range of examples was expanded in a jointly published monograph by the two authors, in which much of the specialised literature on mechanisms is presented to a wider audience, beyond the philosophy of science circles; see Craver and Darden (2013).

23 Forerunners in framing discovery processes as community projects are, e.g., Graßhoff (1998) and Bailer-Jones (2000).

1.2 Heuristics and Strategies in Photosynthesis Research

Modelling a mechanism consists to a large extent in the search for causal relationships between the (potential) elements of the mechanism, and in the attempt to spell out the nature of these causal interactions in detail.[24] In the case of the photosynthesis mechanism, it was mostly the reactions of organic compounds in a cellular environment that were under study. The question was how these compounds interacted with each other in order to give rise to the next step of the metabolic chain. Causal relevance of some compound A for the formation of another compound B can experimentally be established if it is possible to create situations which are almost identical; while in only one of these situations compound A is present. If only then compound B is formed, while it is not formed if A is absent, one may conclude that under the chosen circumstances compound A is causally relevant for the production of compound B. This experimental design is referred to in this book as being a "difference test" (since it is an elaborated version of John Stuart Mill's method of difference), which comprises the ideal basis for sound causal inferences.[25] The causal link between these compounds A and B would then become part of the body of established knowledge—although, by the second half of the nineteenth century, when the structural composition of chemical compounds had become part of the discipline, (bio)chemists were no longer satisfied with this behavioural, phenomenological description of the process. They wanted to know by which concrete interactions or transformations the one was converted into the other. In the case of more drastic conversions, as in photosynthesis, they wanted to know the intermediates and the relationship of this conversion to other processes in the system. The conceptual leap from the outcome of a difference test to the composition of a more comprehensive, explanatory model of the underlying mechanism was far from trivial. It is this step in the process of elucidating a mechanism that this study pays particular attention to; and it is here that heuristic strategies exert their strongest influence.

Speaking of the "heuristics" of science implies venturing into another ill-defined field of debate. William Whewell has been credited with the introduction of the term into the philosophical study of science, when he famously announced in 1860: "If you will not let me treat the Art of Discovery as a kind of Logic, I must make a new name for it. Heuristic, for example". It has rarely been acknowledged, however, that Whewell continued on a far less optimistic note: "Only that, as you know, I do not assert such an art to exist".[26] In this study I do not attempt to prove Whewell wrong and delineate a normative programme of how to make a successful discovery. The aim is to look for patterns of action—strategies in Darden's terminology—that characterised

[24] Vice versa, mechanisms also guide further causal inferences; as the authors of a recently published "manifesto" claim: "It seems that mechanisms are of interest to every aspect of thinking about causality". Illari et al. (2011, p. 16).

[25] See Graßhoff et al. (2000) for an introduction to the analysis of experiments from the point of view of causal reasoning along these lines.

[26] Todhunter (1876, Vol. 2, p. 418).

photosynthesis research at different stages of the longwinded investigation of its mechanism. The similarities to the work done by Bechtel, Richardson, Darden and Craver are obvious, although my project, in contrast to theirs, is entirely descriptive. Yet several of the strategies that they characterised can be observed also in this episode, although slight modifications seem to be in place. Take for example the following quote, written in 1962 by two well-known figures in twentieth-century photosynthesis research:

> We can summarize the over-all conversion of light energy into chemical energy in the form of carbohydrate and oxygen by several steps. First, the light energy absorbed by chlorophyll and related pigments is converted into the high chemical potential energy of some compounds. Second, these compounds react with water and produce oxygen and good reducing agents as well as other cofactors containing high chemical potential energy. Finally, these reducing and energetic cofactors react with carbon dioxide and other inorganic compounds to produce organic compounds.[27]

This is a wonderful example of the *functional decomposition* of photosynthesis into several partial components, in the sense of Bechtel and Richardson, while the components additionally were suggested to work in a temporal sequence (as indicated by the use of the words *first*, *second*, *finally*). In particular, the decomposition of the mechanism into "light reactions", that is, photochemical processes, and "dark reactions", that is, thermochemical processes, was highly influential. It is nowadays assumed, in line with the quotation above, that the splitting of water is a photochemical, light-dependent process; while the reduction of carbon dioxide also proceeds in the dark, provided the necessary reducing agents and ATP molecules are available. However, for a long time the functions ascribed to light and dark processes in photosynthesis had been very different. Up to the 1930s, it was mostly taken for granted that the light reaction in photosynthesis was the reduction of carbon dioxide, possibly in a complex binding with chlorophyll molecules; while the dark reaction was considered to be the polymerisation of the resulting carbon moiety to sugar. This fundamental conceptual change ought to remind us that the functional decomposition of a mechanism is far from trivial; and that in metabolism studies it has been particularly difficult to "carve Nature at its joints". As the plant physiologist Walter Stiles pointedly wrote in 1925: "In considering the stages in the process of photosynthesis, the first question which arises is to determine where the process begins and ends". The limits that were usually drawn at the time, Stiles thought, were "really a matter of convenience".[28] How the scientists dealt with this situation will be one of the recurrent themes of this book.

The importance of strategies that resemble Darden's notion of *schema instantiation* is also evident in the elucidation of the photosynthetic mechanism—not only as a strategy that was used in particular situations, but also in a more general sense. In the history of photosynthesis research two very basic assumptions prevailed, which formed the foundation for much of the research covered in this book. *First*, there

[27] Calvin and Bassham (1962), p. vi.

[28] Stiles (1925, pp. 179, 180).

was the phenomenological description of the process, which was conceived of as the conversion of carbon dioxide and water into carbohydrates and oxygen. This was taken to be a useful, if simplistic, starting point for all further investigation, which could be modified and expanded in more detail in order to arrive at a complementary mechanical description. Very soon these attempts were guided by a *second* widespread assumption. Photosynthesis, thus described, was strikingly similar to the process of respiration, only in reverse: photosynthesis consumed carbon dioxide and produced oxygen and sugar, while respiration consumed oxygen and sugar and produced carbon dioxide. This was highly suggestive and gave rise to the suspicion that the reciprocity of these phenomenological descriptions also extended to their mechanisms. Thus, the investigation of these two fundamental life processes, photosynthesis and respiration, became closely entangled. Since respiration research was part of medical physiology, which attracted far more attention than the functioning of plants, photosynthesis research usually lagged behind, so that respiration research became a popular source of mechanism schemes that were then transferred to photosynthesis.

However, the transferred elements usually were much less comprehensive than the term "schema instantiation" suggests. In many cases, it was rather a transfer of knowledge concerning particular causal links and potential types of interactions. On the other hand, it was not only explanatory schemes that were instantiated or transferred. If one widens the perspective, beyond the mere conceptual elements, it becomes clear that also other pieces of knowledge—"building blocks", as I chose to call them—were picked up elsewhere: methods were imported, experimental set-ups adapted, terms were borrowed and approaches employed that had their origins in a wide range of fields. The pertaining processes will be observed in several chapters of the book, and the same holds true for procedures similar to Darden's *modular subassembly*—although also these received a slightly different interpretation. Observing these strategies, however, hinges on an actor-centred perspective, which deserves an explanation in the following section.

1.3 The Individual and the Community

Scientific research in this study is conceived of as being largely, albeit not exclusively, driven by the scientists' physical and cognitive actions, the reasons for which mainly lie in the actors' goals and purposes, in their beliefs about how to reach these goals, and in their priorities of what to do first.[29] This differs, for example, from the approach taken by Hans-Jörg Rheinberger who chose to focus on the experimental system as the driving force of twentieth-century biomedical research.[30]

---

[29] This very general notion draws on widespread assumptions in action theory without, however, relying on any one particular theory of action.

[30] See Rheinberger (1997) for the classic formulation of the approach.

Rheinberger and others who adopted this approach have convincingly demonstrated that, frequently, it is not a theory that experimental research in biology starts with, sometimes not even a clear-cut research question; but the exploration of the properties and capacities of an experimental system, which consists of a specific set of methods and materials, instruments and, perhaps, model organisms. However, the weaknesses of this approach when it comes to questions on the dynamics of research and its conceptual development, which are the main points of interest here, have repeatedly been emphasised.[31]

Understanding how the model of a mechanism is constructed, requires an actor-centred perspective. In this study it is assumed that the overarching (epistemic) goal of the community of photosynthesis researchers was to explain how plants were able to transform carbon dioxide and water into carbohydrates and molecular oxygen.[32] However, achieving this goal was clearly impossible for any single researcher. One had to find out, for example, how chlorophyll absorbed light and channelled it into chemical reactions, one had to solve the puzzle how carbon was reduced to the stage of sugar at room temperature, one had to work out how photosynthetic oxygen was released, and so on. Explaining photosynthesis required that all these problems, and many more, be solved—thus, there was a wealth of, so to speak, "subordinate goals" (or subgoals for short), the achievement of which was crucial for the actual, superordinate goal to be attained. The definition of these subgoals coincided with the functional decomposition of the photosynthetic mechanism, which was mentioned earlier. If they were still too complex to be accomplished, they had to be divided again into further subgoals—and the mechanism into further functional subunits, in a hope that, at a later point of time, the separate parts of the model would be put together again.

The latter could only be achieved, as everybody was keenly aware of, within the community at large. Thus, the researchers in pursuit of the photosynthesis mechanism were, to a certain extent, obliged to cooperate with each other, if only on a minimal level. "To cooperate" might merely imply taking note of the findings of fellow scientists and letting them know about one's own results, frequently even ahead of publication; yet cooperation might, of course, also imply more, such as commenting on each others methods and data, arranging for concerted research efforts, exchanging collaborators and PhD students, and so forth. All of these action patterns can be observed in the history of photosynthesis research, as will be demonstrated in later chapters. Over the past decades, and in particular since the 1980s, science as a social activity has primarily been described as being dominated by competition

[31] See, e.g., Weber (2005, pp. 143–149); a concise, although slightly dated critical analysis of Rheinberger's approach is also provided in Burian (1995).

[32] The approach to analyse scientific research from the perspective of hierarchies of goals and actions in *epistemic systems* was first developed by Gerd Graßhoff in the 1990s, see Graßhoff (1994). For its application to historical cases, see also Graßhoff (1998, 1999), Graßhoff et al. (2000) and Graßhoff and May (2003).

and the often ruthless pursuit of power, influence, reputation and credit.[33] Although one cannot deny that competition is a widespread element in scientific research, it seems that this element has been over-emphasised at the expense of other aspects that are equally omnipresent. I want to argue that a complex process of investigation as it is analysed in this study can to a large extent be described as (mostly informal) cooperative efforts.[34]

Given this situation, the community of photosynthesis researchers becomes the relevant frame of reference for all attempts to explain the choices between alternative actions.[35] As will be shown at examples, within those groups of scientists pursuing a certain subgoal, such as finding the path of carbon dioxide reduction, direct competition, that is, two teams trying to reach the same goal by the same means, was a rare phenomenon. Rather, the different research teams chose alternative options to tackle the pertinent problems. In some cases, this was attained by explicitly dividing the labour between a number of heads and hands (as around 1950 in the Berkeley-based research group headed by Melvin Calvin and Andrew Benson). But often there is no evidence of open agreements along these lines, although even then photosynthesis researchers were obviously trying to avoid duplicating the efforts of others. They rather strove to find complemental angles to solving research problems. One can nicely observe, for example, how members of the (rather loosely associated) community created for themselves appropriate niches, characterised by specific methods and research questions. On the one hand, they divided by this means effectively the labour between them; on the other hand, this also resulted in minimising competition with others.[36] Furthermore, this coordination of research efforts resulted in covering a broad range of alternative approaches, particularly in situations of prevailing uncertainty, in which no obvious choice of action prevailed: a rather effective strategy for the community as a whole, since chances were high that at least one of these options would, at some point, lead to promising results.

The situation resembles Philip Kitcher's concept of a cognitive division of labour within a community.[37] However, Kitcher thought that this was unrealistic as it required some members of the community to consciously (and altruistically) choose and pursue the less promising options, for the sake of retaining multiple alternatives. As the history of photosynthesis demonstrates, the latter is by no means inevitable: more often than not it was impossible to predict which approaches were more promising than others. Miriam Solomon's concept of social empiricism, which

[33] Cf, e.g., Callon (1994), Bourdieu (1991), Latour and Woolgar (1982).

[34] On this point, see also Nickelsen (2014).

[35] The methodology of scientific communities has not yet received much interest in the growing field of "social epistemology"; see, e.g., the introductory accounts of Schmitt (1994), Goldman (2010) and Goldman (2011).

[36] See on this strategy also Edge (1990) who argues that scientists tend to choose "topics sufficiently close to mainstream concerns to ensure recognition, but sufficiently distinct to prevent duplication, and to ensure that the work will be perceived as significant" (p. 214).

[37] See Kitcher (1990).

she developed as a normative framework of the social epistemology of science, provides a helpful perspective on this "principle of plurality", as one may want to call it, that prevailed among photosynthesis researchers.[38] According to Solomon, it is entirely rational for a community to maintain a variety of theoretical approaches to a problem, as long as these approaches are empirically successful (broadly conceived). It seems that the actors' choices in photosynthesis research were frequently well in line with Solomon's idea of "rationality".

The division of labour in photosynthesis research went hand-in-hand with an increasing degree of differentiation and specialisation in twentieth-century science. Many photosynthesis researchers tended to work in a rather limited domain: they focused on a certain subgoal (such as finding the rate of photosynthesis under different conditions), used a certain technique (such as manometry) and persevered with these for a rather extended time span. Which subgoal and which method an individual scientist chose for herself can often be explained by examining a scientist's education and career—their "investigative pathways", in Frederic L. Holmes's terminology.[39] Many of the actors in this study were introduced by their supervisors and mentors not only to a certain discipline but also to a certain order of priorities and a certain method (the importance of which is, of course, well-known to post-Kuhnian history of science). All the aspiring young scholars who spent some time in the laboratory of the German cell physiologist Otto Warburg, for instance, used manometric techniques for the rest of their working lives. Since they had learned to master one of the best techniques available at the time, there was little incentive to try out alternative techniques. Which themes these scientists chose to investigate and which conceptual approaches they used, frequently also depended on their earlier experiences. Few of the actors examined in this study changed radically in the course of their professional careers their methods or their principal notion of how photosynthesis had to be addressed. James Franck, for example, who had come from quantum physics of atoms and molecules, always remained reluctant to accept that the photochemical process in photosynthesis diverged from everything he had learned in his study of inorganic processes. But despite this blind spot of his, Franck nevertheless contributed, in his specific way, to the development of a mechanical model of photosynthesis that included a hitherto unheard-of photochemical process. Within a community, as Solomon has argued, conceptual biases and ill-founded decisions of individual actors are easily tolerated and eventually balanced by others.[40]

The aforementioned division of labour also mirrors the fact that many of the major players ventured into photosynthesis research from a different field of study, in many cases with the intention to merely launch one contribution to the field, from their respective disciplinary background of concepts and methods. They planned, in other words, to tackle a specific subgoal that seemed attainable with the package of competencies they had and then, without spending too much time and resources,

---

[38] See Solomon (2001).

[39] Holmes (2004).

[40] See, e.g., Solomon (1994a).

return to their "actual" field and research goals. In the period under investigation here, photosynthesis research, which developed as a field between the established scientific disciplines, was particularly prone to this widespread pattern of behaviour: to "research opportunism", as it is called in this study (not without a pinch of irony). It was to the great benefit of the field that a number of those who just came by for one contribution stuck around and would make photosynthesis research their investigative focus of many years to come, as, for instance, Otto Warburg, James Franck and Martin Kamen. But this may suffice as a sketch of the systematic approach taken in this study and as an introduction of some of its main threads. I shall close the introduction with an overview of the chapters of this book and their content.

1.4 Survey of the Book

The story begins in chapter 2 with the first attempts made in nineteenth-century Europe to reconstruct the mechanism of photosynthesis. It were particularly German chemists who concerned themselves with this project, including eminent figures such as Justus von Liebig, Adolf von Baeyer and Emil Erlenmeyer. None of them worked on photosynthesis for a long time—the whole period is characterised by "research opportunistic" contributions, along the lines introduced above. The points of particular interest in this chapter are: how, in a situation of utmost uncertainty concerning the processes in the living cell, these chemists came up with a variety of models of the mechanism; how these proposals were evaluated; and how they were related to each other. The chapter ends with the model presented by Richard Willstätter and his collaborator Arthur Stoll in 1918, which became the standard for many years to come. Chapter 3 then studies the work of one central actor, the cell physiologist Otto Warburg, in more detail. Warburg turned to photosynthesis in 1919, and it would remain part of his research interests up to the end of this professional career. His individual research strategies are examined in relation to his biographical background and earlier research preferences, which allows the characterisation of the various "building blocks" that Warburg availed himself with for his photosynthesis work. His main contribution to the field was a methodical revolution through the import of manometric techniques in combination with the study of the unicellular freshwater alga *Chlorella* as the experimental organism of choice.

This approach clearly dominated photosynthesis research in the 1930s, which is the subject of chapter 4. In this decade, a wealth of exciting new findings concerning the photosynthetic mechanism were amassed, while it proved extremely difficult to integrate these into a coherent whole. Researchers of very different disciplinary backgrounds had turned to the study of the photosynthetic mechanism, including William Arnold, Robert Emerson, James Franck, Charles Stacy French, Hans Gaffron, Robin Hill and Cornelis B. van Niel. The chapter tries to highlight the different sources of their work on photosynthesis by taking a group-biographical approach: the investigative pathways of the different scientists are followed in order to clarify why they chose their particular focus and method of photosynthesis research and how they (frequently research-opportunistically) stumbled into the field. The productive line of research in this decade was slowed down by the development of a long and acrimonious controversy on the maximum quantum yield of photosynthesis, which is analysed in chapter 5. The debate started when, in the late 1930s, American research groups found a minimum requirement of 8–12 quanta for the photosynthetic production of oxygen, while the standard value of 4–5, which had been proposed by Otto Warburg back in 1923. The resulting controversy lasted until the mid-1950s, and it is a fine example of how a community reacted, first, if incompatible experimental results were found by teams known to be equally knowledgeable in the field; second, if one of the participants—Warburg—constantly violated unwritten conventions.

Chapter 6 then turns to the investigation of the mechanism of carbon reduction in photosynthesis (which, by then, was known to be the "dark" reaction of the process). This was achieved in Berkeley, in a large research team headed by Melvin Calvin and Andrew A. Benson that made skilful use of radiotracer techniques and paper chromatography. With recourse to published sources, correspondence and the extensive oral history interviews of former participants, the main body of this chapter examines the modelling strategies of the group and the criteria according to which they modified their proposals. The final chapter 7 then looks at how the "light" reactions in photosynthesis were elucidated, culminating in the two photoreaction, two pigment system model, which to this day dominates the field. The introduction of new spectroscopic methods was enormously important in this context, since for the very first time it became possible to nail down the changing redox states of individual molecules, in response to, for example, the onset of illumination. Furthermore, it was in this period, that people also tried to localise the partial processes in the cellular environment. In contrast to the study of the carbon reduction process, several research teams independently worked on this question, and almost simultaneously came to very similar conclusions. The eventual model subsequently became known as the "Z–scheme" (after a standard form of representation).

In terms of explaining photosynthesis at the level of molecular details, the Z–scheme was only the beginning; for this study, however, it marks an appropriate endpoint, which was also felt by the actors themselves. On a certain level of decomposition, a model of the photosynthetic mechanism had been established that was able to describe a complete sequence of reaction steps; and this is what the major players in this episode had been in search for.

Chapter 2
In Pursuit of a Pathway (1843–1918)

> If we try to nail down the present state of our scientific views on the assimilation of carbon in the chlorophyllous parts of plants, we are forced to confess that [...] in this fundamental question of plant physiology we are still at the stage of discussing the possible and the probable.[1]

The quote above succinctly characterises the state of photosynthesis research in the second half of the nineteenth century. The author of this quote openly acknowledged that so little was known about photosynthesis that the formulation of "hypotheses", which in this case were taken to be equivalent to "speculations", seemed to be the only, and, therefore, justified, resort. The body of knowledge at the time, concerning organic reactions in general and the processes in the living organism in particular, was scant and the methods available went hardly beyond input–output measurements without any means to access the stages in-between. This situation was not to change for decades to come. Even as late as 1925 the British plant physiologist Walter Stiles still maintained, in his monograph on carbon assimilation in plants, that "the nature of the intermediate substance or substances formed in photosynthesis is a subject on which [...] our real knowledge is practically negligible".[2]

Yet, photosynthesis still attracted the attention of a number of scientists, some of which concerned themselves with possible mechanisms and pathways of photosynthesis, as will be introduced in this chapter. The period analysed stretches from 1843, the year in which the German organic chemist Justus Liebig brought forward a rudimentary idea of the photosynthetic mechanism—the first attempt to account for photosynthesis in terms of a chemical pathway—until 1918, the date of publication of the voluminous monograph compiled by the German organic chemist Richard Willstätter together with his Swiss collaborator Arthur Stoll. Finding the biochemical pathway for the reduction of carbon dioxide to the stage of carbohydrates in plants was no trivial task given the situation described above. The main topic of this chapter is to elaborate the characteristic features of the models that chemists came up with and to spell out the chemists' strategies to deal with the enormous methodical

[1] Reinke (1882, p. 289).

[2] Stiles (1925, p. 193).

K. Nickelsen, *Explaining Photosynthesis*, History, Philosophy and Theory of the Life Sciences 8, DOI 10.1007/978-94-017-9582-1_2

difficulties. The highly critical attitude held by several plant physiologists in view of this methodology provides evidence for the fact that the chemists' strategies were deeply embedded in a specific research context and community that complied to conventions which were far from universally accepted.

2.1 The Nineteenth-Century Conception of Photosynthesis

"The leaves and other green parts of a plant absorb carbonic acid, and emit an equal volume of oxygen", the German organic chemist Justus Liebig, a towering figure at the time, maintained in 1842.[3] He thereby captured one of the central features of photosynthesis that had been generally known since the work of Joseph Priestley and others in the eighteenth century: the gas exchange that took place in the green parts of plants. Liebig went on to explain in some detail how the assimilated carbon then might be used in the plant to synthesise a wide range of compounds, primarily, he thought, carbohydrates. The latter was confirmed in the early 1860s by the German plant physiologist Julius Sachs, who identified starch—a polysaccharide—as the first distinctly recognisable product of photosynthesis.[4] It was also clear that sunlight was related to this process: in 1845, the German physician Julius R. Mayer described the plant as a reservoir of "solar force", which the plant absorbs and then transforms into a different type of force that Mayer called "chemical potential" ("*chemische Differenz*"). The latter would then be used, Mayer suggested, in the plant's growth and metabolism.[5] The resulting, widespread notion of photosynthesis—or "carbon assimilation", as it was usually referred to at the time—was extremely stable, up to the late 1920s.[6] A characteristic formulation is provided in the following lines, which were published in an encyclopedia for the German educated middle class of 1907:

> Assimilation in the botanical sense of the word is the formation of carbohydrates from carbonic acid and water while oxygen is released. This process is limited to the chlorophyllous assimilation system (assimilation tissue) and requires the involvement of sunlight.[7]

---

[3] Liebig (1842, p. 24). In 1845, Liebig was ennobled and, henceforth, became known as Justus *von* Liebig. Further literature on Liebig will be cited below, when his specific photosynthesis model is being discussed.

[4] See Sachs (1862, 1864) for the original papers; the most informative biographical work on Sachs still is Pringsheim (1932).

[5] See Mayer (1845, particularly pp. 37–42). On Mayer's contribution to the discussion of physical forces ("*Kräfte*"), see Caneva (1993); Smith (2003).

[6] See, for example, the definition in the textbook by Heinrich Schroeder in 1928: "Carbonic acid assimilation (photoenergetic assimilation) [...] is the specific ability of the chlorophyllous plant to reductively synthesise organic compounds—in the first step carbohydrates—from carbonic acid (carbon dioxide and water) by using radiant energy, while, at the same time, releasing oxygen". Schroeder (1928, p. 653); paragraph 1a. All translations into English, if not otherwise declared, are by the author (K.N.). See also Gest (2002) for a brief history of the term "photosynthesis" and its definitions.

[7] Anonymous (1907a).

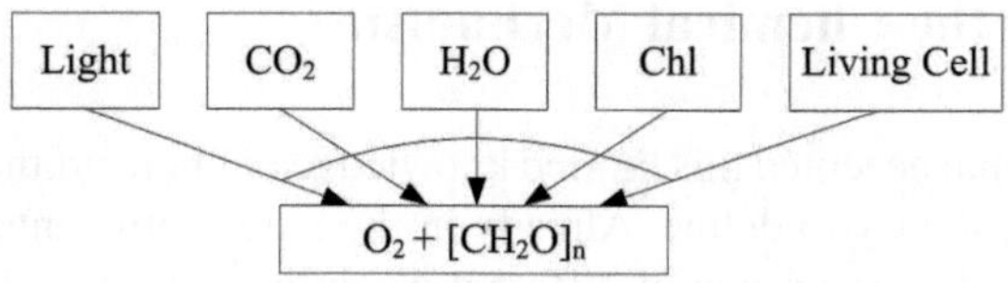

Fig. 2.1 The elementary one-step model of photosynthesis: the basic consensus on the causal factors of the overall process from the mid-nineteenth century until the beginning of the twentieth century.

Carbon dioxide (mostly taken to enter the reaction in the form of carbonic acid) and water, absorbed from the atmosphere and the soil, were taken as the starting materials of the process, which, in the green cells of plants, were then converted, under the influence of light, into carbohydrates and oxygen. From the work of Henri Dutrochet, Julius Sachs and others, chlorophyll pigments were known to play a crucial role in this process, although the precise character of this role was highly disputed.[8] Equally disputed was the reason for the fact that photosynthesis stopped as soon as the cell was damaged; the "living cell", or some specific aspect of it (usually suspected to be either some part of the protoplasm or a structural component), also seemed to be a necessary factor.

This (rather limited) body of knowledge on the mechanism of photosynthesis can be conceived of as a simplistic, one-step model, which is visualised in graph form in Fig. 2.1.[9] Molecular oxygen and carbohydrates, the general chemical formula of which is $[CH_2O]_n$, were taken to be the effects of a process in which carbon dioxide and water, light, chlorophyll ("Chl" in the figure) and the "living cell" acted as causally relevant factors—yet, how these factors interacted with each other was highly debated. Given the complexity of the process, everybody knew that, of course, a great many different steps were required to reach the final stage of photosynthesis, passing through a wide range of intermediate compounds. Photosynthesis researchers agreed neither on the order of the processes involved nor on the question as to which of these processes were light driven and which were not. Yet, everyone was aware that, in whatever way the more complex model would be drawn up, the basic causal links that are represented in the basic one-step model had to be accounted for some way or another. This made it the starting point for all investigations of the photosynthetic process in the period under study and, therefore, also for the subject of this chapter.

---

8 Dutrochet (1837); Sachs (1864).

9 This notation is largely in line with the wide-spread conventions of representing causal graphs; see, e.g., Pearl (2000); Baumgartner and Graßhoff (2004). However, in contrast to the more rigid use of this notation, in this book no metaphysical commitment is implied concerning the factors included in these graphs. Specifically, while causal graphs usually name "events" as relata of the causal links, the adequacy of this interpretation for representing biochemical mechanisms, which rather outline the interaction of concrete entities, is doubtful.

2.2 Finding the Chemical Mechanism

The community that accepted this limited knowledge of photosynthesis as a common denominator is not easy to define. Already in the nineteenth century, photosynthesis research spread over a range of different disciplines, as was observed in one of the first monographs on the subject matter, published in 1925. The author, Walter Stiles, maintained that "the processes taking place in the green leaf, which involve the absorption of carbon dioxide from the air and the manufacture of carbohydrates from it and the water supplied by the soil [. . .] are among the very few problems of botany which have attracted the attention of workers in other fields".[10] The methods of research taken to be necessary to make any progress at all in elucidating photosynthesis included microscopical anatomy, experimental physiology as well as analytical and theoretical chemistry.[11] This range of techniques corresponded to the range of very different aspects of the process to be considered. The structural prerequisites of photosynthesis, for instance, were studied by plant anatomists and morphologists, who tried to explore the close relationship between the morphology of the leaf and its functional requirements;[12] others investigated the development, organisation and distribution of chloroplasts in the green organs of plants.[13] Plant physiologists concentrated on the influences of several macro-parameters on the process, such as light intensity, temperature and carbon dioxide concentration; while the effects that the incident illumination had on the pigments were mainly explored by physicists, who were interested in the mechanism of light absorption and the chemical efficiency of rays.[14] The question of the (bio)chemical mechanism of photosynthesis, finally, was predominantly, if not exclusively, the domain of chemists; and this is the aspect of photosynthesis research that I shall focus upon in the following sections.

It was primarily a number of German nineteenth-century chemists who felt attracted to the problem of the photosynthetic mechanism and who tried to elucidate as far as possible the course of events by which in the leaves of plants carbon dioxide (in its dissolved condition, that is, as carbonic acid) was worked with water into sugar.[15] From the point of view of chemistry, there were two key questions to be answered in this search for the pathway of photosynthesis: (1) How was carbon dioxide, a highly stable compound, reduced? This question was closely related to the release of oxygen, as reduction still was mostly conceived of as oxygen disposal. However, the second question immediately followed: (2) How were the one-carbon units that

---

[10] Stiles (1925, p. 1).

[11] Cf. Reinke (1882, pp. 290–291).

[12] See, e.g., Haberlandt (1881, 1884).

[13] See, von Mohl (1837) for a seminal contribution along these lines.

[14] These early photochemical studies were greatly inspired by the new technique of photography; see, e.g., Boberlin (1993).

[15] On the development of physiological chemistry, or: chemical physiology, in Germany and elsewhere, see Höxtermann (2007b); Holmes (1985); Kohler (1982); Fruton (1972a). On the general situation of chemistry in Germany around 1900, see Johnson (1990).

(presumably) resulted from the carbon dioxide reduction joined together to form large molecules such as sugars?

Above all, it was the first question that puzzled the chemists, as carbon dioxide was one of the most chemically inert molecules known to exist: How could this molecule be made to undergo complex reactions without exposing it to extremely high temperatures or atmospheric pressure? In the following sections of this chapter, the main approaches developed by the nineteenth-century chemists involved in photosynthesis research will be described in terms of background, content, evidence and their relationship to each other. Although some of these approaches are well-known—notably Adolf von Baeyer's formaldehyde model—no in-depth comparative analysis has yet been undertaken, so that some detail is required in order to understand the dynamics of this line of research.[16]

2.2.1 *Justus Liebig and the Organic Acid Model*

It was the aforementioned Justus Liebig who first put forward a possible pathway for the process of photosynthesis.[17] Liebig started to think about this theme around 1840, that is, when he began to consider the impact of chemical knowledge to problems within the domain of life processes: an interest that was at least partly stimulated by the increasing demand at the time in Germany to improve the foundations of agriculture.[18] One of the first results of this line of thinking was the highly influential book *Chemistry in its Applications to Agriculture and Physiology* (1840), which was translated almost immediately in several languages and went through numerous editions.[19] Therein, Liebig emphatically propagated that it was high time to integrate chemical methods and concepts into the study of plants and their internal functioning, among others: his method of studying metabolic changes in terms of input–output balances.[20]

Liebig found the investigation of plants as it was practiced (or so he thought) by his colleagues in the botany departments deeply unsatisfying: "In botany the talent and labour of inquirers has been wholly spent in the examination of form and structure: chemistry and physics have not been allowed to sit in council upon the

[16] See, e.g., Florkin (1977, pp. 147–151), for a discussion of Baeyer's formaldehyde hypothesis. The latter is also treated in Rabinowitch (1945, pp. 255–260), in which Baeyer's approach is compared with Liebig's point of view.

[17] For Liebig's biography, see Brock (1997).

[18] Cf. Allen (1975, pp. 154–157). On Liebig and his influence on agricultural science, see also Rossiter (1975).

[19] The original German version was published as Liebig (1840).

[20] On the integration of chemical and physical methods and concepts into botanical research during the nineteenth century see, e.g., the contributions to Hoppe (1997b) and Janko and Strbánová (1991).

explanation of the most simple processes"[21] Liebig complained. Yet, the botanists were unable to solve the central issues of inquiry, Liebig went on, because they lacked the skills to perform the necessary experiments, "it being an art which can be learned accurately only in the chemical laboratory".[22] Liebig's arrogance in this matter, in combination with his far-reaching ignorance of the actual state of plant physiology at the time (he seems to have taken his knowledge of the field from one single textbook), did not go unnoticed, as is documented in the reactions by, for example, the renowned plant physiologist Matthias J. Schleiden, who was never shy of polemics, and the equally distinguished Hugo von Mohl.[23] If Liebig chastised the plant physiologists for their lack of knowledge in chemistry, Hugo von Mohl wrote, Liebig's lack of knowledge of plants and their organisation surely would have to be considered equally disadvantageous. Mohl was ready to admit that the study of plants, insofar as it was of an experimental nature, was "more in the sphere of the chemist than of the botanist". However, Liebig's own suggestion to solve the pertinent problems presented "splendid evidence for the proposition that chemistry has not yet found out much more than nothing at all about the chemical processes in the interior of plants".[24]

In fact, Liebig's plea for experimental research in physiology and his pride on the chemists' achievements in this respect should not be overrated. Liebig was able to quantitatively determine what the plants took in and what they gave off; yet, like everybody else in the field, he lacked the techniques to investigate the processes that went on within the organism. Liebig nevertheless developed a proposal, which was first brought forward in a publication of 1843, while he repeated the principal idea in a number of other places, among those, several editions of his own chemistry textbook.[25] (A slightly simplified reconstruction of Liebig's model, in the form of a graph, is given in Fig. 2.2) This is how Liebig introduced his suggestion:

> If one considers that unripe fruit, for example, grapes, cannot be enjoyed due to their high acid content; that in sunlight these fruits behave in the same way as leaves, namely, that they are capable of absorbing carbonic acid and releasing oxygen; that at the same time as the acids decrease, the sugars increase: in view of these points, one cannot reject the idea that the carbon of the organic acids in unripe fruit becomes part of the sugars in ripe fruit; that, therefore, the acid is transformed into sugar, effected by the release of oxygen and the components' absorption of water.[26]

[21] Cited: 2nd edition, Liebig (1842, pp. 37–38).

[22] Liebig (1842, pp. 39–40).

[23] See Werner and Holmes (2002) for a detailed analysis of this controversy; Werner (2001) describes how Alexander von Humboldt tried to resolve the disagreement.

[24] Cited Werner and Holmes (2002, pp. 436–438). Mohl's defence was published as von Mohl (1843).

[25] For the first version, see Liebig (1843); see also Florkin (1977, p. 147); Stiles (1925, p. 194); Schroeder (1917, pp. 2–3), and Rabinowitch (1945, p. 255).

[26] Liebig (1843, pp. 61–62).

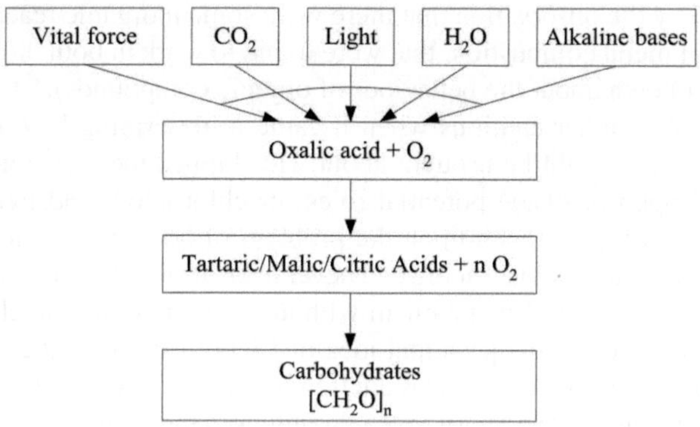

Fig. 2.2 The different processes involved in photosynthesis according to the organic acid hypothesis, which was originally proposed by Liebig (1843). The precise sequence of organic acids was unclear.

Starting from the observation that fruits gradually sweeten as they ripen, Liebig surmised that the tartaric, citric, malic, etc. acids in fruits might be the intermediates on the pathway from carbonic acid to sugar. And since all these acids were usually found in leaves in the form of their ions combined with the ions of alkalis, such as potassium or calcium, to a salt, Liebig concluded that these alkalis played a crucial role in the process too. (He took this to be the reason for the fact that plants would not grow without a minimal amount of these alkaline substances being available, either in the natural soil itself or added in the form of artificial fertilisers).

Liebig's account of the actual sequence of acids in the pathway of photosynthesis was rather vague, although he suggested that initially oxalic acid might be produced from carbonic acid by the release of oxygen—given the presence of an alkaline base, light and some hypothetical vital force. The latter was his interpretation of the unspecified "living cell" factor that was mentioned earlier, since the vital force was thought to fade away when the living organism was damaged or destroyed.[27] Liebig thought that, in later stages, oxalic acid might be reduced to tartaric, malic or citric acid, from which carbohydrates were then formed, thereby releasing additional oxygen. Thus, Liebig postulated a stepwise path from carbonic acid to carbohydrates via compounds that became increasingly poor in oxygen and rich in hydrogen. Some empirical support was taken from the fact that, in the presence of alkali and at high temperatures, the decomposition of oxalic, tartaric and citric acids to carbon dioxide had been observed in the test tube by the French chemist J. L. Gay-Lussac; and in view of this finding, Liebig considered it entirely feasible that the reverse reaction could take place in plant cells.[28] (Note that this assumption of reversibility was, at the time,

[27] See Schroeder (1917, p. 2). Although, in many instances, Liebig rejected the practice of using a vital force as an explanatory factor, he still acknowledged that there were some phenomena that could not be explained without this force. On Liebig's position between reductionism and vitalism, see Lipman (1967) or Hall (1980). See also Caneva (1993) on this point.

[28] Liebig (1843, p. 63).

only supported by the observation that there were some inorganic reactions, notably in the context of metal combustion, that were found to work in both directions, while very little was known about the behaviour of organic compounds).

Liebig was also rather cautious when it came to describing how this sequence of reaction in plants might be brought about. He skipped the questions of possible sources of hydrogen or of the potential roles for chlorophyll and light in the process, and he did not even touch upon the problem of how carbohydrates might be formed from the organic acids. Liebig had never been afraid of formulating sweeping hypotheses on the course of metabolism without going into any much detail.[29] Yet despite the lack of detail, the principal idea of Liebig's model—that organic acids were the intermediates in the gradual reduction of carbon dioxide to carbohydrates—was still being debated in the 1920s, even though both the vital force and alkalis had by then been abandoned as relevant factors of the process. The main points in favour of Liebig's model were: *first*, that it was, in fact, possible to construct a stoichiometrically plausible pathway from carbon dioxide to carbohydrates through the stages of various organic acids; and, *second*, that this approach provided an explanation for the fact that organic acids were found in surprising abundance in all parts of the plant, while nobody had been able, up to then, to identify their physiological function.[30]

2.2.2 *Adolf von Baeyer and the Formaldehyde Model*

One could hardly say that Liebig's proposal aroused either passionate interest or decisive rejection among his contemporaries. In fact, it was only in 1870 that a serious alternative was being advanced by the German organic chemist Adolf von Baeyer. In essence, Baeyer's model comprised the assumption that the first reduction product of photosynthetic assimilation was formaldehyde: a small (and highly noxious) organic molecule, which resulted, Baeyer surmised, from the photolysis of carbon dioxide in the presence of water, light and chlorophyll, while at the same time oxygen was released. As we shall see in later chapters of this book, even far into the 1930s, parts of this hypothesis were still counted among the most promising candidates for a photosynthesis model.[31]

It is worth taking a quick look at Baeyer's general preoccupations at this time.[32] He is, of course, particularly renowned for his research on the plant dye indigo: Baeyer

[29] See Werner (2001). Werner and Holmes (2002) provides an analysis of the dispute between Liebig and Matthias Schleiden and Hugo von Mohl on this point.

[30] Many of these acids are central intermediates of cellular respiration, namely, of the citric acid cycle, as is well-known today. As Liebig rejected the thought that there was respiration in plants, this explanation was not even conceptually available to him.

[31] See on Baeyer's model and its broad reception also Nickelsen and Graßhoff (2011).

[32] See Baeyer (1905) for his autobiography. Further information on his life and work is provided by Klemm (1953) and in Baeyer (1966).

successfully synthesised this important dye in the test tube in 1880, and by 1883 he had completely elucidated the molecule's structure. (Baeyer was awarded the 1905 Nobel Prize in Chemistry, in part because of these achievements.) However, around 1870, Baeyer was also interested in condensation reactions, and he achieved a major breakthrough in 1872 when he succeeded in carrying out the poly-condensation of phenol and formaldehyde. Formaldehyde had been discovered in 1855 by the Russian chemist Alexander M. Butlerov and had since become a product of central interest in the field of organic chemistry. Baeyer based his photosynthesis model on empirical evidence that Butlerov had presented in 1861: on heating trioxymethylene (a condensation product of formaldehyde which today is known as 1,3,5-trioxane) in an alkaline medium, a viscous fluid was produced, which seemed to have some of the properties of sugar.[33] Baeyer took this as the starting point for his proposal of how carbohydrates were synthesised in living plants. In a short paper devoted not even entirely to the problem of carbon assimilation Baeyer made the following argument:

> The general assumption in regard to the formation of sugar and related bodies in the plant is that, under the action of light, carbon dioxide is gradually reduced in the green parts [of a plant] and by subsequent synthesis is converted into sugar. [...] Butlerov's discovery provides the key [to the alternative assumption that sugar is formed directly from carbon dioxide], and it is indeed surprising that it has up to now been so little utilised by plant physiologists.
>
> The similarity that exists between the blood pigment and the chlorophyll has often been referred to; it is also probable that chlorophyll as well as haemoglobin binds carbon monoxide. Now, when sunlight strikes the chlorophyll, which is surrounded by CO_2, the carbon dioxide appears to undergo the same dissociation as at higher temperatures: oxygen escapes and carbon monoxide remains bound to the chlorophyll. The simplest reduction of carbon monoxide is to the aldehyde of formic acid—it only needs to take up hydrogen, $CO + H_2 = COH_2$. Under the influence of the contents of the cells, as well as through the alkalines, this aldehyde is then converted into sugar. [...] Glycerol could, in addition, be formed by the condensation of three molecules and the subsequent reduction of the thus formed glyceric aldehyde.[34]

According to this proposal, the carbon reduction in photosynthesis consisted of several processes, which are reconstructed in Fig. 2.3. First, carbon dioxide binds to the chlorophyll, which is shown as [Chl–CO_2] in the figure; in this state and under the influence of light the carbon dioxide is reduced to carbon monoxide, upon which oxygen escapes. Baeyer justified the assumption of this step by referring to the structural similarity between chlorophyll and haemoglobin: since the latter was known to bind carbon dioxide, it was reasonable to assume, he thought, that chlorophyll could do so as well. The carbon monoxide is then reduced further to formaldehyde by the bonding of either molecular hydrogen or two atoms of hydrogen from other sources (which were not specified). Thus, in contrast to the conceptualisation of the

[33] See Butlerov (1861). The episode is also discussed in Stiles (1925, p. 194); Florkin (1977, p. 147); and Rabinowitch (1945, p. 255).

[34] Quoted in Stiles (1925, p. 194); also in Florkin (1977, pp. 147–148). Translation provided by Jörgensen and Stiles (1917), with minor changes introduced by the author, K.N. For the German original, see Baeyer (1870, pp. 67–68).

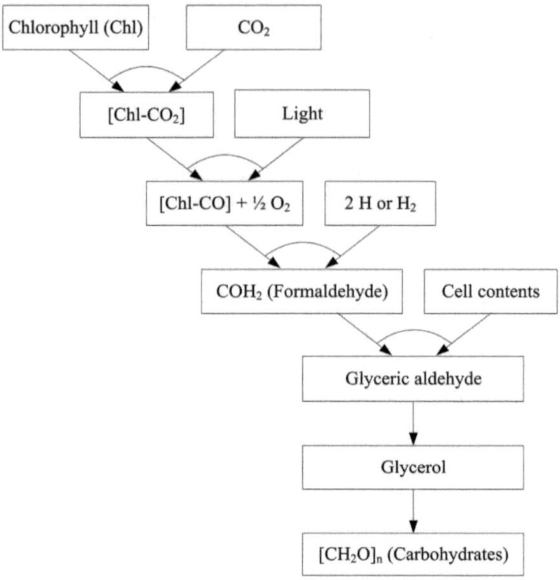

Fig. 2.3 The processes involved in photosynthesis according to Baeyer's formaldehyde model (1870).

process in the organic acid hypothesis, the actual reduction of carbon dioxide to the oxidation state of sugars (which was instantiated in formaldehyde) was assumed by Baeyer to occur without the formation of any intermediates. In subsequent reactions, the formaldehyde was then thought to produce carbohydrates—a process that was presumably promoted somehow by the contents of the cell. Baeyer hypothesised, for example, that the first sugar product might still be associated with the components of the cell, and that it would only later be released as sucrose, starch or cellulose.[35] However, before Baeyer presented his own point of view, he dismissed the organic acid hypothesis:

> The intermediate steps [of the gradual reduction process] have been sought in the organic acids—formic acid, oxalic acid, tartaric acid, etc.—which can be regarded as the reduction products of carbon dioxide. According to this opinion, at those times when the green parts of the plant are most strongly subjected to the action of the sun's rays, a strong accumulation of acids should take place, and these should then gradually give way to sugar. As far as I know, this has never been observed, and when it is remembered that in the plant sugars and their anhydrides are found under all circumstances, whereas the presence of acids varies according to the type of plant, the particular part and its age, then the opinion already often put forward, that the sugar is formed directly from the carbon dioxide, increases in probability.[36]

[35] Baeyer (1870, p. 68).

[36] Quoted in Stiles (1925, p. 194); also in Florkin (1977, pp. 147–148). Translation provided by Jörgensen and Stiles (1917). For the original German text, see Baeyer (1870, pp. 67–68).

Hence, Baeyer's main objection to the organic acid hypothesis was that one of its (conjectured) empirical consequences, namely the accumulation of intermediate products at times of strong photosynthetic action, had, as yet, not been observed. Furthermore, Baeyer pointed out that the acid content of a plant was strongly dependent on parameters that were probably not connected to photosynthesis—such as the species, the part of the plant, the time of year, and so on—which did not tie in with the assumption that these acids were the intermediates of the general photosynthesis pathway. At the same time, Baeyer stressed that he was able to propose a much easier and more direct pathway than Liebig had done: "Indeed, it would be difficult to attain the goal so easily through a gradual synthesis following the other theory!"[37] As a matter of fact, his proposal does seem pretty straightforward if one writes it down as a formula:

$$CO + 2\,H \longrightarrow COH_2 \ (= \text{formaldehyde})$$
$$n\,[COH_2] \longrightarrow [CH_2O]_n \ (= \text{carbohydrates})$$

Put into prose: if carbon monoxide is formed, you only need to add two atoms of hydrogen to arrive at formaldehyde. The latter is already very close to the basic unit of carbohydrates (which is $[CH_2O]$), so that in order to form carbohydrates the formaldehyde only needs to be slightly rearranged and its units multiplied (in condensation reactions); and finally the resulting glyceric aldehyde would be transformed into a sugar—although Baeyer never explicitly discussed this additional complication. In fact, after this short contribution, barely fleshed out on a couple of pages, Baeyer never again returned to the subject matter.

2.2.3 *Testing and Modification*

While the reception of Liebig's model had been rather lukewarm, Baeyer's contribution undoubtedly sparked off a lively discussion. Over the decades between the two proposals the audience for the theme had dramatically multiplied: physiological (and agricultural) chemistry, or biochemistry as it was later called, had become a field of growing interest.[38] In 1861, the first independent institute of physiological chemistry had been founded in Tübingen, Germany, headed by Felix Hoppe-Seyler, who, in 1877, had started the first journal of the field, the *Zeitschrift für physiologische Chemie*. In the first issue Hoppe-Seyler confidently stated that "biochemistry [...] has grown to a science that has not only placed itself on a par with biophysics, but in activity and success competes with it for rank".[39] The importance of metabolic

[37] Baeyer (1870, p. 68).

[38] On the uneasy relation between "pure" and "applied" chemistry (such as, e.g., agricultural and animal chemistry) in the decades around 1900, see Johnson (1990, pp. 25–27).

[39] Quoted in Fruton (1972a, p. 8). On the history of metabolism studies in the early twentieth century, see Holmes (1986). An exemplary reconstruction of the roots of (plant) biochemistry in botanical research is given in Höxtermann and Sucker (1989).

studies was increasingly recognised, and slowly research themes also beyond the processes of digestion, respiration and fermentation, which had been the first to attract the chemists' interest, were being investigated. Owing to these developments, also the processes of photosynthesis received more attention. It was more and more considered a question not only of scientific interest but also of economic importance: first, in view of its value for the fields of agriculture and horticulture but, second, also in view of the fact that a pathway might emerge that enabled scientists to artificially synthesise sugar. Thus, interest in elucidating the photosynthetic mechanism was high, and Baeyer's suggestion was followed by a period of intense investigation that put the models to the test.

However, while Liebig's organic acid model continued to be debated—scientists tried to find out more about the conversion of one acid into another, about the function of acids in plants and about a possible pathway of carbohydrate formation—many of his contemporaries came to regard Baeyer's model as the most promising proposal to explain carbohydrate synthesis, inside and outside the living plant. This was due to the fact that, under certain conditions, formaldehyde was repeatedly found to be formed in artificial systems that contained carbon dioxide, water, and sometimes chlorophyll—among these instances were, for example, the reduction of carbon dioxide caused by magnesium or by silent electric discharge.[40] The conditions in question were usually very different from those predominant in the plant; most of the time, they were, in fact, extremely unfavourable for any life-sustaining process to occur. Yet, the results still seemed to endorse the assumption that there was, in principle, a pathway from carbon dioxide to formaldehyde. This was complemented by findings which demonstrated the occurrence of the second step: the formation of sugars from formaldehyde.[41] The final triumph came when, in 1890, one of Baeyer's former students, the German organic chemist Emil Fischer, succeeded in demonstrating that formaldehyde was, indeed, a possible starting point for the synthesis of the two hexoses, which were thought to be among the major products of photosynthesis (*d*-glucose and *d*-fructose). At the same time Fischer demonstrated that glyceric aldehyde, the possible importance of which Baeyer had already hypothesised, and glycolic aldehyde, which can also be derived from formaldehyde, were potential intermediates.[42] In view of these findings, even the eminent German plant physiologist Wilhelm Pfeffer, who was the author of the standard plant physiology textbook of the time, admitted that the formaldehyde model of photosynthesis was "very appealing".[43]

---

[40] See, e.g., Fenton (1907) (magnesium); Löb (1906) (electric discharge). Among the most influential contenders of these findings are, e.g., Spoehr (1913); Warner (1914); Ewart (1915); Spoehr (1916); Spoehr and McGee (1923).

[41] See, e.g., Loew (1886, 1887, 1888, 1889); Fischer (1888, 1890a); Fischer and Passmore (1889); Euler and Euler (1906a, b) and particular Nef (1910, 1913).

[42] See Fischer (1890b, c). For further discussion of these achievements see also Schroeder (1917, p. 20, pp. 59–60 and p. 67). Fischer was deeply influenced by his teacher's work and explicitly related his study of sugar synthesis to Baeyer's formaldehyde hypothesis. Fischer himself later summarised his achievements in this field, see Fischer (1909, p. 22).

[43] See Pfeffer (1897, p. 339); Pfeffer used the attribute *"sehr ansprechend"*.

Nevertheless, at the same time various aspects of Baeyer's model were being challenged and several modifications were put forward. The possibility of a direct reduction of carbonic acid was discussed; methane was surmised to be an intermediate product between carbon monoxide and formaldehyde; and very soon it was suggested that hydrogen peroxide was also formed in the process, although it was supposed to be immediately removed by the action of the enzyme catalase.[44] Some scientists thought that the reduction of carbonic acid was brought about by hydrogen, either from the decomposition of organic compounds or from a splitting of water by light action.[45] Thus, while many chemists thought Baeyer's idea that carbohydrates were formed through the condensation of formaldehyde was a promising approach, they were not so convinced by Baeyer's assumption that the reduction of the carbon moiety occurred in one single step. They preferred to look for other solutions—possibly incorporating a Liebig-like mechanism via organic acids. Thus, a significant number of scientists started looking for variants that recombined what were considered to be the respective strengths of the alternatives: a highly interesting move in a situation that, according to traditional philosophy of science, would have called for the application of some rational criterion of theory choice.

Figure 2.4 tries to capture schematically the situation up to this point in terms of the goals pursued and the actions undertaken.[46] The chemists' part in the attempt to elucidate the processes of photosynthesis was to find the biochemical pathway (or mechanism), from the raw materials to the end products. The two main problems have already been mentioned: The reduction of carbon dioxide and the synthesis of carbohydrates from the reduced one-carbon units. The two first alternatives, developed by Liebig and Baeyer, were attempts to provide a solution to these problems, which were vigorously debated in the last third of the nineteenth century. Liebig thought that the carbon was reduced gradually in a process that consisted of a number of small steps and which assumed that carbohydrates were formed via organic acids (alternative A in the figure); while 30 years later Baeyer proposed that the reduction occurred in one single step and that carbohydrates were formed via formaldehyde (alternative B). These two alternatives then were being tested and, if considered appropriate, modified. However, at the same time a new goal emerged that required the scientists to combine the strengths of the alternative options, hence, to find a model that included: (1) a path in which carbon dioxide was reduced via organic acids to formaldehyde; and (2) the synthesis of carbohydrates from formaldehyde. In the following decades, a number of suggestions were put forward that tried to meet these criteria, four of which received particular attention at the time and are discussed in

[44] See, e.g., Reinke (1881a, b) (for the direct reduction mechanism); Maquenne (1882) (methane); Bach (1893) (hydrogen peroxide); Usher and Priestley (1906b) and Usher & Priestley (1906a) (catalase).

[45] See, e.g., Pollacci (1902a, b) and Stoklasa and Zdobnicky (1911) (hydrogen from organic compounds); Löb (1906) and Kimpflin (1908) (hydrogen from the splitting of water).

[46] Arrows leading from one level to the other denote the order of hierarchy; potential relevant influences from bottom to top have been neglected for the moment.

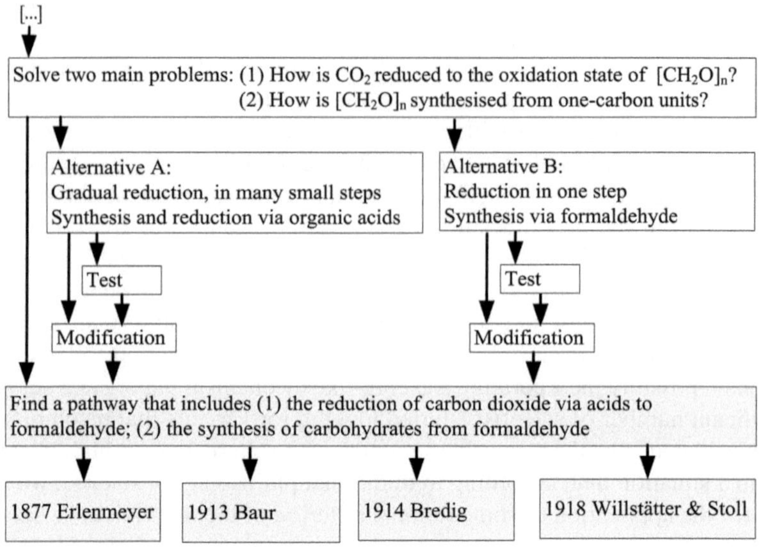

Fig. 2.4 The chemists' goal after 1870: find a model that preserves the strengths, but not the weaknesses, of the two pioneering suggestions by Liebig and Baeyer.

more detail in the following sections: these are the formic acid model, proposed in 1877 by Emil Erlenmeyer; the organic acid-formaldehyde model, advanced in 1913 by Emil Baur; the water cleavage model proposed by Georg Bredig in 1914 and the chlorophyll complex model of 1918, developed by Richard Willstätter and Arthur Stoll.[47] This list of well-known names, which includes the most eminent chemists of the time, makes it clear that the elucidation of photosynthesis had developed into a problem that was being taken seriously, and that many of the period's best chemists were intent on solving it.

2.2.4 The Formaldehyde Problem

However, before the four hybrid suggestions are presented, it may be helpful to clarify the status of these models and their evidence. Photosynthesis researchers of later periods—up to today—have been rather quick to dismiss the nineteenth-century approaches as being based on nothing but speculation. The formaldehyde model, in particular, suffered not only from a lack of positive evidence but also from a wealth of

[47] The titles of these models were not used in the discussion on photosynthesis research in the nineteenth century but dubbed only in Stiles (1925).

negative findings: despite almost innumerable attempts, with the most refined techniques, nobody was able to detect substantial amounts of formaldehyde in the green parts of plants. Not only were the ashes of plants meticulously scrutinised; scientists also attempted to "feed" plants with formaldehyde via the atmosphere or an aqueous medium. But neither approach conclusively demonstrated that formaldehyde occurred in plants or detect that supply of formaldehyde had a stimulating effect on photosynthesis—every positive finding was countered by an equally convincing rejection.[48] Yet, the principal idea that photosynthetic carbohydrate formation went through the stage of formaldehyde remained part of the standard account of photosynthesis for the next decades—it was taken for granted as late as 1938.[49]

It may be tempting to assign the reluctance to drop the formaldehyde model either to psychological immobility or to reverence of Baeyer's authority on part of the scientists of succeeding generations. However, this seems very implausible in view of the fact that the crowd of supporters included scientists such as Emil Fischer, Emil Erlenmeyer and Richard Willstätter, none of whom one would reasonably attest a lack of critical thinking or scientific originality. The problem rather lies in the very nature of the model, which can be conceived of as a complex causal hypothesis. A typical test of hypotheses on causal relevance is an experiment designed along John S. Mill's experimental methods, in particular the "method of difference", in a more elaborate version also known as "difference tests".[50] This method requires to look at differences in two situations: one in which a factor is realised, the causal relevance of which is under investigation; and another in which this factor is absent, while all other relevant circumstances are kept the same. If now the effect only emerges in the situation in which the test factor was present, one can conclude that it was causally relevant for the effect to be brought about.[51] A slightly modified variant of this method was Mill's method of "concomitant variation": effects that were brought about by some key factor of the environment were expected to intensify if the latter factor was present to a higher degree. The various "feeding experiments" attempted by many nineteenth-century chemists were set-up precisely along these lines: given the fact that formaldehyde was thought to be a key intermediate in photosynthetic assimilation, one could investigate whether the rate of photosynthesis changed, when

[48] Experimenters who record the identification of formaldehyde in green leaves after illumination include Pollacci (1902a, b); Grafe (1906); Kimpflin (1907); Pollacci (1907); Gibson (1908), Angelico and Catalano (1913); Chodat and Schweizer (1915). Schroeder (1917) provides a forceful rejection of these findings: he argued that they were either obtained by using flawed methodology or, at the very least, in themselves inconclusive. This perspective finds support by Mazé (1920); Rouge (1924) and Sachs (1862).

[49] Cf. Manning (1938, p. 272).

[50] See on Mill's methods and their application in nineteenth century science, e.g., Scholl (2013).

[51] See on this method of experimental design Graßhoff et al. (2000); Baumgartner and Graßhoff (2004); Weber (2005, 2012); Graßhoff (2011). On the condition of homogeneity given in the two situations, see also Hofman and Baumgartner (2011).

an additional supply of formaldehyde was provided.[52] What was found was that without formaldehyde, the rate of photosynthesis remained normal, which was not surprising; yet, even with a surplus of formaldehyde the rate did not change.

An allegedly obvious conclusion from this experiment would be that formaldehyde had no relevant influence on the rate of photosynthesis and the formaldehyde hypothesis ought to be dropped. However, this would have been a fallacy. It is one of the uncomfortable features of causal reasoning that it is impossible to conclusively infer the *causal irrelevance* of any factor.[53] If the result of an experiment is negative—in the sense that the situations with or without the testing factor do not differ in outcome—then the following conclusions are possible: (1) formaldehyde is, indeed, causally irrelevant; (2) the detection method is flawed or inappropriate; (3) formaldehyde is causally relevant but was unable to exert its influence, because at least one necessary cofactor was not realised in the test situation. From this follows that even if all the aspects of the experimentation were carefully designed, set-up and carried out, one could not conclude from an indifferent result, not even from a consistently indifferent one, that the respective test factor was irrelevant. This explains why the (few) pieces of positive evidence were considered so much more important. Notably Fischer's in vitro experiments of 1890 seemed to provide good reason to believe in the existence of a pathway from carbon dioxide to carbohydrates via formaldehyde. At the same time, of course, chemists felt the need to account for their failure to either detect the alleged key intermediate or demonstrate its relevance. Thus, it was soon agreed among nineteenth-century chemists that formaldehyde was, most probably, processed very swiftly by plants: too swiftly to be captured by the chemists' crude methods. This was made even more plausible by the fact that formaldehyde was such a strong cell poison; surely, plants would have developed mechanisms to prevent it from freely floating about.[54] The Swiss chemist Walter Löb even suggested that formaldehyde was perhaps never actually released as such; he surmised instead that the formaldehyde's constituents (C, OH_2) immediately condensed to sugar.[55]

2.3 Reconciling the First Approaches

While these attempts to account for the formaldehyde problem left many questions open, the majority of chemists still chose to include the formaldehyde pathway into their model suggestions, albeit in a modified version and contextualised in a different

---

[52] This general approach was practised by Baker (1913); Boitreux (1920); Grafe and Vieser (1909, 1911); Jacoby (1919, 1922); Moore and Webster (1913); Sachs (1862).

[53] See Nickelsen and Graßhoff (2011) for more details.

[54] See Schroeder (1917, pp. 8–12). Manning (1938, p. 122) underlined this point, stating that, even if photosynthesis was running at a maximum rate, formaldehyde "concentrations higher than a few hundredths of a per cent are distinctly toxic".

[55] Löb (1906).

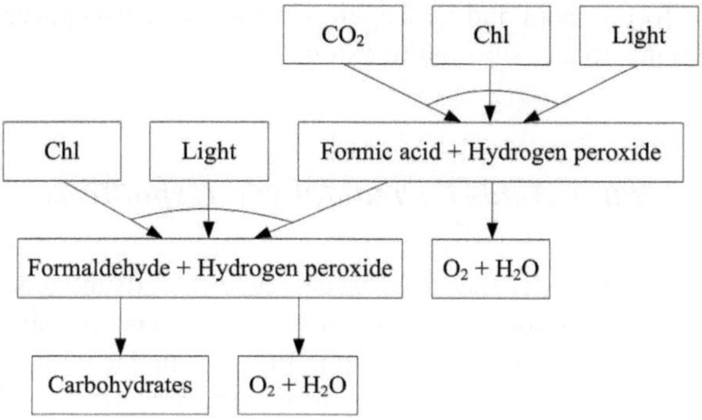

Fig. 2.5 The processes involved in photosynthesis according to the formic acid hypothesis, which was proposed by Erlenmeyer (1877).

manner. The aim of the following sections is to introduce the four most important attempts to reconcile the approaches to the photosynthesis mechanism put forward by Liebig and Baeyer.

2.3.1 The Formic Acid Model

One of the attempts to integrate the advantages of the earlier models by Liebig and Baeyer, while at the same trying to avoid their shortcomings, was the assumption that the first reduction product was formic acid, which scientists believed could be further reduced to formaldehyde—the supposed precursor of carbohydrate formation. The earliest proponent of this point of view was the German chemist Emil Erlenmeyer. He suggested, in 1877, that, in the process of photosynthesis, carbonic acid was first reduced to formic acid; this would yield hydrogen peroxide as a by-product, which would then be immediately decomposed into water and molecular oxygen.[56] A graphical reconstruction is given in Fig. 2.5. This proposal was mainly based on experiments that Erlenmeyer had carried out with glycolic and lactic acids: both these acids were readily decomposed following the pattern described above, so that Erlenmeyer believed he could postulate that carbonic acid reacted in the same way (although he had not been able to test it in the laboratory). As Erlenmeyer wrote, he was convinced that, in view of the ready decomposition of hydrogen peroxide, this path was the most obvious way to explain the liberation of free oxygen in photosynthesis. In subsequent steps, Erlenmeyer assumed that formic acid would, under the influence of light and chlorophyll, be further reduced to formaldehyde and then polymerise to form carbohydrates.[57] (Unfortunately, along with formaldehyde,

[56] Erlenmeyer (1877).

[57] Erlenmeyer (1877, p. 634).

the presence of formic acid and hydrogen peroxide was also never detected in the green parts of plants to any substantial extent.[58])

2.3.2 *The Organic Acid–Formaldehyde Hypothesis*

An alternative hybrid model that tried to combine the advantages of Liebig's and Baeyer's approaches was proposed in 1913 by the Swiss physical chemist Emil Baur.[59] A graphical reconstruction is given in Fig. 2.6. In line with Liebig's hypothesis, Baur argued that it was highly improbable that the reduction of carbon dioxide or, rather, carbonic acid, was accomplished in one single step, as Baeyer had postulated. Considering the respective oxidation states of the carbon atom, Baur thought that several potential intermediates might be formed on the path from carbon dioxide to carbohydrates; and since chemical processes almost always include the formation of intermediates, as Baur pointed out, he preferred to assume that they did, in fact, occur.[60]

Baur was convinced that oxalic acid was the first product of photosynthesis, and that it was produced after the carbon dioxide had interacted with the pigment, which then absorbed and utilised the light energy—two different processes, neither of which Baur discussed in any detail.[61] In the later stages of the gradual reduction of carbonic acid, which, most probably, involved the formation of glycolic and formic acids, oxygen would be released. And, although Baur found it highly improbable that formaldehyde was the first reduction product, he nevertheless believed that the final stage of photosynthesis in which carbohydrates are formed was reached via formaldehyde.

2.3.3 *The Water Cleavage Model*

A third hybrid variant was proposed by the German physical chemist Georg Bredig in 1914,[62] while this suggestion was subsequently supported and elaborated by the (less prominent) chemists Karl August Hofmann and Karl Schumpelt.[63] A reconstruction

---

[58] See Stiles (1925, p. 199).

[59] Emil Baur is not to be confused with the German geneticist *Erwin* Baur, who was one of the three co-authors of a notorious German textbook of genetics and race hygiene in the 1920s.

[60] See Baur (1913, p. 474). To substantiate this point, Baur also cited H. Euler, *Pflanzenchemie*, 1909, 3rd part, p. 183 and 266; as well as his own monograph, *Cosmografia Chimica*, Milan, 1908, p. 207.

[61] Baur (1913, p. 475).

[62] Bredig (1914b); see also Bredig (1914a) and Bredig (1915). See also Czapek (1913), p. 524, for a review of this theory in the standard plant physiology textbook of the time.

[63] See Hofmann and Schumpelt (1916).

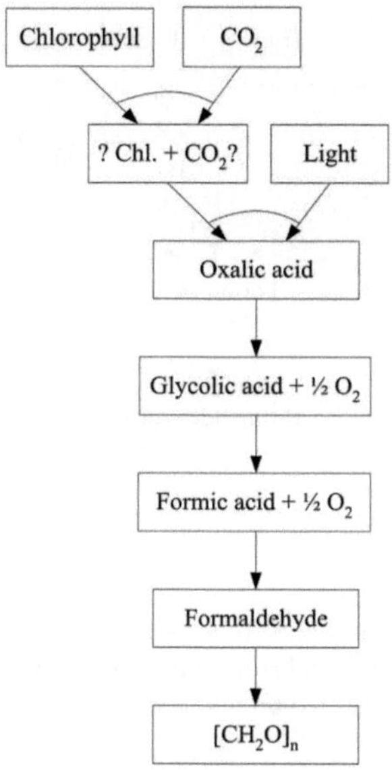

Fig. 2.6 The processes involved in photosynthesis according to the organic acid hypothesis, which was advanced by Baur (1913).

is given in Fig. 2.7. In line with Erlenmeyer, Bredig proposed that formic acid was the first product to be formed, which was supported, he believed, by his recent finding that, under moderate pressure, the salts of formic acid were produced from the salts of carbonic acid in the presence of a surface-providing catalyst, such as palladium. This also tallied with his earlier discovery that, under the influence of surface-providing catalysts, hydrogen was removed from organic substances and transferred to other molecules.

Bredig suggested that in nature the catalytic function could be ascribed to the chlorophyll, while the hydrogen came from water cleavage: it had been observed, after all, that, under the influence of ultraviolet light, water decomposed into an explosive mixture of molecular oxygen and hydrogen (oxyhydrogen gas, in German called *Knallgas*, i.e. detonating gas). In plants, the catalysing agent took the hydrogen from the decomposition of water and used it in the reduction of carbonic acid, whereby oxygen was released. Thus, Bredig was one of the few scientists of these decades to address explicitly the question of the origin of the reducing hydrogen equivalents; and he was one of the few scientists at the time to consider water as a possible source of hydrogen and, at the same time, oxygen. (Walter Löb and some other contemporary chemists also cautiously held this view, which turned out to be correct.) However, Bredig admitted that water might be replaced as the source of hydrogen in plants by other substances that would remove the hydrogen under the influence of sunlight.

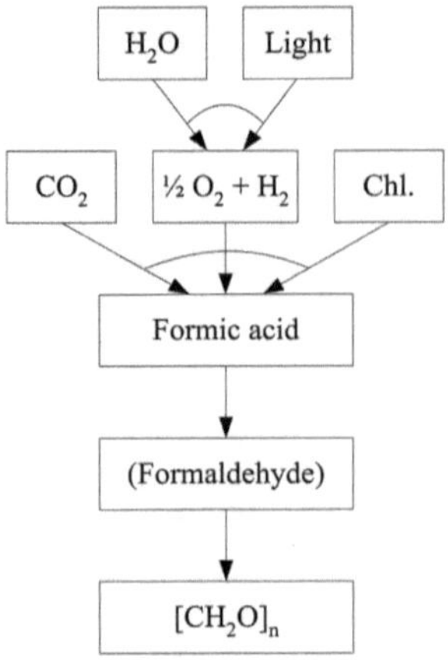

Fig. 2.7 The processes involved in photosynthesis according to the water-cleavage hypothesis, which was proposed by Bredig (1914) and Hofmann and Schumpelt (1916). While Bredig doubted the role of formaldehyde, it was reinserted by his successors; therefore the factor is put in *brackets*.

Bredig seriously doubted the validity of the formaldehyde hypothesis: apart from the fact that there was no convincing evidence that formaldehyde occurred in green plants, he pointed out that "the formaldehyde, in any event, is the one reduction product of carbonic acid, the production of which would require *the highest energy input* by light; and for this reason alone it is not very likely that nature should have chosen this detour".[64] Yet, no alternative explanation was given by Bredig to account for the subsequent stages from formic acid to carbohydrates—so that Hofmann and Schumpelt tried to amend this gap by the suggestion that formic acid was further reduced to formaldehyde (even without the presence of hydrogen peroxide), which then served as a starting point for the synthesis of carbohydrates following the well-known sequence.

2.3.4 *The Chlorophyll Complex Model*

The last model that should be presented in this context is the sophisticated suggestion advanced by Richard Willstätter and Arthur Stoll, which emerged as a result of their comprehensive 1918 monograph on the role of chlorophyll in photosynthetic assimilation.[65] A reconstruction of this model is given in Fig. 2.8. From their experimental findings, Willstätter and Stoll concluded that, once carbon dioxide had found its way into the plant's green cells, the first stage of photosynthesis consisted of a dissociable

[64] Bredig (1914b, p. 363); emphasis in the original.

[65] Willstätter and Stoll (1918).

binding of the gas to unknown, organic constituents of the plant cell, presumably plant proteins or amino acids. It was by this means that the concentration of carbon dioxide within the plant cells was thought to increase, so that photosynthesis operated more efficiently. (After all, the concentration of carbon dioxide in the air is rather low.) Willstätter and Stoll surmised that the carbon dioxide was probably chemically altered in the course of these events, converted either into carbonic acid or into one of the latter's derivatives. The product of this absorption process, which was considered to be a purely chemical, light-independent process, is symbolised in the figure as the (CO_2*–Cell complex), where the asterisk indicates that the original carbon dioxide was added to this complex in a modified form. Willstätter and Stoll believed that the modified carbon compound was then passed to the chlorophyll. From their experimental findings, they postulated that an additive compound of the bicarbonate type was formed by chlorophyll and either carbonic acid or one of its derivatives. This intermediate substance is symbolised in the figure as [Chl–CO_2**], indicating that the original carbon dioxide had undergone a second conversion. The actual photochemical step of the process was believed to be the chemical rearrangement of the carbonic moiety of this intermediate, additive product into an isomer, which was higher in energy and was then reduced in the process that followed (indicated in the figure as [Chl–CO_2***]). Willstätter and Stoll assumed that this product was a kind of peroxide, most probably formylhydroperoxide.

It was thought that the further decomposition and reduction of this compound was effected by an enzyme, which Willstätter and Stoll assumed was the mysterious "living cell" factor (see above). They surmised that, in the course of this catalysed reduction process, oxygen was released and formaldehyde synthesised, and that subsequently carbohydrates were produced via condensation reactions: we have seen this pattern in most of the models discussed so far. However, Willstätter and Stoll found it highly probable that no intermediate product was released before the reduction process of the carbon had been completed, so that the stages of this reduction process were very hard to establish. In Fig. 2.8, the result of these reactions is, therefore, symbolised only vaguely by the expression [Chl + 'C' + O_2], which is the last intermediate before the final carbohydrate stage $[CH_2O]_n$ is reached.

In contrast to the other scientists mentioned so far, Willstätter and Stoll considered chlorophyll to be the central factor in the whole process of photosynthesis: in addition to its capacity to absorb light and make it chemically available, chlorophyll was also assumed to be the actual *site* of carbon reduction, which involved the formation of an intermediate additive compound. The action of light was thought to be effective only in interaction with this additive compound; that is, the light did not act on the chlorophyll, as one might assume, but on the carbonic moiety, which was thereby converted into one of its isomers. This differed sharply from the widely held view at the time that chlorophyll only acted as a sensitiser in photosynthesis, in that it transformed rays of shorter wavelengths into more efficient rays of longer

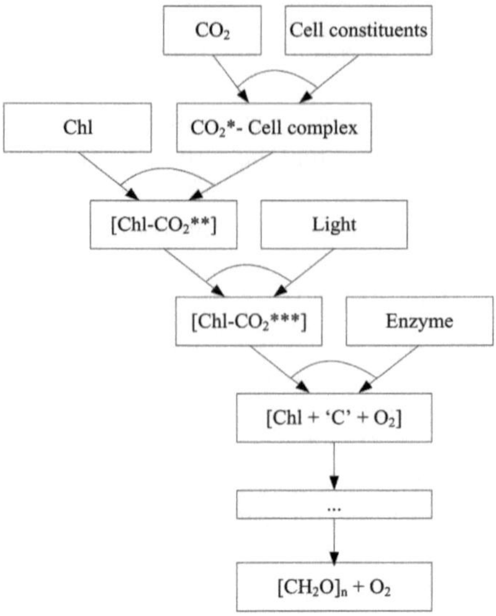

Fig. 2.8 The processes involved in photosynthesis as conceptualised by Willstätter and Stoll (1918).

wavelengths—although it was unclear which effects these rays would then bring about.[66]

2.4 Features Common to the Models

The previous sections had the purpose to introduce the range of models that were under debate by 1918. Despite their divergences in terms of content, all these models shared a number of structural features. The aim of the following sections is, therefore, to spell out these common features, reflect on the type of mechanism that was being suggested and the evidence that was invoked. This will provide us with a reasonable basis from which then the chemists' strategies and methodology of model building in this period will be discussed.

2.4.1 Simplicity and Simplification

The most obvious feature that all the photosynthesis models had in common was their simplicity: none of them included more than a very limited number of factors and

[66] See, e.g., Czapek (1913, p. 614), for a review and emphatic endorsement of the "sensitiser" position.

reaction steps. The conscious simplification of the process to be explained is a typical, well-known and even indispensable aspect of the construction of scientific models.[67] In the case of photosynthesis, it is striking, for instance, that almost all the scientists limited their choice of "root" factors, that is, the raw materials but also the chlorophyll and other preconditions, to those of the previously discussed one-step model. There were only two exceptions: *first*, Liebig, who postulated the involvement of the vital force and of some alkaline bases, both of which were immediately dropped by his successors; *second*, Willstätter and Stoll, who believed that an unknown enzyme was an additional root factor, which, they assumed, catalysed the final reduction steps. Along with the root factors, the actors also limited the range of end products under consideration—these were taken to be quite definite, as one could read in a review of 1916 by the plant physiologists Ingvar Jörgensen and Walter Stiles: "The substances which are known to be produced as a result of carbon assimilation are oxygen and carbohydrates".[68] It is remarkable that hardly anyone played with the option that some of the many other compounds which by then were known to be synthesised in plants came out of photosynthesis. The widespread argument for this restriction was that the "assimilatory coefficient" of plants, that is, the quotient of carbon dioxide absorbed and oxygen released, in most cases approached unity, which corresponded well to the assumption that only molecular oxygen and carbohydrates were produced. The formation of, for example, proteins or fats would have favoured different coefficient values. However, one could have constructed scenarios that might have included the formation of a mixture of substances, which would still have produced an assimilatory coefficient of unity. It seems that for those parts of the process that were reasonably established even the chemists were inclined not to introduce additional speculations. (As we shall see in later sections, they were far less scrupulous when it came to the modelling of intermediate stages of the process).

In addition to the selection of factors also the number of reaction steps was minimised to the extreme; this holds true even for the most sophisticated variant by Willstätter and Stoll. And, finally, all the chemists in this chapter shared the implicit assumption that the same mechanism of photosynthesis operated in all species of higher plants. None of the actors mentioned above considered the possibility that in the leaves of trees, for example, the process of photosynthesis might differ from the process of photosynthesis in the leaves of grass. In his widely read textbook on photosynthesis of 1917, also the plant physiologist Heinrich Schroeder believed that one

[67] This has been underlined in many accounts of models and model building heuristics, such as Bailer Jones (2009) or Frigg and Hartmann (2012), although the latter rather speaks of "idealization", which partly, but not fully coincides with the "simplification" in our case.

[68] Jörgensen and Stiles (1916a, p. 186). One may want to add, however, that a wide range of different carbohydrates was considered as potential photosynthetic products, including polysaccharides, disaccharides, hexoses and pentoses; see, e.g., the synopsis given in the cited review. The authors also maintained: "We do not wish it to be supposed that we [...] support the view that glucose is the first sugar of carbon assimilation. We hold that the data so far produced from analyses of carbohydrates in leaves and from microchemical examination provide insufficient evidence in favour of or against either theory".

could safely assume that the same mechanism was valid for all higher plants, because "given the same raw materials (CO_2 und H_2O), the same products are found, while the circumstances as well as other resources of the transformation are also the same".[69] Given the many morphological and physiological differences that were known to distinguish different species of plants from each other this declaration of like circumstances was not entirely beyond doubt. (In fact, as it turned out much later, in the 1960s, the assumption was inaccurate: grasses do have a different photosynthesis mechanism).

Resorting to assumptions like these, however, was inevitable. The mechanism of photosynthesis was so complicated that none of the scientists working around 1900 could possibly have hoped to achieve more than a rudimentary understanding of it. (Although they unquestioningly shared the principal assumption that there was a linear, decomposable pathway to be elucidated.[70]) In view of this situation, they seemed to have agreed that it was inadvisable to make things even more difficult by taking into consideration more factors than was absolutely necessary, not to mention possible detours of the pathway or variants in different plant species. Methodologically speaking, the chemists (and even most plant physiologists) shared the working hypothesis that photosynthesis acted as a mechanism of the same type in all higher plants; thus, if the causal relationships were established in one instance, the same should be true of all others. "Construction assumptions" of this type, as I would suggest to call them, were an essential part of the modelling process: for the sake of constructing the model, the chemists assumed that the process had certain properties, even if they knew, or strongly suspected, that this was not actually the case (which certainly was true for the number of reactions steps involved). Whether these assumptions were deemed permissible or not, however, was dependent on the context—in the aforementioned, plant physiological review of 1916 by Jörgensen and Stiles, for instance, the authors expressed their astonishment that most of the chemists so easily restricted the range of internal factors to the plant's chlorophyll: recent findings in plant physiology had demonstrated, they thought, that there were significant influences connected with the properties of the cell and their structural environment that deserved closer inspection.[71] The contrast between the different communities will become even more conspicuous in a later section of this chapter, when I turn to the heuristic strategies that chemists utilised in their work.

---

[69] Schroeder (1917, p. 58).

[70] See Bechtel (1986a) on the underlying conception of metabolism, which was the dominant view up to the 1930s.

[71] Jörgensen and Stiles (1916a, p. 176).

2.4.2 *Incomplete Empirical Support*

The second common feature of the chemical models that I wish to highlight is the fact that all the photosynthesis models were dramatically underdetermined by empirical data. In stark contrast to the conservative attitude towards introducing new root and end factors to their models of the mechanism, the chemists in the period under study apparently did not mind postulating completely new *intermediate* factors. This is true, for example, not only of all the assumptions about the products of the gradual or immediate reduction of carbon dioxide but also of the formation of a complex made of chlorophyll and either carbon dioxide (Baeyer) or an unknown derivative of carbonic acid (Willstätter and Stoll). None of the actors seemed too concerned about invoking either unheard-of reaction mechanisms or the occurrence of reverse reactions the existence of which was yet unknown. Still, none of these models was discarded on the grounds of incomplete experimental foundation—at least not by their chemical colleagues; while the plant physiologist Hermann Spoehr sarcastically commented this situation in a review, when he wrote: "It can safely be said at the outset that, when critically considered from a physiological view point, none of the existing theories [on the mechanism of photosynthesis] is even moderately well established by observations of facts".[72]

Take the formaldehyde model as an example. Baeyer's assumption that carbon dioxide formed a complex binding with chlorophyll was exclusively based on the observation that chlorophyll was structurally similar to haemoglobin; and since the latter was known to bind carbon dioxide, the former was assumed to do the same. All inherent differences were silently considered irrelevant. On the other hand, the hypothesis that formaldehyde was formed in living leaves and that the formaldehyde molecules subsequently combined to larger units was based on the observation of an artificial system, that dramatically differed from the conditions in real plants. Likewise, the evidence for the formic acid hypothesis was solely based on the observation of test tube experiments, which were then transferred, by Erlenmeyer and others, to specific life processes. This type of reasoning was omnipresent: the chemists tended to back up their models with evidence that had been taken from observations outside the organism.[73] Even Willstätter and Stoll who actually measured the gas exchanges of living material used empirical data that were unrelated to their physiological observations when it came down to the mechanism. Again, this was a practice which appears to have been perfectly acceptable among the chemists, but was sharply criticised by plant physiologists: "So long as our knowledge of the heterogeneous system in which these [processes] take place is so incomplete, it is impossible to draw conclusions from experiments in which the conditions are clearly so different", Jörgensen and Stiles underlined in their review.[74]

---

[72] Spoehr (1916, p. 2).

[73] The in vitro vs. in vivo debate has vexed biochemistry ever since. See Jacob (2002) for a philosophical perspective on this point.

[74] Jörgensen and Stiles (1916b, pp. 79–80).

2.4.3 Selected Focus and Modules

Closely related to the principle of simplicity is the third observation which I would like to draw attention to: all the chemical models, to a greater or lesser extent, were inherently inconsistent in their level of detail and explanatory scope. All of them treated at length one specific aspect or one particular process of photosynthesis, while other, equally relevant, sequences of events were merely summarised or even totally ignored. Take, for instance, Liebig's model. The key feature was the assumption that the formation of carbohydrates occurred through the formation of organic acids that became increasingly poor in oxygen and rich in hydrogen. Liebig failed, however, to explain how these acids were thought to be then converted into carbohydrates. This neither implies that he considered this reaction step unimportant, nor that he thought that this step had been adequately treated in the model. The synthesis of carbohydrates simply lay outside Liebig's focus of investigation. Baeyer's focus was the introduction of formaldehyde as the central intermediate on the path of carbon from carbon dioxide to carbohydrates. He also hypothesised about the complex of chlorophyll and carbon monoxide, yet remained silent on all the details. Willstätter and Stoll concentrated on the first stages of photosynthesis, which involved the function of chlorophyll, although when it came to the actual carbon reduction they vaguely postulated that some intermediate substance, which was peroxide in nature, was formed, and that it would eventually pass through the stage of formaldehyde and polymerise, as in Baeyer's model.

To some extent, incompleteness is a consequence of simplification: if one chooses to model only those aspects of a problem that are considered of central importance, one will inevitably present an incomplete account of the whole—yet, from this perspective one would expect the same level of simplification for all parts of the mechanism, while simplification alone does not explain the fact that selected aspects were treated in considerably more detail. Additionally, one might argue that different models frequently are constructed to cover one and the same phenomenon in view of different aspects of the phenomenon and different explanatory functions of the model. This is well-known to have happened in physics quite regularly, for example, in the case of the wave–particle dualism, when more than one model was found to be necessary to cover the different aspects of the phenomenon.[75] However, in the case of photosynthesis the function and intended scope of the models were fairly similar to each other, yet the scientists still presented widely divergent accounts. It is significant in this context that the chosen focus of research corresponded so closely to the authors' general knowledge and skills: Willstätter and Stoll, for example, were experts in the field of chlorophyll research, while they had only a limited knowledge of the chemistry of carbohydrates. I shall come back to this observation in a later section of this chapter.

[75] See on this case Wheaton (1983). Daniela Bailer-Jones coined the term "*Aspekthaftigkeit*" ("aspectivity") for this wide-spread feature of models; Bailer-Jones (2009).

2.5 Model-Building Heuristics

In summing up the observations made so far, it would seem that the process of modelling the mechanism of photosynthesis had taken a quite peculiar path: even the most promising models dealt with a highly simplified notion of photosynthesis; all the models were strikingly incomplete and focused on selected aspects only; and, finally, none of the available models was more than partially supported by empirical evidence, while even the available data were mostly gathered from artificial systems, which had little in common with the living plant. It is easy to understand that these deficiencies were sharply criticised by plant physiologists. However, within the group of chemists that form the focus of this chapter these procedures seem to have been part of the generally accepted practice. In the following sections I shall try to extract from the available evidence some of the principles of model-building heuristics that nineteenth-century chemists in search of a photosynthetic mechanism applied.

2.5.1 Extending a Prototype

At the beginning of the chapter, I gave an outline of the body of generally accepted knowledge of photosynthesis. The standard raw materials and end products of the process had been well established by earlier generations of scientists and were never seriously questioned—this is why I referred to this early, phenomenological description of the process as the "prototype model", although as such it was never defended by one particular person. One then could interpret the photosynthesis models introduced in this chapter as extensions and modifications of this first model into different directions.[76] This does include local condensation of model suggestions, as, for instance, the dropping of the vital force and the alkalis as factors from Liebig's model. Baur's version, for example, at the same time extended Liebig's proposal (by adding the synthesis of carbohydrates via formaldehyde) and modified it (by introducing a different sequence of acids). The important observation here is that the modelling always took off from existing knowledge that served as a starting point.[77] Yet, according to which lines of reasoning was this prototype extended? Given the fact that the actual empirical knowledge was scant, owing to a lack of appropriate methods, other criteria and strategies had to be followed. One of the most important among them was the transfer of (mostly: causal) knowledge from one subject to another, which the following section is concerned with.

---

[76] See Graßhoff (1998) for a discussion of how models can be extended, condensed or modified at the example of the modelling of the astrophysical object SS433.

[77] This is a typical procedure for any kind of causal reasoning; see on this point Graßhoff (2011).

2.5.2 *The Transfer of Causal Knowledge*

"With the development of organic chemistry it was but natural that attempts should have been made to explain the mode of reduction of carbon dioxide and the course of sugar synthesis in the green leaf", the plant physiologist Heinrich Spoehr wrote in his review of 1916. However, in doing so "some have fallen into the error of reasoning, that if they can produce sugar, or the substances closely related to sugar, from carbon dioxide and water, by almost any means, that this is necessarily also the process taking place in the leaf".[78] Spoehr found the chemists' practice to base their photosynthesis models on evidence that emerged in the test tube unacceptable. Yet, data that were gathered from the plants themselves were unavailable: more direct access to the intermediate steps of the process only became possible much later, with the introduction, in the late 1940s, of radioactive tracer molecules in metabolic studies. Thus, one could either drop the theme altogether, that is, stop thinking about the photosynthetic mechanism until more appropriate methods were developed; or try to transfer the established knowledge of biochemical reactions from test-tube situations to the photosynthesis processes in the living organism. The former was the plant physiologists' advice; the latter was what the chemists chose to do.

Take Baeyer's procedures as an example: around 1870 the polycondensation reactions of formaldehyde were the focus of his research; and he knew, from Butlerov's investigation, that, under certain circumstances, a syrup-like substance was a product of this type of reaction. Baeyer knew, too, that carbohydrates were structurally composed of a series of molecular units [CH_2O], which were identical to formaldehyde in terms of atomic composition; and although they differed in their structural arrangement, the conversion of one into the other did not seem too difficult. A similar rationale explains how Baeyer came up with the earlier stages of his model. It was well known at the time that, under certain circumstances and under the influence of strong light, carbon dioxide could be reduced to carbon monoxide. Starting from here, Baeyer then assumed that this carbon monoxide would initially form a complex with chlorophyll. In 1870, nobody knew which complexes chlorophyll was able to form because chlorophyll had proven very elusive and impossible to isolate.[79] Yet it was well-known that chlorophyll was structurally similar to haemoglobin, and that the latter easily bonded carbon monoxide. Thus, Baeyer's complex hypothesis was based on the (chemically speaking: quite reasonable) assumption that molecules which are structurally similar undergo the same chemical reactions. The resulting transfer, in this case, closely resembles Darden's notion of the instantiation of a mechanism scheme.[80]

---

[78] See Spoehr (1916, pp. 1–2).

[79] "In the early years of the twentieth century the literature of the pigments of the chloroplast thus showed that knowledge of them was in a very confused state"—this is how the British botanist Walter Stiles characterised the state of affairs in his textbook on photosynthesis of 1925; cf. Stiles (1925, p. 19). Crucial parts of the long-winded history of chlorophyll research are described in Höxtermann (1991).

[80] Darden (2002).

The underlying heuristic assumption in these inferences—from haemoglobin to chlorophyll; from the test tube to the organism; and, as Schroeder explicitly stated: from one plant to another (see quote above)—can be put in a more general form, stating that similar effects under similar circumstances can be taken to be brought about by similar causes and mechanisms. In this form, it is easily recognisable as the content of Sir Isaac Newton's second Rule of Reasoning, which he formulated in his *Principia Mathematica* (1687). (Newton presents his "Rules of Reasoning" at the beginning of the Third Book of the *Principia*. The first rule states: "We are to admit no more causes of natural things than such as are both true and sufficient to explain their appearances"; to which the second rule then adds: "Therefore to the same natural effects we must, as far as possible, assign the same causes". While the first rule defines a minimising approach to the set of factors to consider, the second rule additionally postulates a minimum of explanatory schemata or mechanisms to account for natural phenomena.) Of course, this strategy is fallible; but it has a prolonged and well-established history as a useful heuristic principle. The same assumption that the (unknown) processes inside the organism should operate according to the principles that were known of the processes that occurred outside the organism had governed chemical investigation into life processes since the time of Lavoisier and others, who, for example, conceived of respiration as an ordinary combustion process.[81]

The resulting way of thinking can be categorised as "analogical reasoning", in line with Mary Hesse's classic work on models and analogies: a source system (i.e. processes in the test tube) is explored in order to get to know more about the actual target system (i.e. the organism) which is structurally similar to the source system but might not be accessible to the same degree.[82] Philosophers of science have been quite unenthusiastic about analogies and their epistemic value—not only because they are by necessity uncertain, but also because the term is so vague. However, the "analogical reasoning" in this case can be explicated by framing the situation in terms of the underlying causal reasoning: the chemists' crucial assumption was that the process under study (the reactions steps of the photosynthetic mechanism) fell into the same class of events as other, already well-known processes—that is, for example, the reactions of chlorophyll were taken to be the same as the reactions of the structurally similar haemoglobin. As the relationship of causal relevance in chemistry (and elsewhere) concerns *types* of events and their regularities, and not only individual *tokens*, the grouping together of processes into the same class implies that they are part of the same causal relationships.

Of course, this assumption hinged on the fact that not only input and output but also the conditions in the two cases were sufficiently similar. This was exactly the point most sharply criticised by plant physiologists. The aforementioned Spoehr specified his comment to the chemists' work by emphasising that, in general, they "show a lamentable lack of knowledge of the conditions under which photosynthesis takes

[81] See Holmes (1985); Underwood (1943).

[82] Cf. Hesse (1963). For a recent treatment of analogical arguments, see, e.g., Gamboa (2008) and, in particular, Bartha (2010).

place, as well as of the physiology and structure of the chlorophyllous cells".[83] In order to make up for this lack of knowledge, but, at the same time, account for these differences, chemists had introduced additional factors for modelling the processes in the cell. Liebig had turned to "vital forces" to explain how photosynthesis operated, while Baeyer assigned a special function, perhaps of a catalytical nature, to the material constituents of the cell. These factors filled the explanatory gap (as did the assumption that water was one of the raw materials), while they did not satisfy the plant physiologists' critique. The latter asked for more than merely the causal factors; they required an explanation that referred to the cell structure and the interaction of several cell components: a *mechanistic* model that went beyond the mere existence of causal links in a chain or network.

2.5.3 The Building Block Strategy

Transferring causal knowledge from one situation to another was, to some extent, connected to a second equally widespread heuristic strategy. When one examines the different photosynthesis models it is clear that, although they diverge, most of them were not completely disparate. Rather, one finds "modules" that reappear in different combinations, such as the "formation of a chlorophyll-carbon dioxide complex", the "reducing of carbon dioxide via organic acids", the "formation of carbohydrates from formaldehyde", and so on. These were frequently interpreted as functional subunits, resulting from the (conceptual) decomposition of the of the mechanism. In the graphical notations they may correspond to one "branch" of the graph (or to one section of a longer branch).

Perhaps the most striking example of this practice is the assumption that carbohydrates were formed through the condensation of formaldehyde, which was integrated into most of the later models. Similarly widespread became the assumption, first formulated by Erlenmeyer, that a formic acid derivative and some peroxidic compounds were involved in the process, the decomposition of which gave rise to the photosynthetic oxygen. This module constantly reappeared in later model suggestions, even though the rest of the models did not resemble Erlenmeyer's original concept in other respects. Willstätter and Stoll, for example, ingeniously recombined this module with the chlorophyll complex and the formaldehyde module, added some causal hypotheses from their own field of expertise, notably the enzyme hypothesis, and thus presented a completely new amalgamation of ideas that had, in fact, been around for decades. I shall refer to this practice as the "building block strategy": the chemists carefully examined their predecessors' results and then integrated into their own work whatever they found useful and acceptable. In this respect, the chemists' procedure resembles In this case, the strategy resembles Lindley Darden's notion of

[83] See Spoehr (1916, p. 2).

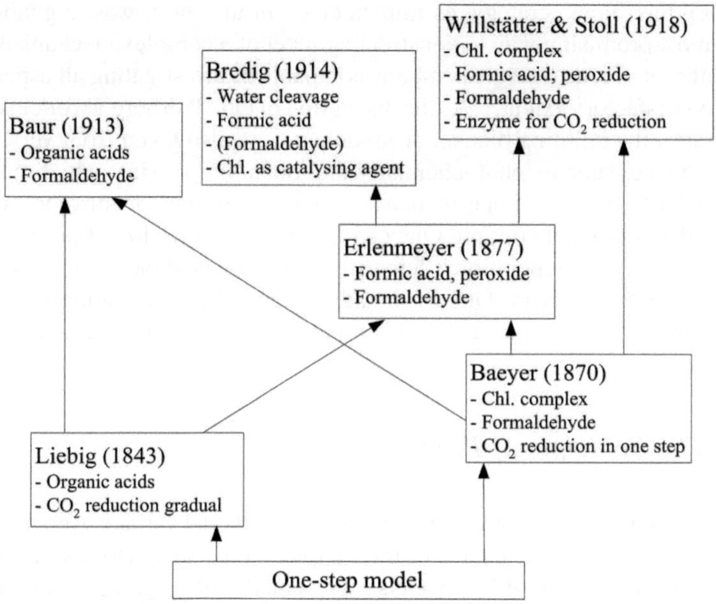

Fig. 2.9 The models and their "building blocks" shown in sequence.

"modular subassembly": one proceeds from known or suspected components of the mechanism under study and tries to rearrange them in a fruitful manner.[84]

Figure 2.9 is a schematic representation of the models discussed in this chapter and shows which modules were reused. The arrows from *x* to *y* denote the relationship "modules of model *x* were integrated into model *y*". The one-step model is taken as the starting point for the two first major alternatives that emerged in the nineteenth century: those proposed by Liebig (1843) and Baeyer (1870). Seven years later, in 1877, came Erlenmeyer's suggestion, which was influenced, on the one hand, by Liebig's proposal of a stepwise carbon reduction via acids (although Erlenmeyer believed that formic acid was the central intermediate, something that Liebig had not mentioned), and on the other hand by Baeyer's suggestion, as Erlenmeyer also included carbohydrate formation via formaldehyde in his model. In 1913, Baur published his proposal, which also attempted to synthesise the different approaches; this was followed in 1914 by the model advanced by Bredig, who, like Erlenmeyer, favoured the path via formic acid (and was rather sceptical about formaldehyde); finally, in 1918 the Willstätter–Stoll model was put forward, which was influenced both by Erlenmeyer's and Baeyer's suggestions particularly the latter's assumption that a complex of chlorophyll and carbon dioxide was formed.

[84] Cf., e.g., Darden (2002); Darden and Craver (2002).

The building block strategy, as implemented in this case, was, arguably, at the time the most promising way to construct a model of a complex mechanism such as photosynthesis, since no single person was capable of investigating all aspects of the process to satisfactory extent.[85] As the discussion of Otto Warburg's work in chapter 3 demonstrates, the building blocks, or modules, were also taken from very different areas of science, such as photochemistry and quantum physics, which one person alone could not possibly hope to master: using "plug-ins" from other scientists' work was the only viable option. One can interpret this division of labour as a very impersonal form of "cooperation": scientists complemented their own accounts with the achievements of others. One of the advantages of reformulating the chemists' models in the form of graphs is that this aspect model dynamics becomes obvious.

2.5.4 *The Principle of Plurality*

Figure 2.9 could give the impression that the earlier model variants were replaced by succeeding ones during the course of this chapter's time span. However, this was by no means the case. None of the models represented in the figure was dropped until well into the 1920s—not even Liebig's and Baeyer's earliest, original ideas. After 1918, the model that Willstätter and Stoll proposed was considered by many scientists to be the most promising option—not least, because it successfully integrated a fair portion of the modules under discussion. Yet, it certainly had its weaknesses too,[86] so that none of the concurrent alternatives was completely abandoned. Every now and then selected modules of the earlier models were revived and re-examined so that at any point of time a range of different suggestions were debated. I call this practice the "principle of plurality", which is a typical feature of ongoing modelling processes.

The strategy to keep as many models as possible under investigation (even those that are thought to be less promising than others) is well-founded if one regards the model-building process as a cooperative enterprise of the whole community. The philosopher of science Philip Kitcher explored these situations from a more formal, normative perspective. The question he rose was whether there are "conditions under which, in light of our goals as an epistemic community, we ought to want to maintain cognitive diversity". Kitcher differentiated between personal and impersonal epistemic intentions, which, at times, might be in conflict: under certain conditions it could be rational for someone "to assign herself to the working out of ideas that she (and her colleagues) view as epistemically inferior".[87] Examples that seem to point in this direction include, according to Kitcher, Wegener's suggestion of continental

[85] Comparable practices were revealed in the construction of botanical illustrations (i.e. representations of species models) around 1800, where elements from earlier images were extensively copied but at the same time modified and adapted to the new context. See Nickelsen (2000, 2006a, b).

[86] A fair number of these were pointedly exposed in Jörgensen and Stiles (1916b, pp. 84–90).

[87] See Kitcher (1990, p. 6 and 8).

drift which, initially, was discarded by the majority of geologists, while, luckily, some individuals continued to advocate and explore it—up to its "revival" in the form of plate tectonics. Hence, Kitcher argued, although the continuous pursuit of Wegener's theory might have been an irrational decision from the point of view of the individual, it turned out for the best from the point of view of the community. However, Kitcher thought that this required the existence of a special type of "altruistically rational agents" in science: "Altruistically rational scientists are those who are prepared to pursue theories that they regard as inferior when, by doing so, they will promote achievement of the goals of their own (and their colleagues') impersonal epistemic intentions". And Kitcher immediately conceded that this "raises an even more bloodless ideal of scientific rationality than that criticised by historians and sociologists of science".[88]

However, in the situation that I sketched earlier, there was no need for any individual scientist to deliberately devote herself to a clearly inferior theoretical approach. Although there were degrees of preference to one or another option, it was impossible to determine conclusively whether carbon reduction was achieved through a series of intermediates, which possibly included organic acids of one kind or another, or whether carbon dioxide was directly converted into formaldehyde or some other compound with the same oxidation state in a complex binding with chlorophyll. Concurrently pursuing alternatives, or, as Kitcher put it, stick to a broad cognitive division of labour, as long as some uncertainty prevailed (which in science is more often the case than not), seemed a reasonable course of action, without assuming any unrealistically high degree of altruism among the chemists. The philosopher of science Miriam Solomon even made a normative postulate of this, claiming that "consensus is not normatively appropriate unless theories show clear and substantial differences in degree of empirical success".[89] Under given circumstances, the plurality of approaches was not only advantageous for the community of photosynthesis research as a whole but also a useful strategy from the perspective of the individual actors: the chemists working in the period under review were struggling to establish causal relationships on the grounds of very insecure data, based on highly fallible assumptions. It was not at all improbable that an outsider module might prove, in the course of time, to be the better horse to bet on (and we know today that the water cleavage hypothesis was exactly of this kind).

It may be helpful to remember that, at the time around 1900, it was not even remotely plausible that any one of the model alternatives under discussion was "accurate" in the full sense of the word. The actors were keenly aware of this situation and repeatedly pointed to the fact that, for the time being, only preliminary hypotheses were brought forward. Willstätter and Stoll's model had some explanatory value, from the chemists' point of view, and explained relevant sets of data with recourse to established chemical knowledge; but it still seemed rather odd to assume that

[88] Both quotes: Kitcher (1990, p. 9, Footnote 5).

[89] Solomon (1994b, p. 339).

highly poisonous compounds such as peroxides and formaldehyde could be regular intermediates of this fundamental process of life. On the other hand, given the body of knowledge of chemistry at the time, the occurrence of the whole process of photosynthesis appeared highly unlikely, if not virtually impossible: both water and carbon dioxide were known to be extremely stable, inert molecules, so that it was hard to believe that they were able to decompose at all at room temperature. Yet, since the actual existence of photosynthesis could not be disputed, the chemists felt entitled to invoke even improbable mechanisms to explain the process.

2.6 Collective Versus Individual Goals

It was mentioned earlier that each of the models discussed in this chapter had a specific focus: that is, each of them was particularly detailed in some respect, while other parts were treated more superficially. Thus, none of the models was intended to grasp the biochemical pathway of photosynthesis in all its complexity. On closer inspection, these differences in focus can be explained by turning to the chemists' individual interests and skills. It is striking that none of the actors discussed in this chapter studied the process of photosynthesis for any lengthy period of time. Rather, all of them made only a limited contribution to the field, which was frequently presented in one single paper, and they then moved on to other concerns: more precisely, they returned to their original, main, research goals. I will elaborate on this aspect in more detail in this section.

2.6.1 Photosynthesis as a Side Issue

I shall start by taking another look at Liebig. At the time that he formulated his model, Liebig was involved in the general (and rather ambitious) project to explain *all* agricultural processes from a chemical point of view, not only in order to gain fundamental knowledge but also because he had specific applied and utilitarian purposes in mind. Among other things, Liebig was concerned with enhancing crop production and, thus, of ensuring adequate food supplies.[90] This was the context in which Liebig started to think about photosynthesis. From this larger perspective, it is clear that he regarded the task of "explaining photosynthesis" at best as a lower-level goal: if one wanted to enhance crop production, for example, by developing an efficient fertiliser (which was one of Liebig's objectives) or by advising farmers how to grow their plants, it was obviously advantageous to have some knowledge about photosynthesis, the source of all plant growth.

[90] See Brock (1997, chapter 6).

The motivation behind Baeyer's contribution was briefly mentioned earlier. Around 1870, Baeyer was studying condensation reactions, including the condensation reactions of formaldehyde; and his formaldehyde model of photosynthesis was directly related to this work. Indicative of this relation is also the title of the paper in which the model was published: "On dehydrogenation and its meaning for the life of plants and for fermentation processes".[91] Chemically speaking, the "dehydrogenation" of compounds is one of the effects of those reactions that, following Baeyer's suggestion in the paper, came to be called "condensation reactions". Indeed, Baeyer's famous and influential model of photosynthesis only appeared in this paper to illustrate one of the types of condensation reactions that Baeyer was investigating. Another type of these reactions, according to Baeyer, was central to fermentation processes. Thus, the modelling of photosynthesis was not even one of his subgoals. Rather, Baeyer probably realised that he could make a contribution to photosynthesis research only while he was working on the more general phenomena dealt with in the paper. One may call this an "incidental goal", since it was the by-product of work done while attempting to attain other superordinate goals. Typically, incidental goals of this type are very limited and specific, and can quickly be reached on the basis of immediately available knowledge and skills. Having reached the incidental goals (that is, in the case of Baeyer, having published his thoughts on how the condensation reactions of formaldehyde might help to explain photosynthesis), Baeyer immediately went back to his original work; and even though his photosynthesis model was the subject of protracted discussion, Baeyer himself never again returned to it.

The other cases examined in this chapter were very similar. Erlenmeyer was, around 1877, primarily working on the structural elucidation of organic molecules and on the chemical properties of double and triple bonds. In the one paper of relevance here, photosynthesis was mentioned only in passing, as a possible application of Erlenmeyer's thoughts on "Water as an oxidising and reducing agent" (as the title of the paper reads).[92] In 1913 on the other hand, Baur published a whole series of papers on the topic of photolysis, including the photolysis of formaldehyde. This presumably led him to think about other problems related to photoreactions and formaldehyde, including photosynthesis. As in Baeyer's case, it did not cost Baur much in terms of resources to make a quick contribution to photosynthesis before returning to his original line of research. Bredig, meanwhile, was particularly interested in the effects of surface catalysis, and since he assumed that chlorophyll acted as a surface catalyst (like palladium or platinum), he followed this approach when framing his photosynthesis model. The latter was not an obvious topic of interest for Bredig, who was, after all, a physical chemist. The findings he presented provide some indication as to how he chanced upon the subject: while working on formates (the salts of formic acid), Bredig had found that, under the influence of a catalysing

[91] Baeyer (1870, p. 64). The original German title of the paper reads: "Über die Wasserentziehung und ihre Bedeutung für das Pflanzenleben und die Gährung".

[92] "Das Wasser als Oxydations- und Reductionsmittel"; see Erlenmeyer (1877).

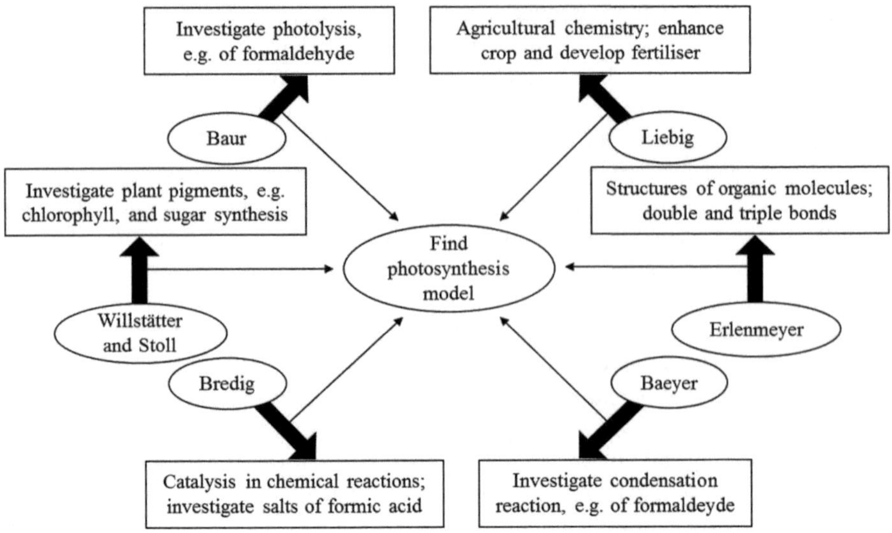

Fig. 2.10 The actors and their goals: diverging individual (superordinate) goals; and the subgoal or incidental goal of finding the photosynthesis model. The *thick arrows in bold* indicate the relationship "*X* pursues the superordinate goal *y*". The *thin arrows* indicate that, in the course of pursuing the superordinate goal, an incidental goal or, less frequently, subgoal (of contributing to finding the photosynthesis model) was reached.

agent, they could be oxidised to the salts of carbonic acid. Thus, an obvious conjecture to make from this was that this reaction might also work in reverse, that is, reduce carbonates to formates, and that this could then be assumed to be one of the processes involved in photosynthesis (Fig. 2.10).[93]

Finally, Willstätter and Stoll had already been working on the chemistry of plant pigments, in particular chlorophyll, for more than ten years when they published their model (they had, for example, established that magnesium was an integral part of chlorophyll and they had been the first to find a practicable method for isolating chlorophyll from plants).[94] It is not too surprising then that this work prompted them to make contributions to the mechanism of photosynthesis—and neither does it come as a surprise that the focus of their model was the role of chlorophyll. Yet, even then, "finding the photosynthesis model" was evidently not their first and primary goal, as the authors made clear in their introduction:

> Even though our experiments may contribute to describing more precisely the processes of [carbon dioxide] assimilation, at the same time they clearly give a negative answer to the question as to whether it is already possible to realise this assimilation outside the living cell. It is too early for experiments of artificial assimilation under the influence of chlorophyll.

[93] See Bredig (1914a, b, 1915).

[94] See their first monograph: Willstätter and Stoll (1913).

> This is not really a negative conclusion, it is a positive finding that ought to inspire and point the way to new work.[95]

This rather defensive-sounding paragraph (in which the authors appear to be trying hard to raise their spirits in view of their failure) reveals that Willstätter and Stoll's primary goal was to realise artificial photosynthesis: to find a way of reducing carbon dioxide to carbohydrates in a light-driven reaction under the influence of chlorophyll, but without the rest of the plant cell. And it was only in order to reach this goal that Willstätter and Stoll found it necessary first to clarify how the process operated in plants. "Finding the (natural) photosynthesis mechanism" was for them, too, a subgoal in their quest to achieve artificial photosynthesis.

2.6.2 Constructive Research Opportunism

Thus, it seems that the work carried out on photosynthesis by nineteenth-century chemists, as it was presented in this chapter, can be interpreted as a by-product (or spin-off) of the scientists' work on other projects. At some point in their research, Baeyer and Erlenmeyer among others seem to have realised that, based on what they had achieved so far (in other topics), they could easily make a contribution to photosynthesis research. I shall refer to this behavioural pattern in this study, with a slight twist of irony, as the principle (or maxim) of "research opportunism".[96] Despite the term's usual negative connotation (which is by no means intended here) it nicely captures that this maxim is about taking advantage of situations if the opportunity arises. The underlying rationale seemed to be something like: "Contribute to solving an interesting research problem, whenever you can do so without being distracted for too long from the pursuit of your principal goals". The maxim can marvellously be illustrated by an episode remembered by one of the giants of twentieth-century photosynthesis research, William Arnold, whom I shall return to in later chapters of this book:

> In June [1950], Dr Bernard Strehler came to the [Oak Ridge National] Laboratory [in Tennessee]. Strehler had a brand-new Ph.D., a tremendous amount of energy, and lots of ideas about almost everything. One day he appeared in the laboratory door and said, "Arnold, how would you like to make one of the fundamental discoveries in plant physiology?" My answer was: "OK, if it won't take too long".[97]

This collaboration thus started between Arnold and Strehler would contribute to the discovery that adenosine triphosphate (ATP) is synthesised not only in the mitochondria but also in the chloroplast (for details, see Chapter 7). Here as well as in other instances, it seems that scientists are quite willing to make a contribution to an open question if the opportunity arises, even though they might never develop

---

[95] Willstätter and Stoll (1918, preface, pp. III–IV).

[96] In this I am following a suggestion by Gerd Graßhoff, given in a personal communication.

[97] See Arnold (1991, p. 79).

more than a passing interest in the pertinent subject. They might provide explanatory approaches, introduce new methods or apply concepts from their actual area of research; but soon afterwards, they often return to their original field.[98] At first glance, Willstätter and Stoll seem to have been exceptions to this rule, since they spent so many years working on questions related to photosynthesis. However, if one looks at their case a little more closely, it is clear that most of the time they studied the structure of chlorophyll and its behaviour in different circumstances. It was only in the final chapter of their second book that Willstätter and Stoll turned to the mechanism of photosynthesis—which they only examined in order to find out how to synthesise sugars in the test tube (see above).[99] The "opportunistic" way of picking research problems also explains why Baeyer never turned to photosynthesis again, even though he left so many problems unsolved; and the same is true of Liebig, Erlenmeyer, Baur, Bredig, Hofmann and Schumpelt. Other eminent scientists who, at the beginning of the twentieth century, made a single contribution to photosynthesis research from their own field of expertise include Emil Fischer (see earlier in the chapter), Felix Hoppe-Seyler, Walther Nernst, Walter Noddack, Jacobus H. van't Hoff and Fritz Weigert. It was not that these men deliberately wanted to hold things back; rather, with the one shot they made they had published all they had to say on photosynthesis. Given the limited resources in terms of research time, infrastructure and money, pursuing an incidental goal is only worthwhile from the actor's perspective if it can be reached quickly and with a minimum of additional effort. This explains why all the contributions discussed in this chapter had so different foci (as was explained earlier) which closely corresponded to the chemists' individual, superordinate goals.

Thus, photosynthesis research around 1900 provides a prime example of research opportunism. Although many people were interested in photosynthesis and contributed some findings to explaining the process, hardly anybody at the time made the subject their centre of interest. As was mentioned earlier, the whole field of chemical metabolism studies was highly problematic at the time, since there were no methods through which one could gather direct information on the course of the pathways. Data were scarce and the interpretations thereof disputed. It would have been highly unreasonable to make a theme in which success was so uncertain the sole focus of one's research. Contributing to the subject in an "opportunistic" manner was the best option, both for the individual scientist, whose costs were limited while the potential gains were high, and for the collective of the chemists as a whole, since this was, after all, a viable option for keeping the topic alive until more adequate methods became available.

[98] See Graßhoff (1998) for a case study from astrophysics in which he also discusses this phenomenon.

[99] Prompted by findings that Stoll made, however, Willstätter and Stoll both contributed once again to the field around 1932; see Stoll (1932) and Willstätter (1933).

This moment came with the rise of manometry in photosynthesis research after 1920, which fundamentally changed the field. The cumbersome and largely speculative search for potential carbonic intermediates lost its former attraction (although the approach had its comeback in the 1940s, when radioactive tracer molecules became available; remarkably, still then formaldehyde was among the products that chemists tried hard to find). Yet, even though it then became principally possible—and reasonable—to concentrate one's efforts on photosynthesis, which by this time was no longer considered a high-risk field of study with almost no hope of success, the subject remained a side issue in plant physiology (and other disciplines). The set of people that spent their professional lives on photosynthesis studies remained limited. In addition to the specialists, there always remained a substantial number of "opportunistic" contributors, in the sense outlined in this section. It will transpire in the following chapters of this book that both parties played an important role in solving the problem of how photosynthesis worked.

Chapter 3
Otto Warburg and the Turn to Manometry (1912–1925)

In this chapter the focus shifts from the collective attempts to explain the mechanism of photosynthesis to a close-up of a particular individual, namely the German cell physiologist and biochemist Otto Warburg. His contributions to the field mark a turning point in twentieth-century photosynthesis research: Warburg introduced a number of revolutionising new techniques to measure the rate of photosynthesis (which resulted in the move to kinetic studies of the process), he put forward a new model of the mechanism, and added a completely new perspective to the subject by attempting to establish the efficiency of the process in terms of the minimum quantum requirement of photosynthesis.

The example of Warburg illustrates the enormous potential of research opportunistic behaviour in the sense that was introduced in the previous chapter. The application of one experimental technique, measuring metabolic processes using manometric methods, dominated virtually Warburg's entire career and, beginning with his seminal work on cell respiration, he was constantly searching for the effect of metal-containing enzymes acting on internal cell surfaces. With this limited range of concepts and techniques, Warburg was able to contribute, at the highest level and with great success, to fields as diverse as cell respiration, photosynthesis and cancer research. In this chapter I shall trace the "investigative pathway", a term that was coined by Frederic L. Holmes[1], that brought Warburg to make his contributions to the field of photosynthesis research, including the question which sources he used to compose his model hypothesis and why Warburg chose to measure the quantum efficiency of the process. The general objective of this analysis is to gain a better understanding of the reasons why individual actors pick out certain subgoals within a field of study and how their specific background shapes the contributions they make.

To this end, three different sources of inspiration are explored: Warburg's early research into cell respiration; his father's work on the quantum yield of photochemical reactions in general; and the way in which Warburg reacted to the photosynthesis work carried out by Richard Willstätter and Arthur Stoll (which was summarised in chapter 2). How Warburg ingeniously availed himself of fragments taken from these

[1] See Holmes (2004).

K. Nickelsen, *Explaining Photosynthesis,* History, Philosophy and Theory of the Life Sciences 8, DOI 10.1007/978-94-017-9582-1_3

contexts and recombined them in a new and innovative way is an instructive example of the successful use of conceptual and methodical building blocks.[2] I shall begin by outlining Warburg's career up until 1920, focusing on those details that are of most significance for the aim of this chapter.

3.1 New Materials and Methods

3.1.1 Otto Warburg (1883–1970)

Otto Warburg was one of the most successful and influential biochemists of the twentieth century.[3] Born in a middle-class German family of partly Jewish origin, his father was the experimental physicist Emil Warburg, one of the most eminent scientists of his time, who had converted to protestantism before Otto was born. In 1905, after having held various academic positions (in Strasbourg, Freiburg (Breisgau) and Berlin), Emil Warburg was appointed as the President of the renowned *Physikalisch-Technische-Reichsanstalt* (PTR) in Berlin, where he remained for the rest of his working life, that is, until 1922. This Berlin-based appointment would have a momentous influence on the career of his first child and the only son.

Otto Warburg had a typical middle-class German education, attending a humanistic *Gymnasium*, before going on to study chemistry at Freiburg (Breisgau) in 1901. After having spent some time there, Warburg moved to Berlin, where he continued his studies in the laboratory of the organic chemist Emil Fischer (the same Fischer who had contributed in establishing the path of carbon to sugars via formaldehyde; see Chapter 2). While working in Fischer's laboratory, Warburg earned his doctoral degree in chemistry in 1906.[4] In the years before, Warburg had received additional training in his father's laboratory at the PTR, where he became familiar with, among other things, the vacuum bolometer, an apparatus devised by Emil Warburg to

[2] See also Nickelsen (2009) on this point.

[3] For general accounts of Otto Warburg's life and work, see Krebs (1979); Henning (1987); Höxtermann and Sucker (1989); Werner (1991) and Höxtermann (2001). Selected parts of his sister's personal notes were published in Rüskamp (1989). Warburg's contribution to the theory of cell respiration, as reflected in his correspondence with the physiologists Jacques Loeb, Leonor Michaelis and Otto Meyerhof is treated in Werner (1996), while Kohler (1973a) investigates the background of Warburg's concept of *Atmungsferment*. On Warburg's experimental methods in photosynthesis, see also Hoppe (1997a, pp. 19–20).

[4] It has not escaped the notice of Warburg's biographers that, in his botany examination, Warburg demonstrated only "satisfactory" knowledge of "carbon assimilation", unlike the excellent results he received in all his other subjects. Clearly, the young Warburg had not yet developed a passion for what would later become one of his main research themes. See Höxtermann and Sucker (1989, p. 21), for a facsimile of the exam's documentation; a transcription can be found in Werner (1991, p. 24). The original document is preserved in the Archives of the Humboldt University of Berlin (Phil. Fak. No. 411, folio 210).

measure light intensities.[5] This was one of the measuring instruments that Otto Warburg would later use in his research work on the quantum yield of photosynthesis.[6]

Warburg could then have embarked upon a career as a chemist. Instead, he chose to broaden his education by studying medicine at Heidelberg, with, among others, the well-known physician and physiologist Ludolf von Krehl. He earned a second doctorate in medicine in 1911, and in 1912 attained his habilitation. Warburg stayed with Krehl for 1 more year, and it was at Heidelberg that he began his successful research work on the processes of cell oxidation. Warburg's findings in this field were to bring him his first major breakthrough as a scientist in his own right. And they were to have lasting consequences: Warburg was awarded the Nobel Prize in Physiology or Medicine in 1931, largely because of the studies in cell oxidation, which he resumed in the 1920s.

In 1913, at the age of 30, Warburg returned to Berlin, having being appointed head of his own research department (his first such appointment) in the newly founded Kaiser Wilhelm Institute (KWI) for Biology in the Dahlem district of the city.[7] However, since the institute's new building was not completed in time, Warburg had a gap to fill between leaving Heidelberg and starting at the KWI. He kept himself busy first by working again in the PTR's radiation laboratory, where he undertook some photochemical work, and then in the physico-chemical laboratory of Walther Nernst at Berlin's Friedrich Wilhelms University, where Warburg apparently worked on the oxidation potentials of living cells.[8] Both these topics would lay the groundwork for his later research into photosynthesis.

Warburg's first years in Berlin were interrupted by the outbreak of the First World War in 1914. He immediately volunteered and started serving in the Prussian Horse Guards, who were involved in activities near Germany's Eastern Front. Although Warburg remained in service until 1918, he returned to Berlin before the official end of the war—presumably due, at least in part, to a letter he had received from Albert Einstein, urging him to return home.[9] Warburg apparently agreed to this suggestion, and so his father and the plant physiologist Carl Correns, Warburg's superior at the KWI for Biology, entered upon a lengthy correspondence with the Ministry of the Interior, requesting Otto Warburg's release. In addition to citing general scientific reasons, both Emil Warburg and Correns stressed that Otto Warburg's release would benefit the public, since, they argued, the resumption of his research work would very likely yield results that would help improve the population's nutrition (most

[5] See, e.g., Warburg et al. (1907) and Warburg (1909). On the PTR's history, see Cahan (1989).

[6] For an autobiographical account of this period, see the taped interview with Warburg of 1966, quoted in Krebs (1979, pp. 94–95).

[7] On the early history of the society, see, among many other publications, Vierhaus and Brocke (1990); on the KWI for Biology, see also Sucker (2002).

[8] See Werner (1991, pp. 75 and 113).

[9] Albert Einstein wrote a letter to Warburg, on the initiative of the latter's mother, in which he tried to convince Warburg that he was more urgently needed in Berlin than at the Front; Warburg, it seems, was won over. This letter can be found in Schulmann et al. (1998, pp. 694–697; Nos. 489 and 491). The letter is also transcribed in Krebs's biography of Warburg, which also shows part of it in facsimile; see Krebs (1979, pp. 20–23).

probably alluding to work on photosynthesis in algae).[10] Eventually, the request was successful and in October 1918 Warburg resumed his research in the by-now-completed laboratories of the KWI.

Among the first papers that Otto Warburg published, after having returned to Berlin, were his articles on photosynthesis, the most important of which were two closely related papers, in which he dealt with the general mechanism of photosynthesis;[11] and another two papers on the efficiency of the process.[12] The advances Warburg made in these papers were enormous. He fundamentally changed the field by introducing a number of new techniques that were quickly to become standard practice in photosynthesis research and remained so until the 1970s. These included the use of manometric methods for measuring the rate and the progress of photosynthesis; and to fully exploit the advantages of this new technique, Warburg also replaced the use of leaves and whole plants with the unicellular green alga *Chlorella*, which to this day is a well-known experimental organism in photosynthesis research.[13] In addition, Warburg also employed sophisticated photophysical techniques, such as bolometry, absorption measurements and intermittent illumination by means of rotating sectors, which required not only special skills but also specific instrumentation. He was also the first to use inhibitors systematically in order to discover more about the biochemical process of photosynthesis. From the results of his research, Warburg proposed a mechanism that involved the formation of a "photolyte", a concept that he adopted from contemporary physics, denoting substances that are decomposed by photolysis—indeed, it was his father Emil who had introduced this concept to science.[14] Finally, Warburg brought a new perspective to debates of the period by determining the energetic efficiency and, as a consequence, the minimum quantum requirement of photosynthesis—one of the few parameters at the time that limited the range of possible model alternatives. In view of the enormous impact of these new elements, it seems worthwhile to inspect them in a little more detail.

3.1.2 *Manometry*

Before Warburg introduced manometric methods to photosynthesis research, techniques were employed to determine gas exchanges with sensitivities measured in the

[10] See Werner (1991, pp. 121–122). During this period, German politicians were considering unconventional sources of nutrition in an attempt to counteract the nation's growing problem of undernourishment; one suggestion was to follow the Japanese example of exploiting marine algae, similar to those that Otto Warburg later used in his photosynthesis experiments. Other suggestions of the committee, which was led by Emil Fischer and had been set up to deal with the problem, included using reed or couch grass.

[11] Warburg (1919, 1920b).

[12] Warburg and Negelein (1922, 1923).

[13] See Zallen (1993a) for a thoughtful discussion about the use of *Chlorella* algae (and others) as experimental or even model organisms in photosynthesis research.

[14] See Warburg (1917).

range of millilitres of gas. This meant that, in order to be able to measure the minimum oxygen level in experiments, large areas of plant material had to be illuminated for a relatively long period of time, and one had to use large samples or even whole plants. Warburg's method, by contrast, offered an enormously increased sensitivity, capable of measuring *micro*litres of gas exchange.[15] Consequently, one could not only greatly reduce the sample size and the duration of experiments but also utilise smaller and more manageable light beams, all of which led to a far better control of the whole experimental set-up. The latter was Warburg's main incentive also in other situations, when he turned to designing new techniques and instruments in order to monitor biological processes. "If one finds appropriate reactions specific for the cell component which one wants to analyse, the rest of the cell is part of the test tube", Warburg maintained.[16]

Warburg had started his manometric studies with a basic manometer, which his father Emil had devised in 1900 to measure the velocity of ozonisation processes.[17] However, in March 1912, during a brief visit to the laboratory of the physiologist Sir Joseph Barcroft at the University of Cambridge(UK), Warburg became familiar with a much more sophisticated version of the instrument, which he modified even further for the analysis of either very thin slices of living tissue (for his studies in respiration) or cell suspensions (for his research in photosynthesis).[18] Figure 3.1 shows one of Warburg's own manometers, together with the specific vessel or glass flask that had to be used with it, while Fig. 3.2 is a sketch of the complete measuring device, the "Warburg apparatus".

The principal procedure was straightforward: a suspension of unicellular algae, which had been grown under controlled conditions, was poured into the flasks, which were then connected to a manometer. This combination of manometer and flasks was then mounted on a thermostat (the flasks facing the interior of the thermostat, the manometer the exterior) in order to keep the temperature of the suspension at a constant value. In this position the vessels were shaken to achieve homogenous conditions for all the cells at any time. Illuminating these vessels (with measured light intensities; in this case from below) initiated the different processes of photosynthesis, which gave rise to the evolution of molecular oxygen. The effect—namely an increase of pressure in the system—was measured by noting the changes in the height of the capillary fluid in each manometer. A set of rather simple equations could then be employed to calculate the amount of oxygen produced, taking into account the relative change in manometer fluid, the relationship between the molecules of

---

[15] See Myers (1974, p. 420).

[16] This was Warburg's reply, when in 1928, after a talk on the spectrophotometric analysis of the heme molecule, he was harshly criticised by Willstätter for the application of spectrophotometry to such complex structures as cells. Quoted in Nachmansohn (1972, p. 5).

[17] See Warburg (1900).

[18] A comprehensive account of Warburg's early manometric techniques can be found in his book on tumour metabolism, Warburg (1926). Kok (1960) provides an overview of how the techniques Warburg used in his photosynthesis studies developed over time. For a brief and very accessible introduction, see also Allen (1975, pp. 173–174).

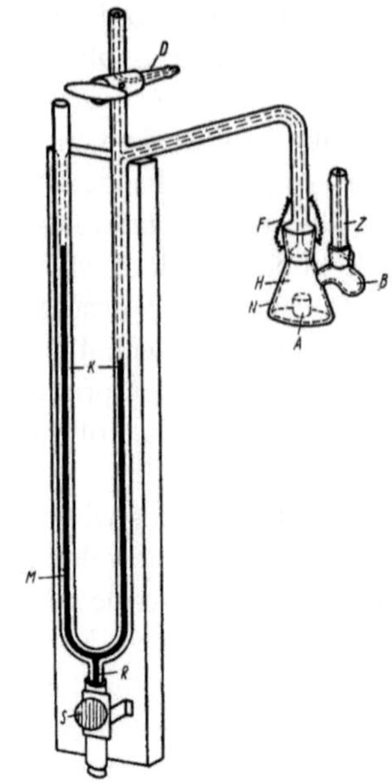

Fig. 3.1 An illustration of a manometer used by Otto Warburg in his photosynthesis experiments. (Reproduced from Warburg (1926, p. 1)).

oxygen evolved and the pressure produced as well as other constants related to the vessel. Almost all of Warburg's path-breaking studies were carried out using this technique, which he continued to develop throughout his career. Its simplicity and its broad range of applications soon made it part of the standard apparatus of every physiological laboratory.

3.1.3 Chlorella

The use of manometric methods led Warburg to reconsider the test organisms suitable for the study of photosynthesis. There were some major disadvantages in using the leaf tissue of higher plants, in addition to the difficulty of ensuring homogenous illumination (see above). First of all, heavy diffusion was a problem. As leaf tissue slices always have several cell layers, the oxygen produced in the interior cells had to diffuse through a great many other cells before it entered the suspension. This resulted in significant inaccuracies in the measurement of the gas exchanges. Second, the parameters of the micro-environment of the interior cells could not be controlled: neither temperature nor carbon dioxide pressure inside the tissue could be taken to be the same as in the rest of the vessel. In his search for alternatives, Warburg finally was successful: "After some preliminary trials I kept to a round immobile green alga

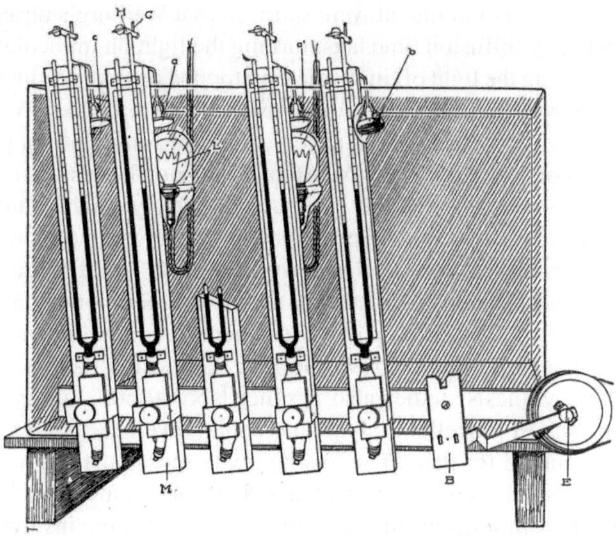

Fig. 3.2 A drawing of the complete measuring apparatus (which became known as "Warburg Apparatus"). The manometers are mounted on a thermostat, so that the vessels can be illuminated from below by light bulbs. A V-belt connected to an electric motor, part of which can be seen on the right of the illustration, oscillates the manometers. (Reproduced from Warburg (1919, p. 245)).

of 3 to 5 micrometres in diameter, which multiplies by successive fission without developing clusters or movable cells, similar to an alga described as '*Chlorella*' in the literature."[19]

As an experimental organism, *Chlorella* had many practical benefits. It is relatively easy to grow in large quantities and the alga's chloroplast occupies half the cell volume, which means that a large proportion of the plant material used is photosynthetically active and the yields are relatively high.[20] For Warburg, however, the main advantage was of a methodological nature. At that time, unicellular algae were the smallest organisms known that were capable of carrying out the full photosynthesis process.[21] (Chloroplasts, which would later become the preferred living structure for testing, were not isolated before the 1930s; and even then most scientists doubted whether the whole array of photosynthetic reactions could be carried out in the chloroplasts). The small size of the algae meant that the paths between the reaction

[19] Warburg (1919, p. 231). Quoted also, originally in German, in Werner (1991, p. 148). It is highly probable that Otto Warburg was helped in his choice of organism by the phycologist Ernst Georg Pringsheim, who was one of the leading algae experts of the time. In an autobiographical account, Pringsheim reported how he was approached by Emil Warburg to participate in photosynthesis experiments being carried out at the PTR. Pringsheim declined the offer but he was still the obvious person to turn to in search for an appropriate single-cell model organism. Pringsheim (1970).

[20] These advantages were still emphasised by Manning (1938, p. 120), Footnote 3.

[21] See Zallen (1993b, pp. 271–273).

sites in the cell and its environment were short, so that Warburg's observations were no longer affected by diffusion time lags: turning the light on immediately produced oxygen, while turning the light off immediately stopped oxygen production. This was a pre-condition for the use of flashing light experiments, in which Warburg studied the effects of light and darkness given at very short intervals. The temperature of the cell interior was practically identical to the temperature of the suspension and, finally, the experiments could be carried out using comparatively small quantities of light: in an algal suspension, light readily penetrates the cells without being absorbed by non-photosynthesising regions or reflected away by a surface. Thus, using *Chlorella* enormously increased Warburg's control of the main experimental parameters, such as temperature, the gaseous and liquid environments, and light intensity.

In the years that followed, *Chlorella* became the most popular experimental organism for photosynthesis studies also in other laboratories—although there were many other species of unicellular algae that could have been used as well.[22] Two main factors accounted for this. *First*, Warburg was one of the internationally leading scientists on his field. Visitors from all over the world came to his laboratory, and then took to using *Chlorella* when they returned to their home institutions. In addition, some of the period's most influential photosynthesis researchers were trained in Warburg's institute, including Robert Emerson and Charles Stacy French, who spread the technique across the USA. *Second*, most of the crucial experiments, in particular those experiments undertaken during the course of the controversy on the maximum quantum yield of photosynthesis (which is the subject of chapter 5 of this book), were carried out on *Chlorella* cells, so that a lot of knowledge on the behaviour of this alga accumulated over the years. Its "representational scope", which Rachel Ankeny and Sabina Leonelli defined as the range of living beings for which an experimental organism was taken to be exemplary, was extremely broad—in fact, for some decades, until people became aware of the fact that there was a variety of photosynthetic pathways, it covered the whole of the plant realm. *Chlorella's* "representational target", however, that is, "the phenomena to be explored through the use of the experimental organism", mostly remained photosynthesis. Initially, *Chlorella* algae were also used for the investigation of other life processes; however, in the course of further research it became obvious that *Chlorella* algae had metabolic peculiarities that were far from generally spread even among freshwater algae. Hence, *Chlorella* never acquired the status of a "model organism", such as *Arabidopsis*, *Escherichia coli* or the mouse.[23]

Notwithstanding these difficulties, *Chlorella* has remained an experimental organism to this day—mostly because of the large amount of information collected

[22] See on this point also Zallen (1993b).

[23] For further reflection on the representational scope and target of an organism and on the distinction between "experimental organisms" and "model organisms", see Ankeny and Leonelli (2011). The quote was taken from p. 315.

about its organismic properties.[24] Starting from the late 1930s it was found that the physiological state of the algae and their growth history strongly influenced their photosynthetic performance. This meant that it was extremely important to grow the same strain of algae under the same (favourable) standard circumstances if one wanted to maintain the experimental conditions as homogenous as possible. Once appropriate culturing conditions had been defined for *Chlorella*, a task which turned out to be far from easy, many scientists were reluctant to change the experimental object again. Any modifications made, either to the conditions or to the organism itself, would have required a lengthy investigation into the comparability of the new situation with the established experimental standard.[25]

3.1.4 Buffer Solution

Using unicellular algae for manometric experiments required finding a more sophisticated solution to the problem of keeping the carbon dioxide concentration of the experimental setting at a constant value. Usually one would have turned to using carbonate–bicarbonate buffers, which were part of the standard equipment of laboratories at the time. However, because of their (slight) alkalinity, the standard buffers of this type, Warburg thought, were potentially harmful to the algae. Therefore, he developed a new buffer with an almost neutral pH and an extremely low carbon dioxide concentration (consisting of 15 parts one mole solution of Na_2CO_3 and 85 parts one mole solution of $NaHCO_3$).[26] It was only later, in his studies of the quantum yield of photosynthesis, that Warburg started to use acidic, phosphate-containing buffers.

3.1.5 Bolometry

In order to control the light intensities of his photosynthesis experiments, Warburg employed sophisticated bolometry, which he had learned to use in his father's

---

[24] This does not, of course, mean that all the scientists working in photosynthesis research exclusively used *Chlorella*. The species of *Euglena*, *Scenedesmus* and *Porphyridium* were also used as experimental organisms, while in the field of genetic engineering *Chlamydomonas reinhardtii* became the standard; see Zallen (1993b, pp. 275 ff).

[25] On the broader question of how model organisms are being used in research, see, e.g., Creager et al. (2007): a collection of illuminating essays on the "model systems approach" as the editors call it, that is, the practice of using organisms that have become standard test objects, or of referring to case studies that have acquired exemplary status within a discipline. The editors argue that model systems, such as in this case *Chlorella*, are used as "models for" something if the system under investigation is too complex to come up with a "model of" it. The aspect discussed here, how the choice of standard organisms enhances the reliability of causal inferences, complements the analyses given in the volume.

[26] See Werner (1991, p. 148).

laboratory at the PTR. This was rather exceptional: at the time most biologists, and even chemists, were not familiar with this technique. A bolometer is a relatively sensitive device for detecting and measuring radiation intensities. The instruments in use around 1900 consisted of a Wheatstone bridge, the two branches of which were connected to very thin (0.0025 mm) strips of metal (e.g. steel, platinum, palladium), a battery and a galvanometer for measuring electrical currents. When one of the two metal strips was exposed to radiation, the metal heated up, which increased its electrical resistance. Consequently, the galvanometer could detect a certain voltage between the two parts of the system, proportional to the amount of radiation energy incident on the metal. Emil Warburg improved the device further when he developed the vacuum bolometer that was mentioned earlier. By 1907, bolometers could thus detect temperature changes of 0.00001°C—which, of course, exceeded by far the degree of precision that was sensible to ask for when working with living organisms.[27]

3.1.6 Rotating Sectors: The Flashing Light Technique

A second non-standard technique (for biological experiments) that Warburg also had got to know through his father was the use of "rotating sectors" and, hence, intermittent illumination. To this effect, a disc with one or more sections was placed between the light source and the algae, so that part of the light could be screened off (see Fig. 3.3). Rotating sectors were standard instruments in the field of photophysics and, therefore, regularly used in the optical laboratory of the PTR, where Warburg carried out many of his early experiments in photosynthesis.[28] This technique enabled Warburg to study photosynthesis under conditions in which the light reactions limited the velocity of the process, as will become clear in a later section of this chapter.

3.1.7 Inhibitors

Warburg was the first to use biological inhibitors as a means to find out more about the biochemical mechanism of photosynthesis. In particular, he investigated the effect of anaesthetics, such as the different urethanes, and of hydrogen cyanide. Warburg already knew the inhibiting effect of these substances from his studies in respiration:

---

[27] See Warburg et al. (1907) for the description of the vacuum bolometer; cf. also the popular account of this instrument, at Anonymous (1907b).

[28] See Warburg (1919, pp. 235 and 255). Warburg also seems to have been inspired to use rotating sectors by the work of the English plant physiologist Horace Brown and his collaborator Francis Escombe; see Brown and Escombe (1905, p. 38). Warburg acknowledges Brown's influence in Warburg (1919, p. 263).

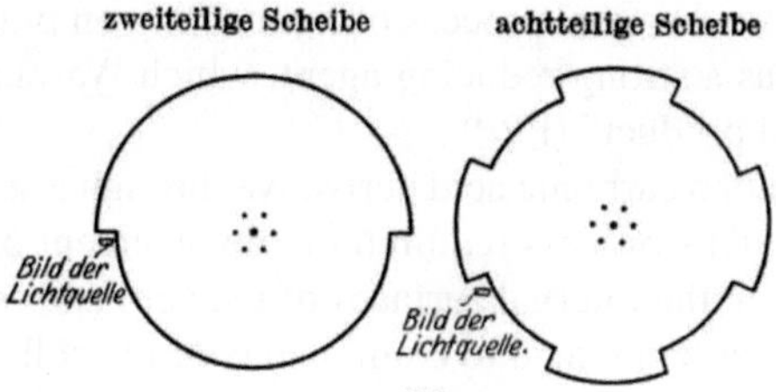

Fig. 3.3 A drawing of Warburg's rotating sectors in Warburg (1919, p. 251), Fig. 9. Two different variants are depicted: on the left, a disc divided into two halves (i.e., two periods: one dark, one light) is shown, while the disc on the right is divided into eight sections (resulting in four dark and four light periods each). The tiny rectangle to the left of each disc marks the position of the light source (*Bild der Lichtquelle*).

urethanes were known to inhibit processes dependent on internal cell surfaces, while hydrogen cyanide blocked the haemoglobin's site of oxygen binding (at the iron component). If the process of photosynthesis, or some of its components, were also inhibited by these substances, one could then draw inferences on the properties of some necessary factors and, hence, on the way the process functions.

3.2 Warburg's Early Photosynthesis Model

Warburg's first two articles were entitled: "On the rate of decomposition of the photochemical carbonic acid". Like most other researchers working at the time, Warburg took it for granted that the decomposition of carbonic acid was the source of oxygen in photosynthesis.[29] His goal was to ascertain "by which means those substances that take part in the assimilation process are rendered reactive in living cells".[30] Why was it, Warburg asked, that, in the process of photosynthesis carbon dioxide decomposes, although under normal circumstances (notably at room temperature) it is usually almost completely inert? His explanation was, in short, that the reactivity was increased by the substances involved binding onto the surfaces of those solid cell constituents that contain heavy metals. Therefore, if these surfaces were destroyed, then the reaction sites were destroyed and, hence, photosynthesis was inhibited. Warburg postulated that three different classes of reaction were involved:

---

[29] Warburg (1919, 1920b). The original German title reads: *Über die Geschwindigkeit der photochemischen Kohlensäurezersetzung*.

[30] See Warburg (1921, p. 354). Original German text: "... wodurch die an dem Assimilationsvorgang beteiligten Stoffe in der lebenden Zelle reaktionsfähig werden".

(i) A primary photochemical process of light acting on pigments. The product of the process was a strong reducing agent, which Warburg called "the primary photochemical product" (PPP).
(ii) The formation of a carbonic acid derivative through a series of ordinary chemical reactions. This process required the involvement of heavy metals, which are embedded in the internal surfaces of the cell, and included the intermediate binding of carbonic acid to components of the cell. Thus, the process was surface dependent.
(iii) Secondary reactions in which the carbonic acid derivative reacts with the PPP, which would eventually lead to the release of oxygen and the synthesis of organic substances. These reactions were also thought to be surface-dependent chemical processes.

In the following sections, I shall present Warburg's main evidence for these hypotheses and reconstruct the course of his argument—unfortunately, no laboratory notes of Warburg survived for that period of his research, so that the following is based on his publications.[31] The model that Warburg developed was not to last for long, yet his argumentation deserves attention for two reasons: *first*, it impressively demonstrates the difficulties that biochemists and cell physiologists at the time were struggling with, as the body of data to infer the course of internal processes was so meagre and only allowed for indirect conclusions. At the same time, *second*, it equally impressively demonstrates the ingenuity of Warburg's work, both in terms of experimentation and interpretation. Warburg really squeezed out as much as possible from the little evidence he had and argued for the legitimacy of every single step—hence, his argument provides an excellent example of the construction of a complex model hypothesis. In order to explain Warburg's inferences from his data—and to appreciate his approach to the problem which fundamentally differed from the chemists' work presented in chapter 2—some technical detail is required, while a more approachable summary is given in section 3.2.2.

3.2.1 Experimental Findings and the Interpretations thereof

3.2.1.1 Carbon Dioxide Concentrations

Warburg began his work in photosynthesis by re-examining the standard parameters of photosynthesis as investigated thus far by plant physiologists—this was inevitable, Warburg explained, as his predecessors had been using unsatisfactory techniques and instruments. (None of the chemists mentioned so far had ever attempted to repeat—and improve upon—the plant physiologists' work; while, even if they had considered this option, they would not have had the methodical skills to do so.) The first theme

[31] A first account was given in Nickelsen (2007).

Warburg revisited was the relationship between photosynthesis and the levels of carbon dioxide concentration, measured at high light intensities. There were no surprises here: Warburg confirmed the findings of the English plant physiologist Frederick F. Blackman and his collaborators, who in 1905 had established the fundamental Law of Limiting Factors, a reformulation of Justus Liebig's Law of the Minimum. This so-called "law" stated that it was not the totality of resources that limited the rate of a chemical reaction (or of a physiological process such as growth) but the availability of the scarcest factor.[32] As Blackman had demonstrated, at low carbon dioxide concentrations the rate of photosynthesis increased in proportion to a rise in carbon dioxide concentrations. However, after a certain point, additional increases in carbon dioxide concentrations no longer promoted an increase in the rate of photosynthesis, until the rate remained constant, notwithstanding any further increases in the gas.

Like Blackman before him, Warburg concluded that, while in the first part of the curve carbon dioxide concentrations limited the rate of the process, in the second part of the curve some other limiting factor must have been present. Yet, Warburg gave the theme a new turn. Since light intensity and temperature were chosen favourably, he thought, the limiting factor in the second part of the curve had to be an additional substance, X, which would react with carbonic acid in the course of photosynthesis. Substance X might possibly be a component of the green cells, Warburg hypothesised, alluding to Willstätter's discovery of the occurrence of this type of reaction.[33] Carbonic acid would react with substance X to make an unknown derivative, and only then could further reaction steps occur, leading to the release of oxygen. A reconstruction of the sequence of events that Warburg proposed is shown in Fig. 3.4.

3.2.1.2 Light Intensity

The second issue that Warburg re-examined was the relationship between photosynthesis and light intensity, measured at high carbon dioxide concentrations. He found that at low light intensities the rate of photosynthesis increased in proportion to the light, while this effect became less prominent at higher light intensities. After a certain point, the rate of photosynthesis reached a plateau and additional increases in light intensity were unable to promote the process any further. Again, the phenomenon itself was familiar (although Warburg's new technique produced a slightly different curve), but Warburg proposed his own interpretation, while he emphasised the similarity of this effect to the one described above:

---

[32] See Blackman (1905). Blackman was the first to investigate, together with various collaborators, the influence of several parameters on the rate of photosynthesis, including light intensity, temperature and carbon dioxide concentrations.

[33] Warburg (1919, p. 253): "We can understand the shape of the curve if we take the rate of assimilation to be proportional to the concentration of carbonic acid and the concentration of a second substance, which reacts with the carbonic acid" (author's translation). Warburg cites Willstätter and Stoll (1918, p. 172, 226 ff.)

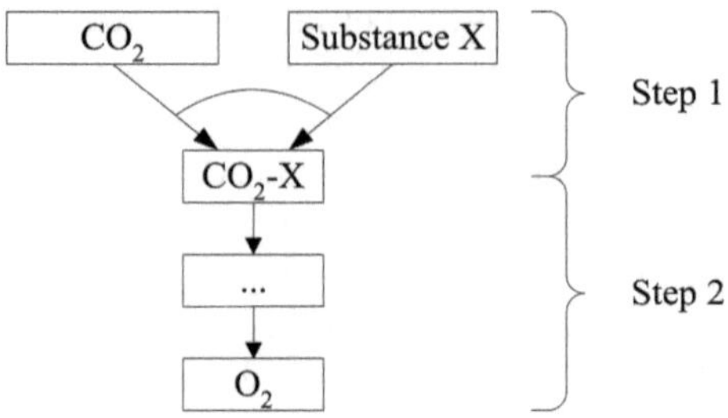

Fig. 3.4 Warburg's interpretation of the carbon dioxide curve. In the first part of the curve, CO_2 itself would be limiting the process, while a second substance, X, was thought to be the limiting factor in the second part of the curve, so that no additional increase in carbon dioxide concentrations would be able to promote the formation of oxygen any further. The formed complex of carbon dioxide and substance X (the "carbonic acid derivative") was assumed to undergo further reaction steps before oxygen could be released.

> The appearance of the curve is very similar to the one that demonstrates the influence of different carbonic acid concentrations at constant light intensity; the "concentration of light energy" operates in this case like the concentration of a chemical substance. This agreement suggests that each light intensity corresponds to a specific concentration of a primary photochemical product, which, according to its concentration, would, in turn, be effective in a chemical reaction. The explanation of the shape of this curve would then have to be similar to the earlier one, by assuming that the rate of assimilation is in proportion to the concentration of the primary photochemical product and the concentration of a second substance, which reacts with this primary photochemical product.[34]

Thus, Warburg thought that also the light curve resulted from two different factors that influenced the rate of photosynthesis under different light conditions. Indeed, this time Warburg went even further, since he not only proposed two different *factors* but also two different *reactions* that would limit the whole process at low or high light intensities.[35] This was the first time that the shape of this light intensity curve, well-known since the time of Blackman, had been explicitly interpreted in this way. If one follows Warburg's argument, a series of at least three reaction steps emerges: In the first stage light reacts with some other substance, Z, to form the PPP, which in the second stage reacts with another substance, Y, to further the process, before oxygen could be released in the final, third stage (see Fig. 3.5).

---

[34] Warburg (1919, pp. 257–258).

[35] Warburg also interpreted the shape of the CO_2 curve to indicate that two different reactions were required to form the carbonic acid derivative. However, he did not elaborate on this point any further and dropped it completely in his 1921 article; therefore I have also omitted it from my discussion. See Warburg (1920b, pp. 210–211).

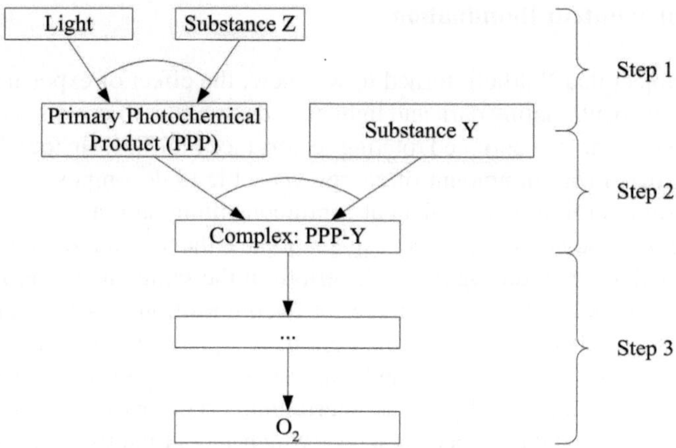

Fig. 3.5 Warburg's interpretation of the light curve. The first step consists of a primary photochemical reaction of light with substance Z, resulting in the primary photochemical product (Step 1: PPP). This product immediately undergoes a reaction with a second substance, Y, and a complex of PPP and Y is formed (Step 2: PPP-Y). The latter is then subject to further reaction steps leading to the release of oxygen (Step 3).

3.2.1.3 Temperature

Finally, Warburg also re-examined temperature, the third classic parameter of photosynthesis. At high concentrations of carbonic acid and at high light intensities, Warburg found, at the standard temperature interval between 15 and 25 °C, a temperature coefficient of about 2 (that is, with a rise in temperature of 10 °C the reaction rate doubled), which was in agreement with the literature.[36] This indicated that under these conditions a thermochemical process was limiting the assimilation rate. At low carbonic acid concentrations and at high light intensities, Warburg found coefficients of 4–5, that is, an even stronger dependence on temperature; again, a thermochemical reaction was, presumably, a limiting factor—this, too, was not a new finding. And, finally, at low light intensities, Warburg confirmed "Blackman's important discovery", as he called it, of a coefficient approaching unity, which would mean that under these conditions the rate of photosynthesis was governed by a process that is practically temperature independent: a photochemical reaction was the obvious answer.[37]

---

[36] Warburg (1919, p. 258).

[37] In his 1921 article, however, Warburg slightly revised this last result by presenting evidence which showed that at low light intensities the coefficient was *negative*, that is, the rate of the process rose as the temperature decreased. This, Warburg argued, indicated that in this process high energy substances, such as PPPs, were the limiting factor. Warburg (1921, p. 355).

3.2.1.4 Intermittent Illumination

The next subject that Warburg turned to was new: the effect of exposing photosynthesising cells to alternating dark and light periods. In order to investigate this effect, Warburg used the aforementioned rotating sectors (see Fig. 3.3). He found that at high light intensities a certain amount of energy was able to decompose more carbonic acid at intermittent illumination than at continuous illumination.

Warburg proposed two alternative explanations: either decomposition of carbonic acid continued to occur during the dark periods at the same rate as before, possibly because of some sort of energy storage; or decomposition was interrupted during dark periods, and then resumed during periods of light at double the rate. Warburg preferred the latter interpretation, and suggested that, while decomposition itself stopped when the source of light was interrupted, other processes would continue until an equilibrium state had been reached (which at continuous illumination would never be attained). Warburg assumed that during these "dark" processes a substance was formed that would be decomposed by light energy. As a higher concentration of decomposable substance would be available after a dark period, light could then act more efficiently—supposing that light of sufficient intensity was available. At low light intensities the products produced during the dark periods would not be properly processed. This interpretation perfectly matched Warburg's interpretation of the light intensity curve: the light-dependent reaction, the primary photochemical process, provided only part of the necessary raw materials for the eventual release of oxygen. The other component was supposed to be an additional substance, Y, which had to react with the PPP (see above and Fig. 3.5). In addition, Warburg now assumed that substance Y was derived from a precursor substance, Y', by way of light-independent chemical reactions. With the resumption of light after a dark period, therefore, the PPP would encounter increased concentrations of Y and the process would thus proceed at a higher rate (see Fig. 3.6), for the extended model).

3.2.1.5 Anaesthetics

The effect of inhibiting substances, especially anaesthetics, on photosynthesis played an important role in Warburg's reasoning (see above). He predominantly investigated the effect of urethanes, in particular phenylurethane, which was known to reversibly inhibit life processes. Warburg confirmed this general finding for green algae and extended it to his conclusion that photosynthesis was far more sensitive in this respect than, for example, respiration. He interpreted this finding following the general mechanism of anaesthesia:

> Taking into account that the effect of anaesthetics is due to changes in the boundary layers, one must conclude that the slightest changes in these layers thus inhibits the process of [photosynthetic] assimilation. This agrees with the experience that, in contrast to other life

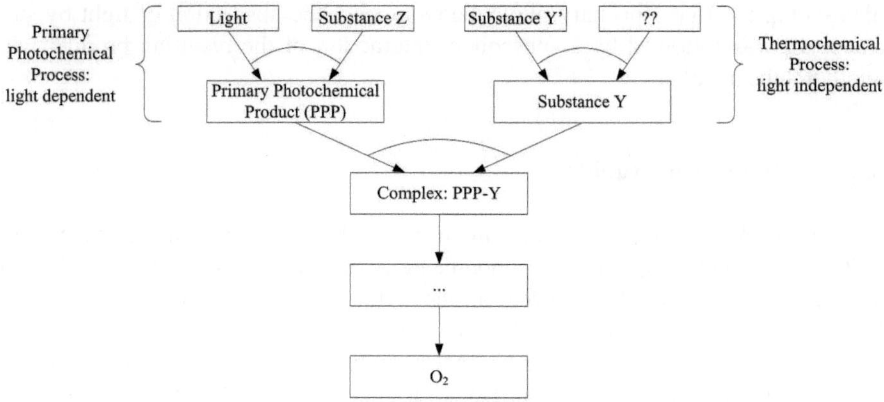

Fig. 3.6 The extended model of Warburg's interpretation of the light curve: whereas the PPP is formed during a light-dependent process, substance Y is produced during a light-independent series of reactions. The former limits the rate of photosynthesis at low light intensities, the latter at high light intensities.

> processes, as, for example, respiration and fermentation, the slightest mechanical change to the cell structure will suspend [photosynthetic] assimilation.[38]

This interpretation matched Warburg's earlier finding that the inhibiting effect of an anaesthetic substance was proportional to its adsorptive capacity, that is, its tendency to adhere to surfaces.[39] Since the inhibiting effects were observed under all circumstances—that is, at low and at high light intensities as well as at different carbon dioxide concentrations—Warburg concluded that all the reactions that limited the rate of the process under different conditions were surface dependent. That photosynthesis is sensitive to anaesthetics at high light intensities and low carbon dioxide concentrations, for example, demonstrated that the limiting process under these conditions (which he considered to be the bonding of carbonic acid to an unknown substance, X) was a reaction that took place on the cell's internal surfaces, presumably on the surface of the membranes.[40] The same applied to the limiting process at low light intensities and at high carbon dioxide concentrations, which also proved sensitive to anaesthetics. According to Warburg, the limiting process under these conditions was the light-dependent stage. As the absorption of light itself was surely not sensitive to anaesthetics, Warburg concluded that a secondary (although indispensable) surface-sensitive reaction must take place. This corresponded well to

[38] Warburg (1919, pp. 265–266). Note that Warburg used the term *Grenzschichten* which was translated here as "boundary layers".

[39] Warburg (1920b, pp. 196–197).

[40] Warburg (1920b, pp. 197–199).

his assumption that a primary photochemical step, the absorption of light by substance Z, was followed by a subsequent interaction of the resulting product with another substance, Y.

3.2.1.6 Hydrogen Cyanide

In addition, Warburg examined the influence of hydrogen cyanide, another substance with inhibiting effects, albeit for fundamentally different reasons. Warburg demonstrated that even at very low concentrations of this substance, such as by an n/10,000 hydrogen cyanide solution, assimilation was reversibly inhibited.[41] By contrast, respiration was not even inhibited by an n/100 solution of hydrogen cyanide, that is, at a 100-fold higher concentration. However, this strong inhibition of photosynthesis could only be observed at high light intensities. Warburg, thus, suggested that hydrogen cyanine inhibited "the ability of carbonic acid to undergo photochemical reactions".[42] This corresponded to Warburg's assumption that carbonic acid had to bind to another substance, X, before the resulting derivative could be decomposed. It was this binding process that Warburg thought would be inhibited by hydrogen cyanide. From other contexts, it was known that (1) hydrogen cyanide mainly acted by inactivating necessary heavy metals and that (2) these heavy metals were usually part of the catalysing enzyme. Warburg, hence, inferred that the reaction in question was an enzyme-catalysed reaction requiring the involvement of heavy metals.

3.2.1.7 Photochemical Induction

Finally, Warburg investigated the phenomenon of "photochemical induction". The principle effect had first been observed in the photochemical reaction between chlorine and hydrogen: if this mixture was irradiated, hydrochloric acid was formed. The rate of this reaction was initially slow, gradually accelerating to a constant final value. As Warburg explained, this delay had been shown by Walther Nernst to be primarily caused by secondary reactions of this chain reaction process rather than by the primary photochemical reaction.[43] A similar phenomenon, Warburg argued, could also be observed in photosynthesis, when studied under intermittent illumination. Only after some minutes of illumination, Warburg reported, would the usual constant value of carbonic acid decomposition be reached. However, this was only the case at high light intensities, as he could not demonstrate any such delay at low light intensities. He suggested the following:

> This phenomenon [i.e. the induction period in photosynthesis] cannot be interpreted by assuming that during the dark periods substances accumulate that would immediately react

---

[41] Warburg (1919, p. 266).

[42] Warburg (1920b, p. 199).

[43] See Warburg (1920b, p. 189).

> with the oxygen that is formed on illumination, so to say, *in statu nascendi* ; in this case the induction period should be longer, the lower the intensity of illumination, while in actual fact the opposite can be observed. Thus, it rather follows from the observations that 1) no oxygen is released in the course of the primary process and 2) no substances are formed in the course of the primary process that would spontaneously (in dark reactions) give rise to oxygen. [...] Points 1 and 2 are all that can safely be said about the primary process; both make it very unlikely that the primary process concerns the carbonic acid molecule.[44]

3.2.2 Photosynthesis Framed as Photolysis

Warburg integrated all these findings into a model of the mechanism of photosynthesis, which is reconstructed in a graph form in Fig. 3.7. Warburg had investigated the different partial aspects of the mechanism of photosynthesis more comprehensively than anybody before him and had tried to collect quantitative data on every single step of the process. Yet, the final model of this mechanism still comprised conceptual steps that went beyond these data. Warburg considered photosynthesis as a complex form of "photolysis", that is, "light splitting"—a concept that had been introduced by his father Emil in the course of his studies in general photochemistry. The substances that were decomposed by photolysis were called "photolytes"—both terms were clearly derived from the words "electrolysis" and "electrolytes". In all such reactions, Warburg explained, one had to distinguish between the primary and secondary processes: "The primary reaction always involves a change in the [light] absorbing molecule, while the secondary reactions take place between the photochemical primary products or between these and other constituents of the photolyte."[45] The latter, that is, the constituents of the photolyte which react with PPPs, would be called "acceptors". (This, of course, fundamentally differs from what today is called an "acceptor" in photosynthesis, mostly used in the context of either electron or hydrogen acceptors in the electron transport chain). However, as Warburg stressed, photosynthetic assimilation was "not a simple photolysis of carbonic acid":

> The primary photochemical process, during which oxygen is released, affects the chlorophyll molecule and leads to the formation of the primary photochemical product. The rate of the formation of the primary photochemical product is in proportion to the amount of radiation absorbed per time unit. The concentration of the primary photochemical product is determined both by the rates of its formation and its consumption. The primary photochemical product reacts with the acceptor during secondary reactions.
>
> The acceptor is not carbonic acid but a derivative of carbonic acid, which is formed in the cell by a chain of chemical reactions. Thus, there is a third class of reactions in the cell, in addition to the primary photochemical process and the secondary reactions: namely acceptor formation. Acceptor formation is a sequence of spontaneous reactions, which, without illumination, would quickly come to rest, due to the accumulation of end products. On illumination, however, the end products—the acceptors—are consumed during

[44] Warburg (1920b, pp. 208–209).

[45] Warburg (1920b, p. 206).

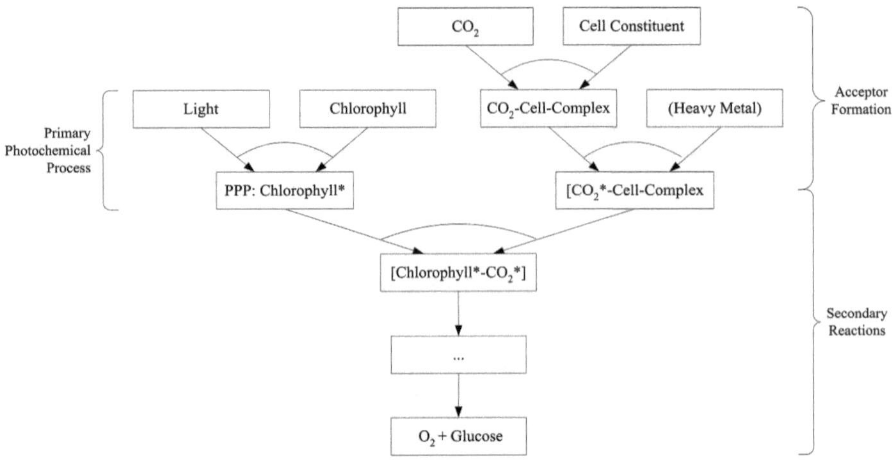

Fig. 3.7 A reconstruction of Warburg's photosynthesis model.

> the secondary reaction, which destabilises the dark equilibrium.l Both the reactions that lead to the formation of the acceptor and the reaction between the acceptor and the primary photochemical product are surface-dependent and, thus, they are extremely sensitive to changes in the surface environment.
>
> In contrast to the secondary reaction, the formation of the acceptor is inhibited by small amounts of hydrogen cyanide. Since the action of hydrogen cyanide probably consists of the transformation of heavy metals from an active form into an inactive complex compound, one should consider the involvement of heavy metals in the process of acceptor formation.[46]

This was the core of Warburg's photosynthesis model. The primary process, as Warburg underlined, was the most elusive reaction of the whole mechanism.The only safe conclusions Warburg felt entitled to draw were that this process did not yet give rise to oxygen and that it involved a change in a light-absorbing molecule.[47] On absorbing light energy, the short-lived PPP is formed, which in 1921 Warburg assumed to be the "isomers of the [light absorbing] pigments, enriched in energy by $h\nu$".[48] The higher energy level of chlorophyll in this activated state is indicated in the figure by an asterisk (*). At the same time, Warburg also held that a second sequence of purely chemical reactions—acceptor formation, as he called it—was necessary if photosynthesis were to continue to occur.[49]

---

[46] Warburg (1920b, pp. 206–207).

[47] To simplify matters, the Fig. 3.7 includes only the chlorophyll molecule, although Warburg acknowledged that in addition to the two kinds of chlorophyll (*a* and *b*), also the xanthophylls and the carotenes contributed to light absorption.

[48] See Warburg (1921, p. 354).

[49] Again, it is important to keep in mind that the current usage of "acceptor" does not correspond to Warburg's notion of the term!

Because of this sequence of reactions, Warburg argued, photosynthesis was highly temperature dependent at high light intensities, that is, when there was plenty of light energy available. In his 1921 article, Warburg used the term "Blackman reaction" for the first time to describe the process that limited photosynthesis under these conditions; it was to become the standard term for this stage of photosynthesis.[50] According to Warburg, it was this class of reactions (which formed an activated carbonic acid derivative) that made carbonic acid susceptible to cleavage. The complete series of reactions was yet unknown, but Warburg considered that at least two steps were necessary: the intermediate binding of carbonic acid to some cell constituent and, subsequently, a reaction step that somehow modified the bound carbonic acid. Since this partial process had proven itself highly sensitive to hydrogen cyanide, Warburg assumed that, in the second step, a heavy metal was involved (presumably iron). This would contribute to converting the carbonic acid into its activated derivative (the activation is indicated in Fig. 3.7 by an asterisk [*]). Furthermore, it was also shown to be surface dependent, given its high sensitivity to anaesthetic substances such as urethanes. In short, acceptor formation was, in Warburg's model, thought to be the result of the catalytic action of an enzyme that contained heavy metals and occurred on internal surfaces. The end product of this reaction was a reactive carbonic acid derivative.

Finally, the PPP and the acceptor—that is, the activated pigment and the carbonic acid derivative—were assumed to interact, whereby the carbonic acid derivative was reduced. Warburg did not go into much detail here, except to characterise these reactions again as surface-dependent, purely chemical processes.

3.3 The Efficiency of the Process

To complement this model of the photosynthesis mechanism, in 1922 and 1923 Warburg carried out an investigation into the efficiency of the process, which he co-authored with his long-standing collaborator Erwin Negelein. The question they hoped to answer was: "Which fraction of the absorbed radiation energy can be transformed into chemical energy in the process of carbonic acid assimilation?"[51] If the absorbed radiation energy is called E and the chemical work accomplished at the same time is called U, then Warburg and Negelein were looking for the quotient U/E. This quotient had been introduced in 1920 by Emil Warburg, who had defined it as the "specific photochemical effect" (*spezifische photochemische Wirkung*), abbreviated to φ, which denoted the chemical work effected by one calorie of absorbed

---

[50] Warburg (1921, p. 355). However, in 1925 Warburg adopted Willstätter and Stoll's notion of the Blackman reaction. They believed that Warburg's "secondary reactions", that is, the reduction of the carbonic acid derivative in the chlorophyll complex, was the Blackman reaction, which limited the rate of photosynthesis at high light intensities. See Warburg (1925).

[51] Warburg and Negelein (1922, p. 235).

radiation.[52] It was known to increase at diminishing light intensities, that is, at low light intensities photochemical reactions tended to be more efficient, until a maximum value was reached close to zero light intensity. It was precisely this limiting case, called φ_0, in which Warburg and Negelein were interested: the "photochemical yield" (*photochemische Ausbeute*) of photosynthesis.[53]

In addition to the theoretical concepts that the authors clearly borrowed from Emil Warburg, they also made use of the latter's facilities. As Warburg and Negelein acknowledged in their article, all the relevant experiments were carried out in Emil Warburg's laboratory at the PTR, where they used the institute's high-quality area bolometer.[54] Warburg and Negelein exposed *Chlorella* to light of wavelengths between 570 and 645 nm, that is, from yellow to orange light. In order to get a reliable value for the amount of absorbed energy, E, Warburg and Negelein used very thick algal suspensions, so that practically all the incident light on the sample was absorbed. By contrast, the chemical work U was measured manometrically, with the measured oxygen release taken as the indicator value.

The results of this study included the important finding that the efficiency of photosynthesis was highly dependent on the conditions under which the algae had been cultivated: the highest efficiency was achieved with cells that had been transferred to low light intensities after having been grown for some time in high light intensities. The efficiency measurements themselves revealed that, on average, an extremely high percentage of between 60 and 70 % of the absorbed radiation energy could be transformed into chemical energy—perhaps even more. This was spectacular, given that the highest efficiency that had ever been measured for chemical reactions (Warburg's father, for example, had measured the efficiency of ozone formation) had been one of 50 %![55]

Although the maximum quantum yield of photosynthesis had occasionally been discussed in the years before Warburg and Negelein turned to the subject, it was far from the "frequently debated" issue that the authors claimed it to be in their introduction. In fact, a review of 1916 had seen a definite need for more research on the "energy relations of the green leaf", to which was added: "This aspect of carbon assimilation exhibits perhaps more than any other an unfortunate isolation of effort in research, the various workers on the subject having generally neglected the results

---

[52] See Warburg (1920a, p. 54).

[53] Warburg and Negelein (1923, p. 205).

[54] Warburg and Negelein (1922, p. 236).

[55] In the second publication of 1923, the average value of 1922 (70 %) was slightly reduced to an average (in red light) of 59 % efficiency, while the maximum value they had been able to achieve was 63.5 % efficiency. This was due to a change in procedure: while in 1922, Warburg and Negelein had determined φ_0 by extrapolating from values at higher light intensities, in 1923 they reconsidered this procedure, since, as they conceded, it was not known which curve the extrapolation should be made to follow. Instead, they measured the efficiency in the lowest possible light intensities, and when no significant increase in value was found, they assumed that this value was the limiting case. See Warburg and Negelein (1923, p. 205).

obtained by others, both along their own and related lines of investigation".[56] Measuring quantum yields certainly occupied the attention of photochemists; but very few people had so far tried to transfer this approach to the study of photosynthesis. The standard estimation had been provided in 1905 by the English plant physiologist Horace Brown and his collaborator Francis Escombe, who had found a maximum efficiency of photosynthesis of no more than 6 %, that is, a fraction of what Warburg and Negelein claimed to be the case. Warburg and Negelein argued that Brown and Escombe's results were invalid: They had used whole leaves and measured light absorbance by observing the weakening of light passing through the leaf. Warburg and Negelein rightly argued that a large part of the issuing light would be scattered by the leaf and would, therefore, remain undetected by the instrument.[57]

The efficiency of photosynthesis was particularly interesting in view of the ongoing search for the underlying mechanism. A simple calculation revealed that reducing one molecule of carbonic acid to the level of carbohydrates required, at the very least, an energy input of 112.3 kilocalories (kcal). From this it followed that, on average, the carbonic acid had to interact with at least three pigment molecules, if each of them absorbed one red light quantum with an average energy of 49 kcal each. Although Warburg and Negelein did not yet dare, in 1922, to draw any concrete inferences from their findings, they did emphasise that in view of the high overall efficiency of the process, the reduction of carbonic acid had to be rather straightforward, that is, without the inclusion of high-energy intermediate reactions.

These general findings were followed in 1923 by an investigation into the influence of different wavelengths on the efficiency of photosynthesis.[58] Warburg and Negelein's most important finding was that φ_0 decreased as the wavelength diminished, that is, the photochemical yield was lower at shorter wavelengths than at longer wavelengths. This finding was in agreement with quantum theory, in particular with Einstein's Law of Photochemical Equivalence, which predicted exactly that. It also agreed with Emil Warburg's measurements of the photochemical yield in the photolysis of hydrobromic and hydroiodic acids. Warburg and Negelein calculated that approximately four light quanta would be required if the algae were illuminated with red or yellow light, and about 5 quanta if they were illuminated with blue, to decompose one molecule of carbonic acid. These results were regarded as the authoritative answer to this question for the next 20 years or so; and it was these figures that sparked off the vigorous controversy on quantum yields and efficiencies between Warburg and the American photosynthesis researchers. This controversy is the subject of chapter 5.[59]

---

[56] Jörgensen and Stiles (1916b, p. 24).

[57] See Warburg and Negelein (1923, pp. 192–193).

[58] Warburg and Negelein (1923). Warburg and Negelein investigated the process at 610–690 nm (red), 578 nm (yellow), 546 nm (green) and 436 nm (blue).

[59] See on this episode also Nickelsen and Govindjee (2011).

3.4 Father, Son and Photosynthesis

Considering the analysis of Warburg's early photosynthesis work from a slightly different perspective, I would like to draw attention to the following three points: (1) Warburg explicitly adopted theoretical concepts—photolysis, the photolyte and the photochemical yield—that his father had used and developed before him; (2) Warburg undertook experiments along the lines of his father's earlier work and stressed that his findings concurred with his father's results; (3) Warburg carried out much of his research work on photosynthesis in his father's laboratory, where he used the PTR's optical infrastructure. In view of these observations, it seems that the connection between Otto Warburg's research and the studies of his father Emil deserves some special attention.

3.4.1 Emil Warburg and Photochemistry

As is well-known, the PTR, to which Emil Warburg was appointed its president in 1905, was a national research institution primarily concerned with careful measurements and the definition of standards.[60] The radiation laboratory, in particular, had a renowned history and a high international reputation. It was in this highly equipped laboratory that the physicist Otto Lummer and his collaborators had started to do research on black body radiation and carried out the measurements that eventually brought Max Planck to advance the existence of a universal energy constant. However, after Lummer had left the PTR in 1904 to take up a professorship at the University of Breslau (today's Wrocław), this instrumentation fell into disuse.[61]

The revival of the radiation laboratory is today considered one of Emil Warburg's most notable achievements as President of the PTR. Already shortly after having taken up this position, Emil Warburg started investigating the energetics of photochemical processes, while after 1911 he made it the focus of his work. Between 1911 and 1919, he published nine important articles on photochemistry, a subject that clearly dominated the latter part of his career.[62] In contrast to the earlier methods of measuring radiation, which for the most part had yielded only qualitative results, Emil Warburg wanted to explain the photochemical energy conversion in quantitative terms. This ambition was sparked by Einstein's seminal publication on the light quantum hypothesis (1905), in which Einstein had postulated that: (1) radiation of

---

[60] See Cahan (1989), which provides an account of the PTR's history, 1871–1918.

[61] Lummer's rotating sectors were used later by Emil Warburg to determine the constant c of the black body radiation law and, eventually, by Otto Warburg for his photosynthesis studies. See on Emil Warburg's work the report on pp. 118–120 in: Tätigkeitsbericht der PTR für das Jahr 1910, *Zeitschrift für Instrumentenkunde*: Heft 4, pp. 106–120; Heft 5, pp. 140–160; Heft 6, pp. 174–195.

[62] On Emil Warburg's photochemical work, see, e.g., Franck (1926, 1931). Brodhun (1913) provides a detailed review of the PTR's then recent activities in optics.

the frequency ν consists of discrete light quanta of the energy h (Planck's constant) times ν; and that (2) matter that absorbs (or emits) radiation will do so in terms of light quanta of this type.[63] Although this hypothesis had many explanatory virtues, it was far from complete and the exploration of its consequences gave rise to a wealth of puzzles to be solved.

Although the role of light in chemical processes had been much debated already in the nineteenth century, the quantitative analysis of these phenomena had remained a problem.[64] On the one hand, methodical difficulties impeded the research, since photometry was still in its infancy around 1900. On the other hand, conceptual problems prevailed. In 1909, Einstein himself summarised one of the main inconsistencies as follows: "Why does the occurrence of a certain photochemical reaction only depend on the colour and not on the intensity of the light? Why are rays of shorter wavelengths generally more chemically effective than those of longer wavelengths?"[65] It did not take Einstein long to answer these questions. As the editors of the fourth volume of *The Collected Papers of Albert Einstein* put it: "The derivation of the law of photochemical equivalence from purely thermodynamic considerations was, arguably, Einstein's major scientific contribution in the years 1912–1914."[66] At the same time, Emil Warburg had turned to these questions, which he approached from an experimental perspective. He had even devised a new type of bolometer for this project, the vacuum bolometer, which allowed for more precise measurements to be taken.[67]

Einstein and Emil Warburg, who were approximately one generation apart, met at the first Solvay Conference (held in Brussels, Belgium) in November 1911, where they discussed photochemistry. It was this discussion that brought Einstein to formulate the aforementioned Law of Photochemical Equivalence.[68] One consequence of this law was the fact that all photochemical reactions would then require the absorption of one light quantum per "photolyte" molecule, that is, per molecule that

---

[63] See Einstein (1905) for the original publication; The paper was reprinted (together with an introduction and additional notes) in Vol. 2 of the *Collected Papers* edition; see Stachel et al. (1989), pp. 134–169.

[64] See the introduction to Einstein's paper in Klein et al. (1995, pp. 109–111). For a review of the problem, see Warburg (1917). See also Boberlin (1993) for an account of the nineteenth-century beginnings of quantitative photochemistry.

[65] Quoted in Klein et al. (1995), pp. 109–110. Original quotation in Einstein (1909), p. 490.

[66] See Klein et al. (1995), introduction, p. xiv. For Einstein's first paper on the subject, see Einstein (1912b), which is reprinted in Klein et al. (1995), Doc. 2, pp. 115–121. For a more detailed study of Einstein's work on photodecomposition, see Bergia and Navarro (1988).

[67] See Warburg et al. (1907) and Warburg (1909) for the announcement of the instrument, while in Warburg (1912, 1913) he presented the first results of the project.

[68] For details, see Klein et al. (1995, pp. 110–111). See also the letter from Einstein to Heinrich Zangger, dated 20 November 1911, and published in Klein et al. (1993), Doc. 309, pp. 352–353: "I also have an issue with Warburg, who was in Brussels. He has proven, wrongly, that there must be a threshold for photochemical excitation. On this occasion I found an interesting thermodynamical proof for the Law of Photochemical Equivalence that Warburg seeks to confirm. 1 molecule is dissociated by ν-radiation at the absorption of an energy $h\nu$." (Translated by the author).

was able to undergo photochemical cleavage.[69] Emil Warburg then took up the task of providing experimental evidence for this law, which proved far more difficult than expected; however, in the end Emil Warburg's research into the photolysis of hydrogen bromide and cyclohexane (*Hexahydrobenzol*) proved fairly satisfying in this respect.[70] This work was facilitated by the fact that Einstein had moved to Berlin in 1914 and had been able to keep in close contact with Emil Warburg since.[71] In 1915 Einstein even worked in the PTR's laboratory (although not, as Emil Warburg might have preferred, with him in the radiation laboratory, but with the Dutch physicist Wander de Haas). Einstein also was a regular visitor to the Warburg household; he became acquainted with Emil Warburg's wife and presumably also heard of Warburg's son, Otto, although they only met after 1918.[72] However, thereafter the two remained in touch. It is reported, for example, that Otto Warburg was occasionally invited to dinner at the Einstein family home.[73]

3.4.2 Influences on Otto Warburg

How is all this related to photosynthesis research? *First*, as was demonstrated earlier, Otto Warburg explicitly adopted the type of questions introduced and defined by his father in the latter's work on photochemistry; *second*, Otto Warburg applied some of his father's photochemical concepts—the "photolyte" and the notion of "photochemical efficiency"—to his own work. *Third*, thanks to his father's position, not only was Otto Warburg able to use the excellent optical instruments at the PTR but was also introduced to photochemical experimentation by the experts in the field as well as given practical support whenever it was needed. If one takes a closer look at Otto Warburg's biography, there were three periods during which he worked in his father's radiation laboratory: in 1905–1906 while studying in Berlin; in 1914 before starting at the KWI for Biology; and, again, in 1918, after having returned from active service in the First World War. It seems that whenever he was in Berlin, Otto Warburg took the opportunity to use the PTR's sophisticated instrumentation. Starting from

---

69 See Einstein (1912a, b). The editorial note to the reprint of the paper in Klein et al. (1995) gives an account of the dispute (and later collaborative work) between Einstein and Emil Warburg on the subject.

70 See Warburg (1917, 1924)..

71 In fact, Emil Warburg had been one of the members of the Berlin Academy of Sciences and Humanities who had been behind Einstein's appointment; see Goenner and Castagnetti (2004) and Goenner (2005).

72 See, on this point, Einstein's letter to Otto Warburg of 23 March 1918, urging him to return to Berlin from the war, in which Einstein wrote: "You will probably be rather surprised to receive a letter from me, for until now we have only circled each other without ever getting truly acquainted"; Schulmann et al. (1998), No. 491 (originally German, translation by the author).

73 The outcome of one of these occasions was the employment of Hans Krebs in Warburg's laboratories in 1926, as Krebs himself reported. Quoted in Werner (1991, p. 137), doc. 45.

1918, the experiments that Otto Warburg carried out at the PTR were documented in the institution's yearly reports, and in his papers Otto Warburg duly gave credit to the support given to him. He even expresses gratitude to the staff for having carried out measurements on his behalf.[74] It would be natural to assume, therefore, that Otto Warburg simply observed his father's activities and then transferred them to a different, namely a biological, field of inquiry.

However, the actual circumstances are more complicated than that. If one examines the PTR's annual public activity report for 1911, that is, for the first year of Emil Warburg's research project on the energetics of photochemical reactions, one finds the following description of this project's aims:

> An important class of photochemical reactions to which belongs, among others, the [photosynthetic] assimilation process in green plants, proceeds with the uptake of energy, which is retrieved from the absorbed radiation and forms a certain fraction of it. We have taken up the task to measure this fraction, the photochemical yield, for a number of cases.[75]

Thus, as early as the first outline of his photochemical research programme, Emil Warburg explicitly mentions photosynthesis as being one of the principal classes of reactions he wanted to study from the point of view of energetics.[76] Emil Warburg's focus of interest was photochemical efficiency: He took it for granted that only a fraction of the light energy absorbed by a molecule was used for the subsequent chemical reactions, and he wanted to find out what this fraction was in particular cases. A second indication that the subject of photosynthesis engrossed Emil Warburg from early on can be found in one of the few surviving letters to his son Otto, dated 9 December 1912:

> Today, I have read in a paper by [Fritz] Weigert[1]—which, by the way, otherwise contains little of interest—that two Englishmen, Brown & Escombe (Int. Trans. Roy. Soc. 183 B 223, 1900; Proc. Soc. 76 B, 1905; Nature, March 1905) have carried out very similar experiments to the ones that we are intending to do. Having scanned Weigert's account, I understand that the process is apparently very complicated, that in particular [photosynthetic] assimilation is largely (1:12) independent of light intensity; this is explained by the fact that the rate of assimilation is determined by CO_2 diffusion.
>
> [1][Footnote:] *ZS für wissenschaftliche Photographie, Photophysik & Photochemie.* Vol. 11, issue 2, p. 381.[77]

The letter is particularly interesting as it refers to experiments that father and son obviously were intending to carry out together. The paper of 1905 that Emil Warburg mentioned is the very article written by Brown and Escombe that Warburg later

[74] See Warburg (1919, pp. 235, 255).

[75] Report of the PTR for the year 1911. *Zeitschrift für Instrumentenkunde* 22 (1911), p. 131.

[76] This view of photosynthesis, as an ideal case study for the laws of photochemistry, was shared by many of Warburg's physicist colleagues; see, e.g., the extensive treatment of photosynthesis in Fritz Weigert's monograph on the chemical effects of light, Weigert (1911).

[77] The original letter is preserved in the Archive of the Berlin-Brandenburg Academy of Sciences and Humanities (Archive of the BBAW) in Otto Warburg's estate, at shelf mark NL Warburg 999. A transcription of the German original can be found in Werner (1991, p. 77); doc. 23.

dismissed as being based on methodically flawed experimentation.[78] Thus, although Otto Warburg only published his first article on photosynthesis in 1919, his father already had been planning to work with him on this theme as early as 1912. One could speculate that they might have discussed this option at an even earlier date, perhaps around the time when Emil Warburg mentioned in his PTR report of 1911 that photosynthesis was one of the reactions that were of particular interest to him. Yet, Otto Warburg was then still deeply involved in the study of respiration, and the planned experiments had to wait.

As no laboratory documentation, neither in Otto Warburg's personal estate nor in the PTR's archives, has survived from this period, it is impossible to tell how far research on the subject had advanced (if at all) when Otto Warburg enlisted for service in the First World War. All we know is that neither Emil Warburg nor Carl Correns hesitated to mention photosynthesis experiments as an argument for calling Otto Warburg back to Berlin. And, as can be taken from Correns's report on the activities of the KWI for Biology from 1 April 1918 to 30 March 1919, photosynthesis was the first theme that Otto Warburg took up after returning from the battlefields:

> Warburg, the head of department, was already back at work in October 1918, but was only able to use his rooms again at the end of the year. In addition to the repairing and renovating of his premises, he was engaged in studies concerning the assimilation of carbonic acid in green cells, in particular attempting to separate the assimilation process from the cell structure and studying the influence of assimilation in living cells.[79]

According to the 1919/1920 report of the KWI, photosynthesis was also Warburg's main research theme in the following year.[80] Thus, on closer investigation, one finds evidence that Otto Warburg, together with his father, was interested in photosynthesis as early as 1912. And although conclusive evidence for Otto's research into this theme is available only for the time after 1918, it is highly probable that this work was influenced, and promoted, by Emil Warburg's earlier research aims, formulated in 1911, and the PTR's facilities. Yet, Otto Warburg did not immediately turn to the reaction's efficiency or its quantum requirement, which was in the centre of his father's interest, but instead presented a comprehensive investigation of the chemical mechanism of photosynthesis. And although Otto Warburg made ample use of his father's methods and concepts therein, an even more important source of motivation, not only for the theme in general but also for its specific treatment, emerges if one takes a closer look at Otto Warburg's research up until 1914, in particular his studies in cell respiration.

---

[78] The references cited by Emil Warburg are Brown and Escombe (1900, 1905); Brown (1905) and Weigert (1912).

[79] Translated from Werner (1991, p. 128). During the war, the rooms of Warburg's laboratory had been used by a team led by Fritz Haber to conduct experiments in poisonous gases. Consequently, to make the rooms amenable to the study of living organisms again, all the toxic residuals from the floors and the benches had to be removed.

[80] Werner (1991, p. 146).

3.5 Studies in Cell Respiration

Otto Warburg's research into cell respiration—or biological oxidation, which is the more general term—has already been the subject of a number of excellent studies.[81] It was for this field of research that Warburg was most famous and for which he received the Nobel Prize in Physiology or Medicine in 1931. However, this section focuses on the early years of Warburg's research, that is, the period 1908–1914, which means that Warburg's final success, his concept of *Atmungsferment*, which was developed in the 1920s, has been omitted from the discussion.[82]

Cell respiration was the first theme that the young Warburg chose to study independently, making it the topic of both his medical dissertation in 1911 and his habilitation in 1912. Warburg's work in these years was strongly influenced by the physiologist Jacques Loeb, then based at the Rockefeller Institute for Medical Research in New York, and his "mechanistic conception of life": the programme to explain life processes in a physico-chemical way.[83] Warburg frequently went to the renowned Zoological Station at Naples (Italy), a research institute that developed into a hub for biological research, where he met the avant-garde of this new research tradition, which included Theodor Boveri, Hans Driesch, Oscar Hertwig and Thomas H. Morgan.[84] However, although Warburg received much inspiration from their work on developmental physiology, he quickly developed his own agenda, which was mainly concerned with the physico-chemical elucidation of energy-producing reactions: a subject that was then extremely controversial.[85]

As later with photosynthesis, Warburg also fundamentally changed research in the field of cell respiration research by introducing both new techniques and new conceptual approaches.[86] Warburg was, for example, the first to study this process in isolated cells. Up to then the experimental organisms used for this purpose had been mice, rabbits and other animals, which made it extremely difficult to control the experimental conditions. Warburg rather chose to investigate sea urchin eggs, which were then the main experimental objects employed in the field of developmental

---

[81] For Otto Warburg's early concept of the *Atmungsferment*, see, in particular, the thorough study by Kohler (1973a). Höxtermann (2007a) provides a complementary explanation of Warburg's understanding of biocatalysis. The latter is also treated in Höxtermann (2001, pp. 265–268) and in Werner (1991, pp. 64–69, 113–118). The introduction of Werner (1996) provides rich historical background to the theories of respiration at the time, while Werner (1997) analyses the controversy that arose between Warburg and Heinrich Wieland on the theory of biological oxidation. More specific references can be found in the above-mentioned publications.

[82] I gratefully acknowledge conversations with Ekkehard Höxtermann, Berlin, on this theme, which helped me improve this section.

[83] Cf. Loeb (1905). On Loeb's influence on Warburg's research, see Werner (1996, pp. 35–47).

[84] See on the *Stazione Zoologica di Napoli*, e.g., Groeben (1975); Fantini (2002) and Groeben (2005). A brief overview is given in Nickelsen (2010). Dohrn (1892) provides a contemporary perspective; see also the excellent collection of essays in Metz and Clapp (1985).

[85] See, e.g., Kohler (1973a, b) and Werner (1996).

[86] Cf. Kohler (1973a, p. 183).

biology. Indeed, it is very likely that also Warburg originally intended to study the problems of embryonic development: in his first publication on the subject, he announced further investigations into the early cleavages of the fertilised egg. The rate of oxidation was only the first parameter to which he turned.[87] In other words, studying the respiratory processes of the fertilised egg initially was only a subgoal, while his actual, superordinate goal was to understand the underlying chemical mechanisms of early embryonic cell cleavage. Yet, the subgoal turned out to be so interesting that Warburg very quickly made it the main focus of his further research.

As Warburg was convinced that the mechanism of cell respiration was the same in all cells, it did not matter to him from which organism the material was gathered. Far more important were methodological considerations: Since it is easier to carry out controlled experiments on simple systems, the preferred choice for quantitative studies were the smallest possible units (we can recognise the similarities to his later choice of *Chlorella* in photosynthesis research).[88] It was also during the course of his studies in cell respiration that Warburg developed his sophisticated technique of manometry by adjusting the manometers that Barcroft had used (see above) to the requirements of his own project. Warburg then explored new ways of using the reversible inhibition of cell processes by anaesthetics as a means of investigation; in particular, he used surface-active substances, such as urethanes, and the heavy metal-binding cell poison, hydrogen cyanide. Indeed, it was Warburg who, through his own work, established how these substances were able to inhibit respiration, either through their adsorptive capacities or through a chemical reaction.[89] Thus, a large part of the equipment and substances that Warburg would later so innovatively introduce to photosynthesis research had originally been developed in his earlier studies in cell respiration: the use of single cells as the experimental object, the technique of manometry and the use of a range of inhibitors, notably hydrogen cyanide and urethanes.

Furthermore, the principal question that Warburg raised in his papers on cell respiration was the same question that he would set himself in his photosynthesis studies: Why are substances, which at room temperature are usually extremely stable, subject to very fast combustion in living cells?[90] Some sort of catalysis, it seemed, had to be involved; but then, how should this catalysis be described? Following the discovery of "zymase" (an intracellular enzyme complex) by the German chemist Eduard Buchner in 1897, two options presented themselves: the decisive factor was either the action of the cell structure (which was the biologists' view) or the action of

[87] Warburg (1908, p. 1).

[88] See, e.g., Warburg (1914, p. 320), where he discusses the advantage of working with cells rather than with living tissue.

[89] On the history of using anaesthetics in respiration studies in general and on the discussion centred around Warburg's use of them in particular, see Werner (1996, pp. 87–95), and Werner (1997, pp. 183–190).

[90] Warburg (1914, p. 314).

an enzyme (which was the chemists' view). Warburg solved this issue by integrating both modes of catalysis into one complex mechanism, as he emphasised in his papers as well as in a speech of 1914:

> I hope I have demonstrated to you today that there is no dichotomy here at all: both ferment chemists and biologists are right. The acceleration of energy-producing reactions in cells is a ferment action *and* a structure action; it is not that both ferments *and* structure accelerate, but that *structure accelerates ferment action*".[91]

Through his use of surface-active inhibitors, Warburg was able to establish over the course of the years that internal cell surfaces were, for the most part, essential for cell respiration. He proposed that a "ferment" was involved, which would accelerate the oxidation processes, and that the action of this ferment itself was greatly accelerated when it was attached to the structural elements of the cell. This conception of the process rested on Warburg's personal notion of what a ferment was: while Buchner and others "considered enzymes to be definite proteins with specific catalytic properties [...] Warburg renewed the colloid chemists' ideas of surface activity as an attractive alternative".[92] This is why Warburg never gave up the somewhat old-fashioned term "ferment", which implied that it was not a single protein that promoted a certain process but the cell as a whole. The fact that respiration was so sensitive to the influence of hydrogen cyanide, which readily binds with heavy metals, was taken by him to be evidence that heavy metals of one kind or another were the active part of this ferment. In 1914, Warburg finally came to the conclusion that this heavy metal was the cell's iron, which acted catalytically to promote oxidation by being reduced from its ferric (Fe^{III}) to its ferrous (Fe^{II}) state.

Warburg was very cautious about the possible presence of intermediates in the process, and for a long time did not even speculate about the elusive substance, called X, which was the first substance to be oxidised. However, in a review article of 1914, Warburg suggested that the relevant mechanism included the "oxidation of lipoids in the presence of iron salt".[93] Warburg concluded this from his experiments with lecithin, which he seems to have taken to be representative of the whole group of lipoids. It was known that the internal cell structures that Warburg considered so important were, in large part, made up of lipoids. In Warburg's model of 1914, these lipoids were part of the structure on to which the iron ferment was adsorbed as well as the actual substances on to which the oxygen was transferred through the action of the iron ferment. Warburg was aware that this was not the final word on the issue. He was, for example, silent on the actual sequence of the initial reaction steps—how the oxygen was brought into contact with the iron ferment (was it first bound to surfaces as well?) and whether there were other components of the structure, in addition to lipoids, that were significant for the full process of cell respiration. But even so, as

---

91 Warburg in 1914; translation taken from Kohler (1973a, p. 190). Emphasis in the original.

92 Höxtermann (2007a, pp. 123–124). For the complicated development of the concepts of "enzyme" and "ferment", see, e.g., Fruton (1972b); Kohler (1973b) and Teich (1981).

93 Warburg (1914, p. 335).

early as 1914, Warburg was able to present an impressively detailed account of cell respiration, based on rich empirical evidence.

The similarities between this model and Warburg's photosynthesis model of 1919 are obvious. The processes were framed in exactly the same way—heavy-metal-catalysed reactions, which occurred on internal cell surfaces—and Warburg used exactly the same techniques to provide evidence for this general view of events. In his 1927 book on the catalytic action of the living substance, a collection of selected papers by Warburg on cell respiration and photosynthesis up to that time, he wrote the following:

> Heavy metal catalysis [such as respiration] is also the Blackman reaction, which is part of the process of photosynthesis, [...]. [...] If I may add that these reactions are also surface reactions, one realises that the most important catalytic actions of the living substance are based on the same principle. The kind of metal and the type of bonding may vary, but the principle remains the same.[94]

The similarities in his conceptions of respiration and photosynthesis were no coincidence. Warburg was thoroughly convinced that the same fundamental principles govern the chemical reactions in all organisms, from bacteria to human beings, which was why he experimented with cells as specific as sea urchin eggs and still did not hesitate to generalise his results to the entire living world. In her diaries, Warburg's sister, Lotte, wrote that, in 1926, Warburg commented on Carl Correns, who was irritated by the fact that, at the time, the field of animal physiology was expanding, thereby displacing plant physiology, his own field of inquiry. Warburg, his sister wrote, considered this point of view ridiculous and narrow-minded: "What, however, is the difference? This all belongs together".[95]

Given this attitude and Warburg's interest in energy-producing reactions (not only in respiration but also in fermentation), his shift to studying photosynthesis is no longer surprising. It rather is another prime example of research opportunism in action. Warburg had been highly successful in elucidating the mechanism of cell respiration: why not examine the second large class of energy-producing reactions, namely photosynthesis, with the same approaches and see whether similar principles held there? From this perspective, it is obvious why Warburg at first turned to the photosynthesis mechanism (which he was able to study with the methods at hand), while only later he decided to study the subject from the point of view of energetics, which also interested his father. Finally, the fact that Warburg had used sea urchin eggs for his studies in cell respiration explains why he did not continue this line of research immediately after 1918, although in 1914 his work was looking very promising. Warburg had either worked in Naples or had sea urchins from the Zoological Station sent to Berlin. Neither of these options prevailed any longer after the war: "The more our currency goes down, the farther away is Naples", Warburg wrote in 1922 in a letter to Reinhard Dohrn, who was then the head of the Naples institution.[96]

---

[94] Warburg (1927, p. 12).

[95] See Rüskamp (1989, p. 252); also quoted in Werner (1991, p. 143).

[96] Quoted in Werner (1991, p. 130); translated by the author.

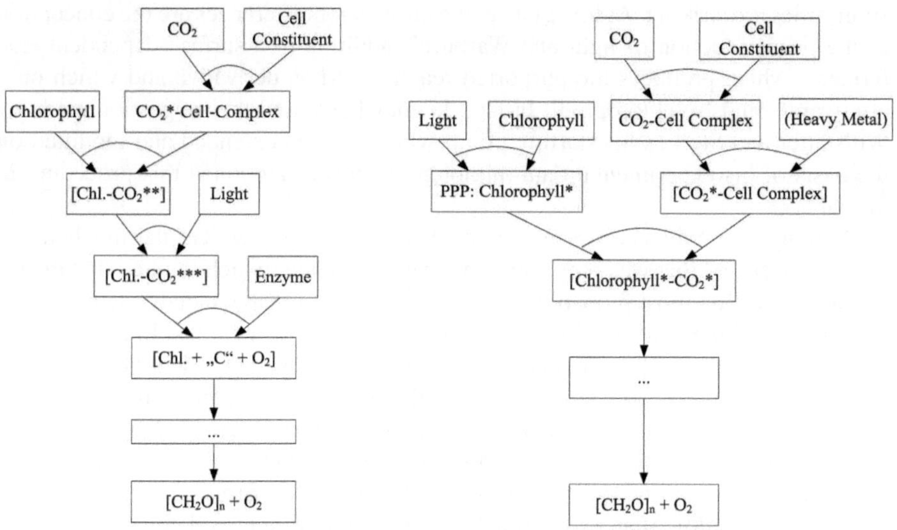

Fig. 3.8 A comparison between the photosynthesis models of, *left*, Willstätter and Stoll (1918) and, *right*, Warburg (1919, 1920).

Thus, Warburg was forced to become involved with research questions that could be carried out using more mundane organisms—such as freshwater algae, which could be retrieved from the nearby Schlachtensee.

To sum up: there is much evidence to support the assumption that Warburg started to contribute to photosynthesis research not because his father prompted him to do so, but mainly because the opportunity arose and he had the means to do so. Perhaps Warburg only chose to embark on this theme in order to make good use of the time during which he could not work on his studies in respiration, because of a poor supply of sea urchin eggs. Warburg then found, however, that with his approaches and techniques he could, in actual fact, make a contribution to the subject; yet, very soon thereafter, he rather turned to the study of cancer. The year 1925 may have marked the end of Warburg's work on photosynthesis—had it not been for the controversy that arose, after 1945, on the maximum quantum requirement, which made Warburg focus his attention on these questions again.

3.6 Comparison with the Chlorophyll-Complex Model

To round off the analysis, I shall now briefly discuss how Warburg's model relates to the lines of research that were followed in chapter 2. In Fig. 3.8, Warburg's proposal has been juxtaposed with the chlorophyll-complex model of Willstätter and Stoll, which was published in 1918 and, as was previously mentioned, was considered to be the most satisfactory photosynthesis model that had so far been developed. The similarities are striking. At first glance, the only major differences are the conception of the primary action of light and Warburg's addition of a surface-dependent iron ferment, which produces the purported reactive carbon derivative and which only afterwards binds to chlorophyll. In fact, he may have used the suggestion made by Willstätter and Stoll as his starting point, which he then extended and modified on the basis of his experimental data (although he never mentioned this procedure in his papers).

Warburg had evidence, for example, that a surface-dependent thermochemical reaction occurred that was sensitive to hydrogen cyanide (which implied that the reaction in question most probably involved the action of heavy metals). Kinetic data also suggested that carbon dioxide (or carbonic acid) was involved in this reaction as well as an additional compound of unknown nature. Warburg identified this compound as part of the cell's constituents, with which the carbonic acid reacted; and he further assumed that heavy metals, which were somehow embedded into the cell surfaces, would activate carbonic acid in this complex binding to one of its derivatives. The latter was simply an extension of Willstätter and Stoll's model hypothesis through the addition of an extra cofactor to the carbonic acid-cell complex module. Willstätter and Stoll had also thought that chlorophyll became part of this complex, upon which the action of light would then activate the carbonic acid derivative again, before the actual reduction took place through the action of an enzyme. Warburg would later identify this complex as the "photolyte": the compound that was the subject of the actual photolysis, that is, light splitting.[97] Warburg again modelled the processes in a similar, but not identical, way. According to his experiments, the primary photochemical process resulted in the production of a strong reducing agent, which he thought was activated by chlorophyll. Thus, the light would only act on the chlorophyll, which then in its activated form would induce the reduction of the carbonic acid derivative. This made the action of an additional enzyme superfluous. The remaining steps to the carbohydrate stage then were the same, since Warburg agreed with Willstätter and Stoll that possibly formic acid and peroxides of some kind were involved.

Thus, although Warburg gathered his data by carrying out completely different experiments, using other methods and a new experimental organism, he modelled his findings very much in line with the standard assumptions of the time—combined with elements that he had taken from his earlier modelling of respiration, such as the involvement of a heavy metal as a catalysing agent. This explains why Warburg's model was seldom regarded as an original contribution to photosynthesis research—even though, as far as their experimental foundation was concerned, Warburg's papers were highly innovative. The fact that most of the factors postulated in the chlorophyll-complex model of Willstätter and Stoll were also inferred by Warburg, albeit from totally different sets of data, was rather taken to corroborate strongly the earlier suggestion, which was well on its way to becoming the new "standard model".

[97] See Höxtermann (2007a) for a comment on Warburg's concept of the "photolyte", which is closely connected to his understanding of biocatalysis and which he developed over time.

3.7 Otto Warburg's Building Blocks

As demonstrated in this chapter, the work that Warburg carried out in the field of photosynthesis between 1919 and 1925 was a direct consequence of his research goals and methods of the years 1908 to 1914—in line with the terminology of Frederick L. Holmes one might speak of a very plausible twist in Warburg's "investigative pathway".[98] Early in his career Warburg chose to focus on the energy-producing reactions of metabolism, in particular those reactions that could be investigated using manometric techniques, that is, gas exchange processes. After achieving considerable success in the fields of respiration and fermentation, the next obvious challenge—given Warburg's general conviction, much the same as Loeb's, that all fundamental life processes were based on similar principles—was the investigation of the curious energy-producing mechanisms of plants. The second line of research that fundamentally influenced Warburg's work in these years were the studies of his father Emil Warburg, which were part of a general attempt being made by physicists working in Berlin at the time to explain natural phenomena in terms of quantum laws. Emil Warburg chose to explore photochemistry in this respect; and photosynthesis, a natural example of a photochemical mechanism, seemed the ideal subject to study. However, as a physicist, Emil Warburg felt that he could not deal with living organisms, so that he tried to convince his son to collaborate with him (after Pringsheim had declined to join this project). In exchange, Emil Warburg could offer Otto the use of the PTR's sophisticated photophysical instrumentation as well as the help of collaborators, who could introduce him to these techniques.

Otto Warburg's unconventional approach to the study of photosynthesis can be attributed to the work that he carried out on cell respiration while in his late twenties. He continued to use manometry as a measuring technique and single cells as experimental objects; and he adhered to the assumption that surface-dependent heavy metal catalysis was the fundamental principle of the energy-producing reactions of respiration, fermentation and photosynthesis. Hence, he exemplifies a knowledge transfer that went far beyond Lindley Darden's transfer of (conceptual) mechanism schemes. And, like his father, Warburg believed that the concept of photolysis was the essential component of photochemical reactions, which also fitted Warburg's own notion of fermentative action. Finally, Warburg used the Willstätter–Stoll model of photosynthesis, with its complex of chlorophyll and a carbon dioxide derivative, as a starting point for his own modelling of the process, to which Warburg then introduced other cofactors and intermediate steps to accommodate his new empirical findings. Thus, Warburg's research pathway not only exemplifies the principle of research opportunism, which was what brought him to study photosynthesis in the first place; it also demonstrates the building block approach that was introduced in chapter 2. Warburg used techniques, instruments and concepts that he had acquired from his own experience as well as from other scientists, such as (1) the standard

[98] Holmes (2004).

body of knowledge of the time; (2) Warburg's own, earlier achievements, albeit not in the same field of research; and (3) the highly successful concepts of his father, which the latter had employed to explain related phenomena.

3.8 Warburg's Impact on Photosynthesis Research

Even though Warburg may have come to study photosynthesis partly by chance, in some ways he provided exactly what the field needed at the time. The extensive review of 1916 by Jörgensen and Stiles that has been cited earlier offered in its final section a vision for the further development of photosynthesis studies:

> This is the prospect that plant physiology is developing into an exact science, utilising the experiences of the fundamental sciences, physics and chemistry, but nevertheless a science, exact and independent, with its own working principles and methods, directing and stimulating the development of the applied sciences, agriculture and horticulture. [...] It is clear that the only way to attain a reasonable rate of progress is to institute a much closer and more intimate cooperation between scientific workers attacking the same problems from different points of view and by different methods.[99]

While Warburg was not the right person to institutionalise cooperation (he always preferred to work by himself), he was one of the few people at the time who combined practical and theoretical expertise from physics, chemistry and physiology. In order to find the photosynthetic mechanism Warburg not only picked up interests and methods from cutting-edge quantum physics and photochemistry, he also availed himself of new biological techniques and employed the methods that he had developed for his physiological work. And by interpreting his measurements of photosynthesis rates in terms of reaction mechanisms (a technique that later became very popular, especially in enzyme studies, but was still a novelty at this time), Warburg ingeniously utilised the progress made in the basic concepts of chemistry. This interdisciplinary mixture of approaches and techniques would soon become characteristic of the field.

The extent to which Warburg influenced photosynthesis research cannot be overestimated. Manometric studies provided information on the chemical and physical details of photosynthesis, which up to then had not been obtainable. The pertinent techniques very soon dominated the field, and by the 1930s the fruitless search for chemical intermediates had largely been dropped. Kinetic studies using gas exchange measurements became the standard approach—at least in those institutions that specialised in photosynthesis research. The German plant physiologist André Pirson wrote in an autobiographical account that, even in the 1930s, the application of manometric techniques was mostly unheard of in general institutes of botany,

[99] Jörgensen and Stiles (1916b, p. 91).

and even a decade after Warburg's first papers had been published, the use of unicellular algae as experimental organisms remained unfamiliar to most botanists.[100] Warburg's (and Willstätter's) findings and particularly their techniques only gradually found their way into university curricula. Photosynthesis remained a side issue, a situation that would not change until after 1945.

[100] See Pirson (1994).

Chapter 4
Struggling with the Standard Model (1930–1941)

In the early 1930s, photosynthesis research was still far from being the popular theme that it would later become. Those working in this field did so in an almost intimate atmosphere, having few potential collaborators—or competitors—as colleagues; and most of them were closely interrelated by friendship, personal collaborations or teacher–student relationships. The main work in photosynthesis research was undertaken in only a small number of places: in Berlin (Germany), in Pasadena and Chicago (USA) and in Cambridge (UK).[1] And it was only a small number of scientists, who studied the subject over a prolonged period of time, among these the main protagonists of this chapter: William Arnold, Robert Emerson, Charles Stacy French, Hans Gaffron and Robert (Robin) Hill. Others became engrossed in the field, although originally they had intended to take a quick research-opportunistic look at the subject before returning to their original interests; these included, most prominently, James Franck, and to some extent Hans Kautsky and Cornelis B. van Niel. The social structure of this group of actors—distributed over different places but in constant communication with each other—makes it very amenable to an analysis of heuristic strategies of a community.

If one were to single out a common feature of the photosynthesis experiments carried out in this period, it would be, in reception of Otto Warburg's work, the application of the technique of manometry to the study of the unicellular green alga *Chlorella*. Many of the scientists listed above, who would become world experts in photosynthesis, spent an extended period of research at Warburg's laboratory in the Dahlem district of Berlin, where they became familiar with the technique and with the model organism. Emerson went to Warburg's laboratory to write his doctoral thesis, and he continued to make use of the technical knowledge that he acquired in Dahlem for the rest of his life. He also transmitted this know-how to all his students, together with the conviction that there was no better alternative. French spent a year of his postdoctoral research in Berlin, having being sent there by Emerson, his mentor at the time. Gaffron worked for several years with Warburg, interrupted only

[1] In addition to these, the universities at Berkeley and at Urbana–Champaign in the USA would soon—in the 1940s—become equally important centres of photosynthesis research.

K. Nickelsen, *Explaining Photosynthesis,* History, Philosophy and Theory of the Life Sciences 8, DOI 10.1007/978-94-017-9582-1_4

by a research stay at Caltech, where he also visited Emerson. Franck never worked directly with Warburg, but they nevertheless knew each other well from the time Franck spent in Berlin, where he had been one of Emil Warburg's students. Thus, out of all the main characters of this chapter only Hill, Kautsky and van Niel had no direct links with Warburg.

As the setting's actors were so closely intertwined, it was difficult to organise the material of this chapter in thematic sections; inevitably a number of arbitrary breaks were introduced. I start with a discussion of the establishment of a new standard model of the photosynthetic mechanism, which became the common reference point of this chapter's players. Its development is closely connected to Franck's entry into the field of photosynthesis research, which was prompted, among other things, by the use of chlorophyll fluorescence as a new way of investigating the photosynthesis mechanism. The contributions Franck made to extending the Willstätter–Stoll–Warburg model of photosynthesis (see Chapter 3) helped make the latter the "received view" of photosynthesis in this decade. The remainder of the chapter then contains the various challenges to the standard model that arose during this decade, and discusses how the different actors reacted to them. Inevitably, again, the description of the different models is rich in chemical detail, while reflective and summarising sections are inserted to help understand the general course of events, even if the technical passages are skipped.

4.1 Fluorescence and the Standard Model

4.1.1 The Kautsky Effect

Beginning in 1931, the German chemist Hans Kautsky, at the University of Heidelberg, began to approach the problem of how photosynthesis works from a new angle: he investigated the fluorescence of chlorophyll solutions, that is, the emission of light after the pigments had absorbed radiation.[2] This first quantitative and systematic study of the fluorescence of chlorophyll was part of a larger project to investigate the energy transformation processes on boundary layers (*Grenzflächen*) of the cell, and had far-reaching consequences on further developments in the field. The results were unexpectedly complex: when photosynthesising cells were illuminated, the fluorescence intensity (starting from a rather low level) rose sharply to a high transient state, and then, after a few seconds, it slowly decreased again until it reached a steady-state level. (This phenomenon would later become known as the "Kautsky effect").[3] Kautsky and his group were the first chemists to interpret systematically the resulting data in terms of an underlying mechanism. This

[2] On Kautsky's life and work, see, e.g., von Gerhard (2004).

[3] See Kautsky and Hirsch (1931) and Kautsky et al. (1932) for the first reports of these phenomena, while Govindjee (1995) provides a historical review of the "Kautsky effect".

approach was based on the observation that, although chlorophyll solutions usually exhibited an intensive and beautifully red fluorescence, this fluorescence was found to decrease (to be "quenched", as it is called today) when the chlorophyll acted as a sensitiser in photochemical reactions, that is, transferred absorbed light energy to other molecules. Accordingly, Kautsky and his co-workers found that the fluorescence of assimilating leaves was comparably low, and that inhibiting photosynthesis with hydrogen cyanide resulted in a strong rise in fluorescence. Thus, fluorescence served as a convenient indicator of the efficiency of the cell's photosynthetic activity: the higher the fluorescence, the lower the utilisation of photons in photosynthesis.

In order to explain the curious rise of fluorescence at the onset of illumination, Kautsky and his co-workers suggested the following sequence of reaction steps: when illumination started, the fluorescence intensity was low because all the absorbed energy could be transferred to an acceptor molecule in the system. However, the concentration of this molecule almost immediately dropped again, which resulted in the peak in fluorescence (since the energy absorbed by the chlorophyll could not be transferred). The rapid increase in fluorescence was neither influenced by temperature nor by the addition of cyanide, and was thus taken to reflect a purely photochemical process. According to Kautsky, the subsequent slow decrease in fluorescence indicated that in this phase the chlorophyll transferred its energy again to an acceptor molecule in the system, the concentration of which rose very slowly. Since in this phase the rate of reaction was strongly influenced by both temperature and cyanide, as well as being linked to a strong rise in oxygen production, Kautsky and his co-workers suggested that, in parallel to the transfer of light energy, a thermochemical catalytic reaction was taking place, which produced oxygen.[4] At the same time, the group thought that oxygen was the molecule to which the chlorophyll transferred the absorbed light energy. A detailed theory of "sensitised photooxidation" was developed, which involved an activated, metastable state of oxygen that was particularly apt to oxidise further molecules in its surroundings. This process was purported to be at the core of the photochemical events that occurred during photosynthesis.[5]

Once it had been discovered that fluorescence studies were related to the photosynthesis problem, another eminent physicist, James Franck, joined the scene and tried to explain Kautsky's observations within the framework of the Willstätter–Stoll–Warburg model of photosynthesis.

Govindjee (2004a) covers the phenomenon of chlorophyll *a* fluorescence from both a historical and a systematic viewpoint.

[4] See Kautsky and Hirsch (1931) and Kautsky et al. (1932).

[5] See Kautsky et al. (1932, 1933, 1935). This proposal was contested by Hans Gaffron, who argued that photosynthesis started without oxygen; see Gaffron (1935). It was presumably through his follow-up of this debate that James Franck first became acquainted with Gaffron's work.

4.1.2 James Franck and Photosynthesis

James Franck is usually remembered for his seminal contributions to physics *sensu stricto* rather than for his work in photosynthesis.[6] Franck started his academic career as a doctoral student of Emil Warburg in Berlin, where he received his doctoral degree in 1906. Despite the difficulties that academics of Jewish origin were experiencing at that time, Franck stayed in Berlin and pursued his scientific interests as an assistant to the experimental physicist Heinrich Rubens, who had succeeded Emil Warburg at Berlin's Friedrich Wilhelm University. In 1911 Franck was promoted, on acceptance of his habilitation thesis, to the status of *Privatdozent* (which is roughly equivalent to the rank of an associate professor but without a proper salary). From 1912 to 1914, Franck collaborated with Gustav Hertz, another of Rubens's assistants and nephew of the renowned physicist Heinrich Hertz. The celebrated paper that arose from their collaborative work confirmed Planck's 1900 quantum hypothesis by showing that electrons scattering on a gas of mercury atoms lost energy only in quantised amounts. This "discovery of the laws governing the impact of an electron upon an atom" earned Franck and Hertz the 1925 Nobel Prize in Physics.[7] However, the beginning of the First World War, which placed other themes on their agendas, brought this fruitful collaboration to an end. In 1921, Franck accepted the Chair of Experimental Physics at the University of Göttingen (Germany), where he spent twelve highly productive years. Franck's focus of interest slowly shifted to the problems of energy exchange in photochemistry, in particular to the phenomena of fluorescence, phosphorescence and chemiluminescence.[8] With hindsight, this work paved the way for Franck's later interest in the physical foundations of photosynthesis. (Incidentally, it was also at this time that Eugene Rabinowitch became Franck's private research assistant: Rabinowitch was another of those physicists who would catch the "photochemical bug" and would eventually be drawn into the world of photosynthesis research).

Franck's happy years in Göttingen abruptly ended after the Nazi Government came to power in 1933. Following the infamous "Law for the Restoration of the Professional Civil Service", issued on 7 April 1933, all persons with at least one Jewish grandparent were dismissed from the civil service, which included university academics. And although Franck, as a First World War veteran, would have fallen under the only exemption clause to this law, he publicly resigned from his

---

[6] On Franck's life and work see, e.g., the biographical memoir by Kuhn (1965) and the tribute by Rosenberg (2004). Beyerchen (1996) analyses Franck's emigration from Germany and its consequences, notably his scientific migration to photosynthesis research. See also the extensive biography by Lemmerich (2007).

[7] Quote taken from the Nobel Prize Announcement at http://www.nobelprize.org/nobel_prizes/physics/laureates/1925/. See Franck and Hertz (1914) for the pertinent publication. Hon and Goldstein (2013) provides a lucid account of the discovery.

[8] This work included Franck's well-known paper on the "elementary processes of photochemical reactions", an analysis of the shape of molecular absorption and fluorescence spectra, which includes what later became known as the Franck–Condon principle; see Franck (1925).

professorship at Göttingen in protest. This courageous step caused an enormous stir, nationally and internationally, among scientists, politicians and the wider public.[9] The consequences were far-reaching. Although Franck had originally intended to stay in Germany, he soon realised that he would be unable to find a new academic post or a position in industry in his home country as long as the political circumstances did not change. Thus, after a short stay at the Johns Hopkins University in Baltimore, Franck spent a year at Niels Bohr's institute in Copenhagen (Denmark). In the meantime a professorship at Johns Hopkins had been arranged for him, which he was able to accept in 1935.

It was during these first years of exile that Franck became interested in the photochemical aspects of photosynthesis. In a detailed study of Franck's emigration from Germany, and his coincidental migration to a different field of science, the historian Alan Beyerchen identified Franck's stay in Copenhagen as the crucial turning point.[10] Franck's role there, as envisaged by Bohr, was to pursue current problems in nuclear physics. However, Franck became increasingly unhappy with this function: he found the field of nuclear physics too crowded, while his access to appropriate resources was too limited for him to be able to compete on an equal footing. Instead, Franck began a project on chlorophyll fluorescence of green leaves with Hilda Levi, a young molecular spectroscopist.[11] In addition, Franck collaborated again with Rabinowitch, who had in the meantime also emigrated to Copenhagen.[12] In an interview with Levi, Beyerchen learned that Franck became involved only because chlorophyll made good fluorescing solutions, which could be used to study the underlying energy exchange processes. At the time these processes, in particular the mechanism of sensitised photooxidation, were the subject of highly controversial debates.[13] However, Franck must have developed a genuine interest in photosynthesis shortly thereafter, since in the very same issue of the journal *Naturwissenschaften*, in which he published his findings with Levi, Franck also published his first conceptual paper on the photochemical mechanism of photosynthesis.[14]

A little later, Franck left Copenhagen and took up the tenured position in Baltimore. However, since he found working in nuclear physics equally unsatisfactory there, Franck continued the line of physico-chemical research that he had begun while in Copenhagen—and he would keep to the photochemistry of green plants for the rest of his working life. In his "Remarks on Photosynthesis" (1935), Franck

[9] An English translation of the pertinent documents (as well as perceptive commentaries and useful background information) can be found in Hentschel (1996, pp. 21–34).

[10] Beyerchen (1996, pp. 77–79).

[11] See Franck and Levi (1935a, b) for the resulting publications.

[12] See Franck and Rabinowitch (1934), in which they formulated the hypothesis of the "cage effect", based on their investigation of the photolysis of different compounds.

[13] Beyerchen (1996, p. 80). Beyerchen refers to an interview that he conducted with Hilda Levi on 12 November 1980 in Copenhagen.

[14] See Franck (1935a).

presented his ideas to the English-speaking world for the first time;[15] while in the following decades Franck published a series of increasingly sophisticated physico-chemical photosynthesis models, which he developed with various co-authors.[16] In view of this new field of research of his, Franck was invited, in 1938, to set up a laboratory dedicated to the study of photosynthesis at the University of Chicago—a project that was financially supported by the Jewish philanthropist Samuel Fels. Franck would direct the Fels Laboratory until his retirement in 1949. Thereafter, he was succeeded by his longstanding co-worker and friend Hans Gaffron, but even though he had given up the directorship, Franck continued to take an active part in the work carried out at the laboratory. Franck had invited Gaffron to come and work with him in Chicago in 1939 and, as Franck's former collaborator and biographer Jerome Rosenberg wrote, "the two constituted an interesting complementary pair, one emphasizing physical mechanisms, and the other comparative biochemistry and plant physiology".[17]

Although Franck started his photosynthesis studies as a typical research opportunist (he intended to have a shot at this theme, based on the expertise he had gathered in other fields, and then move on to other subjects again), events took a different turn. In a talk delivered at the Franck Memorial Symposium in 1966, Gaffron claimed that Franck had admitted that, by opting for photosynthesis, he had got more than he had bargained for: "His fate resembled that of the man who curiously puts a finger on a strip of flypaper, does not succeed in shaking it off and winds up in a terrible mess. In Franck's case this mess was biochemistry".[18] In the same vein, (the aforementioned) Rabinowitch, one of Franck's most ardent admirers, described Franck's entry into the sphere of photosynthesis research:

> He thought that the confusion prevailing in this field was due to [the] lack of precise definition and controlled experimentation by biologists, and that the quantitative approach of a physicist would soon dispel it. But he did not reckon with the complexity of phenomena in living cells. Franck believed that each measurement must mean something in biology, as it does in physics, and can be used as a reliable stone in constructing a mechanism or formulating a theory. The trouble is that in biology, no experiment can be "controlled" in the full sense this term has in physics, because the state and the properties of a living cell depend on its whole history, and thus on more variables than can be reliably controlled.[19]

Franck himself came to acknowledge the unforeseen difficulties: "It differs fundamentally from physics", he wrote to his close friend and former colleague Lise Meitner in 1941; "there, the most simple solution nearly always is correct, but this is

[15] See Franck (1935b).

[16] See Franck and Herzfeld (1937); Franck et al. (1941); Weller and Franck (1941); Franck (1945, 1949). Franck's final attempt to solve the problem was completed shortly before his death: see Franck and Rosenberg (1964).

[17] Rosenberg (2004, p. 73).

[18] Quoted in Beyerchen (1996, p. 82).

[19] Quoted in Beyerchen (1996, pp. 82–83).

absolutely not the case in living material".[20] On the other hand, only a few of Franck's colleagues in the field of photosynthesis research were able to grasp the gist of his contributions—first and foremost because they lacked the necessary background in physics, but also because, at the time, few people were interested in the details of the primary photochemical process to which Franck had turned his attention. In the end, most of his work on photosynthesis was superseded. However, Franck brought more to photosynthesis than his personal theories: He raised questions from the point of view of a physicist that drew attention to lines of research that were not sufficiently appreciated by his fellow biochemists and physiologists. Franck's outspoken goal was to make his colleagues realise that all models of the mechanism of photosynthesis had to meet the fundamental laws of physics—even though this would mean to discard some of their biological pet hypotheses.[21]

4.1.3 The New Standard Model

4.1.3.1 Stoll and Willstätter Again

Not only Franck developed a strong interest in the findings presented by Kautsky and his group but also Arthur Stoll and later Richard Willstätter made another attempt to solve the problem of how chlorophyll acted in photosynthesis, based on Kautsky's observations. Confirming their earlier suggestions,[22] Stoll, in 1932, reported his finding that the hydrogen atoms at position 9 of the chlorophyll molecule were very loosely bound, so that the chlorophyll could easily and reversibly be dehydrogenated. This made it probable, Stoll maintained, that chlorophyll played the role of both hydrogen donor and acceptor in photosynthesis.[23] While Stoll repeated his and Willstätter's earlier assumption that chlorophyll was able to transfer hydrogen to an activated derivative of carbonic acid, bound to the central magnesium atom of chlorophyll, he now considered more precisely the actual origin of this hydrogen: namely water. Chlorophyll, Stoll surmised, might be able to decompose water under the influence of light, possibly according to the equation: $2\ H_2O \longrightarrow 2\ H + H_2O_2$. Stoll suggested that the hydrogen released in this process would hydrogenate the chlorophyll, thereby raising the latter to a higher state of hydrogenation than usual. And in order to prevent the hydrogen peroxide, which was formed during the decomposition of water, from immediately dehydrogenating the chlorophyll again, the peroxide had to be decomposed to water and oxygen. This, Stoll stated, would match the earlier finding by him and Willstätter that a temperature-dependent process

[20] Franck to Meitner, quoted in Lemmerich (2007, p. 238); original German.

[21] Cf. Franck (1935b, p. 433).

[22] Cf. Willstätter and Stoll (1918).

[23] Stoll (1932, p. 957).

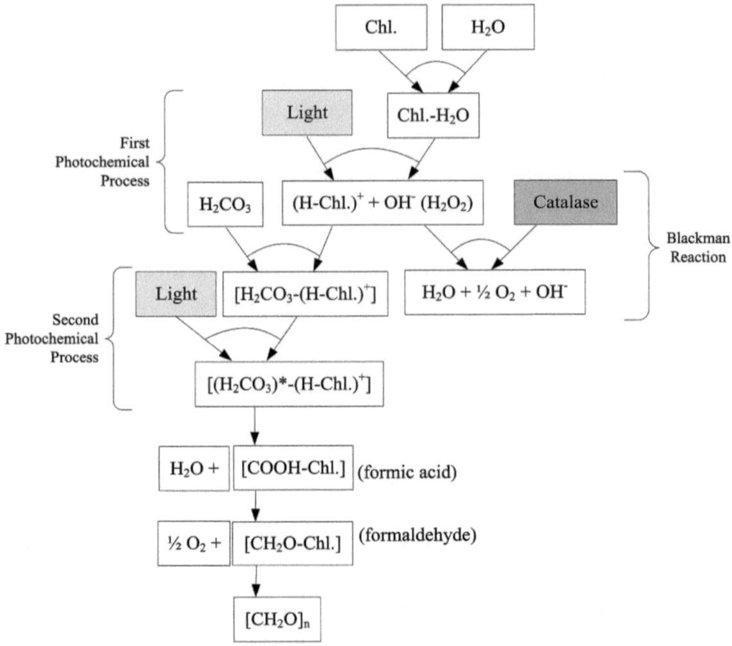

Fig. 4.1 The extended model of photosynthesis proposed by Stoll (1932).

occurred during photosynthesis, which most probably involved an enzyme similar to catalase, and which prompted the decomposition of hydrogen peroxide.[24]

Figure 4.1 shows this new photosynthesis model in a graph form. Water binds to chlorophyll, forming the complex Chl-H_2O. The latter is decomposed under the influence of light, whereby chlorophyll is hydrogenated to H-Chl. This is the first photochemical process. The simultaneously produced OH radicals would, most probably, combine to form hydrogen peroxide, which would immediately be removed under the influence of the enzyme catalase, whereupon oxygen is released. This was interpreted to be the temperature-dependent, enzymatic Blackman reaction. Hydrogenated chlorophyll (H-Chl) then binds carbonic acid (H_2CO_3) to form a complex. Under the influence of light, the carbonic acid in this complex would be activated (which in the graph is indicated by a star) and transformed into a derivative that is susceptible to reduction: this is the second photochemical reaction. A hydrogen then is transferred from chlorophyll to the activated carbonic acid derivative, which yields an unstable intermediate (H_3CO_3-Chl) that immediately decomposes to a derivative

[24] This was in line with the general belief at the time that the Blackman reaction (according to its kinetics) consisted of a reaction between catalase and a peroxide. Warburg and Uyesugi (1924) was particularly influential in this respect. It was only in Emerson and Green (1937) that the supposed similarity between the Blackman reaction and the reaction between catalase and hydrogen peroxide was contested.

bound to chlorophyll (H_2CO_2-Chl) and a hydroxyl radical (OH). The further steps of the process were not entirely clear. Most probably the sequence of first and second photochemical reaction would be repeated: the chlorophyll of the complex would be hydrogenated again and the formyl derivative activated. Thereupon another hydrogen transfer (and loss of oxygen) would eventually yield a formaldehyde derivative chlorophyll complex (H_2CO-Chl), from which the formaldehyde was released that would undergo condensation reactions to form glucose. Stoll argued that this mechanism was in close agreement with the findings of Kautsky and his group: in the dark periods, Stoll assumed, chlorophyll was bound to both water and carbonic acid. At the onset of illumination the chlorophyll was able, in principle, to transfer the activated hydrogen to its acceptor (the carbonic acid derivative), while the latter still had to be formed. This delay in acceptor formation should cause the fluorescence to increase rapidly, until the hydrogen acceptor was available in sufficient quantities. However, Stoll disagreed with Kautsky's verdict that oxygen was the first hydrogen (or electron) acceptor—without, though, fully explaining his objections.

Unlike Stoll, Willstätter, his former mentor, accepted the involvement of oxygen and integrated it into a model, which was published in 1933.[25] In this model, reconstructed in a graph form in Fig. 4.2, oxygen was, in fact, needed for photosynthesis to take place, namely for the dehydrogenation of chlorophyll: oxygen oxidised chlorophyll (that is, it took away one of the chlorophyll's loosely bound hydrogen atoms), which resulted in the formation of monodehydrochlorophyll (Chl^{DH}) and the radical O_2H. However, as this form of chlorophyll was considered unstable, it would be rapidly rearranged to the completely (di-)dehydrogenated form of chlorophyll (Chl^{DDH}) by donating the second loosely bound hydrogen to its central $Mg(H_2CO_3)$ complex (Willstätter also took it for granted that carbonic acid would bind to the chlorophyll's magnesium):

$$\text{Chl.} + O_2 \longrightarrow O_2H + \text{Chl.}^{DH}$$
$$\text{Chl.}^{DH} + CO_3H_2 \longrightarrow [CO_3H_2 + H] + \text{Chl.}^{DDH}$$

The Chl.^{DDH} thus formed was thought to react, under the influence of light, with water, whereby hydroxyl radicals and Chl.^{DH} were formed. Again, the latter donated the loosely bound hydrogen to the centrally bound carbonic acid:

$$\text{Chl.}^{DDH} + H_2O \underset{\text{light}}{\longrightarrow} OH + \text{Chl.}^{DH}$$
$$\text{Chl.}^{DH} + [CO_3H_2 + H] \longrightarrow [CO_3H_2 + 2H] + \text{Chl.}^{DDH}$$

Willstätter identified the latter as the central photochemical reaction, which was repeated three times, so that altogether four hydrogen atoms were transferred to the central magnesium complex. This sufficed for the complete reduction of the carbonic acid molecule. Thereafter (that is, as soon as the magnesium complex was fully saturated and carbonic acid completely reduced), Chl.^{DH} would, in a reaction with water molecules, be reduced to ordinary chlorophyll again. Willstätter did not go into any detail about the fate of the reduced carbon moiety, but one can safely

[25] Willstätter (1933).

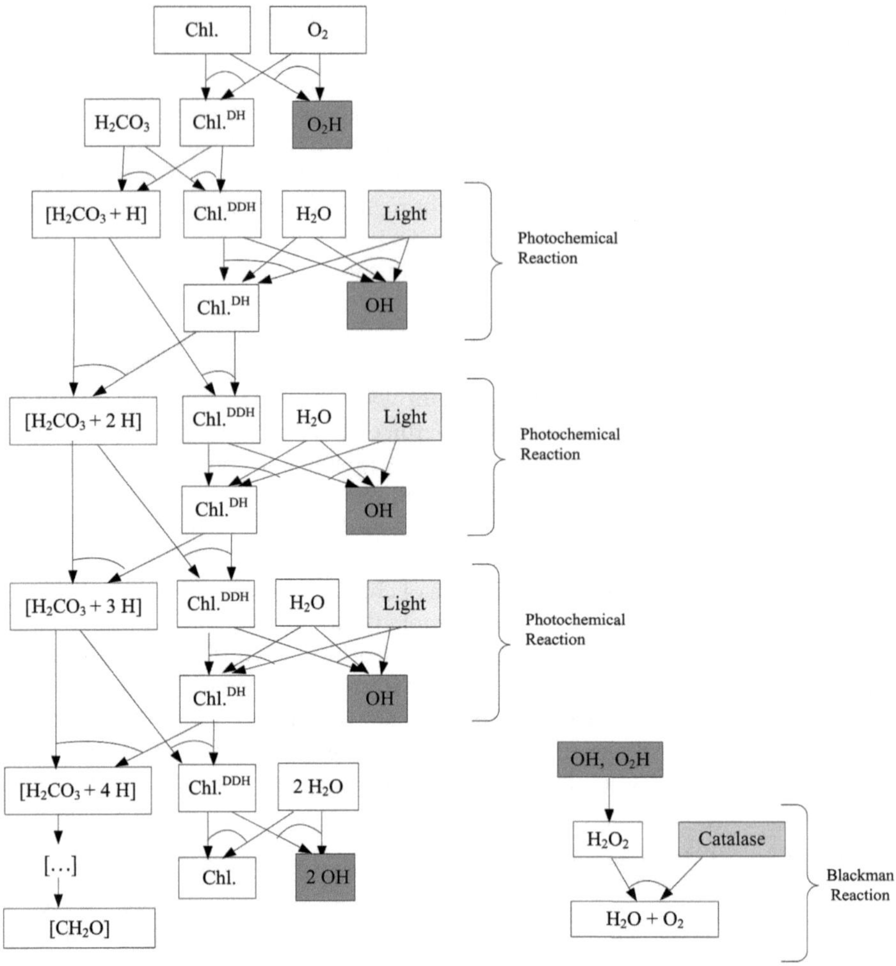

Fig. 4.2 Willstätter's photosynthesis model (1933).

assume that he believed that it was reduced to carbohydrates via formaldehyde in the usual way. All the radicals that were produced in the process (OH, O_2H) were assumed to end up as hydrogen peroxide, which was decomposed in the Blackman reaction through the action of the enzyme catalase.

Thus, in contrast to Stoll, who thought that, in photosynthesis, chlorophyll acted in a *higher state of hydrogenation*, namely as H-Chl., Willstätter assumed that it was the *dehydrogenated forms* of chlorophyll that entered the photochemical reaction. Both scientists, however, introduced the possibility that water might be decomposed in the course of photosynthesis and might donate hydrogen to the chlorophyll, which was then transferred to the carbonic acid bound to the central magnesium atom of chlorophyll. And both of them were convinced that the thermochemical Blackman reaction consisted of the decomposition of hydrogen peroxide through catalase, as a result of which molecular oxygen was released. It is interesting to see, though, that this "module" (since as such it was treated by both) was integrated very differently into the two divergent options. Neither of these suggestions contested the earlier Willstätter–Stoll model; rather, one of the partial processes (in this case, the light-driven reduction of carbonic acid or its derivative by the action of chlorophyll) was singled out and modelled in more detail than before—triggered, among other things, by the new empirical results of Kautsky's group and by Stoll's finding that the structure of chlorophyll has two weakly bound hydrogen atoms. Thus, both suggestions are classic examples of a model being locally "extended".

4.1.3.2 Franck Joins the Field

This was the state of affairs at the time that Franck published his first contribution to photosynthesis studies in 1935.[26] He conceded that, in 1918 and then in their contributions of 1932 and 1933, Willstätter and Stoll had offered:

> ... strong evidence that chlorophyll not only acts as a sensitizer, but that it enters into the course of the chemical reactions. Chlorophyll, having two especially loosely bound hydrogen atoms, is assumed to give off these atoms in reducing carbon dioxide and to regain the hydrogen by dissociating water.[27]

However, given his background in theoretical photochemistry, Franck was not satisfied with the prevailing suggestions for the underlying mechanism. Franck's main argument was that the steps proposed by Willstätter as being the core of the photochemical process were energetically impossible if one took for granted that for each step one quantum of red light was available (as was generally assumed to be the case, based on the findings by Warburg and Negelein of 1923; see section 3.3). Franck was equally dissatisfied with Kautsky's explanation of the course of chlorophyll fluorescence, which, Franck argued, implied assumptions that were at odds with the body of general knowledge of the fluorescence of liquids.

Hence, Franck presented an alternative mechanism, which not only met the energetic requirements but also explained why monodehydrochlorophyll ($Chl.^{DH}$) was necessary for the process to start (on this matter Franck agreed with Willstätter); and why the intensity of fluorescence was such a complicated function of irradiation time. Franck's model suggestion is reconstructed in a graph form in Fig. 4.3. In his paper, Franck emphasised that the following conditions had to be met:

> (i) If 4 quanta are necessary to reduce one carbon dioxide molecule, four different photochemical reactions have to be considered, since storing up energy in the form of the excitation energy of molecules is impossible. Hypotheses about metastable states with a long life time were likewise ruled out because the reactions took place in a condensed system.

[26] Franck (1935a, b).

[27] Franck (1935b, p. 433).

(ii) For each photochemical partial reaction the energy of 1 quantum of red light had to suffice.
(iii) Each individual photochemical step had to take place with a yield of unity, in accordance with the total quantum yield. Therefore, only those photochemical partial reactions could be considered in which at least one of the products was not a radical, so that back reactions would not take place.[28]

Franck believed that the last condition in particular dealt a final blow to Willstätter's 1933 proposal, which required the involvement of several radicals (such as OH, O_2H). This was far too costly, energetically speaking, given that Warburg and Negelein had determined the minimum quantum requirement of the process as four to five. Stoll's assumption that hydrogenated forms of chlorophyll might be involved was likewise refuted by Franck, since he believed that these compounds were too unstable to play a major role. Four photochemical steps had to be found, each of which required no more energy than was provided by one quantum of red light: this was, from Franck's perspective, the principal challenge. As can be taken from his papers, Franck found it almost impossible to devise a photosynthesis pathway that was sufficiently parsimonious in terms of energy expenditure. In his attempt to solve this task, Franck assumed that first monodehydrochlorophyll (Chl.^{DH}) was formed under the influence of light:

$$\text{Chl.} \xrightarrow[\text{light}]{} \text{Chl.}^{DH} + \text{H}$$

(This reaction, Franck maintained, was the reason for the induction period of photosynthesis, which had repeatedly been observed; at the same time it explained the rapidly appearing peak in fluorescence that Kautsky had reported). If no oxygen was present, the initial state of the chlorophyll would be quickly restored by the reverse reaction; while in the presence of oxygen the hydrogen atom would be used up in the following processes:

$$H + O_2 \longrightarrow O_2H$$
$$O_2H + O_2H \longrightarrow H_2O_2 + O_2$$

The hydrogen peroxide (H_2O_2) would then be removed by the action of catalase, in agreement with the earlier suggestion made by other authors. However, according to Franck the main procedure consisted of a series of reactions that occurred in and around the chlorophyll molecule, in which hydrogen atoms were exchanged for OH radicals. Concurring with Willstätter and Stoll, Franck assumed that these exchange reactions took place in a complex of chlorophyll and carbonic acid (or one of its derivatives), which went through the stages of formic acid and formaldehyde. The formaldehyde then was the usual starting point for the formation of carbohydrates in condensation reactions. If illumination was stopped, the monodehydrochlorophyll (Chl.^{DH}) would be restored to the usual form of chlorophyll (Chl.) by taking up a hydrogen atom from formic acid or formaldehyde (which would destroy some of the light reaction products). Franck was also ready to assume that, instead of oxygen, other primary hydrogen acceptors might possibly be involved in the first

[28] Cf. Franck (1935b, p. 436).

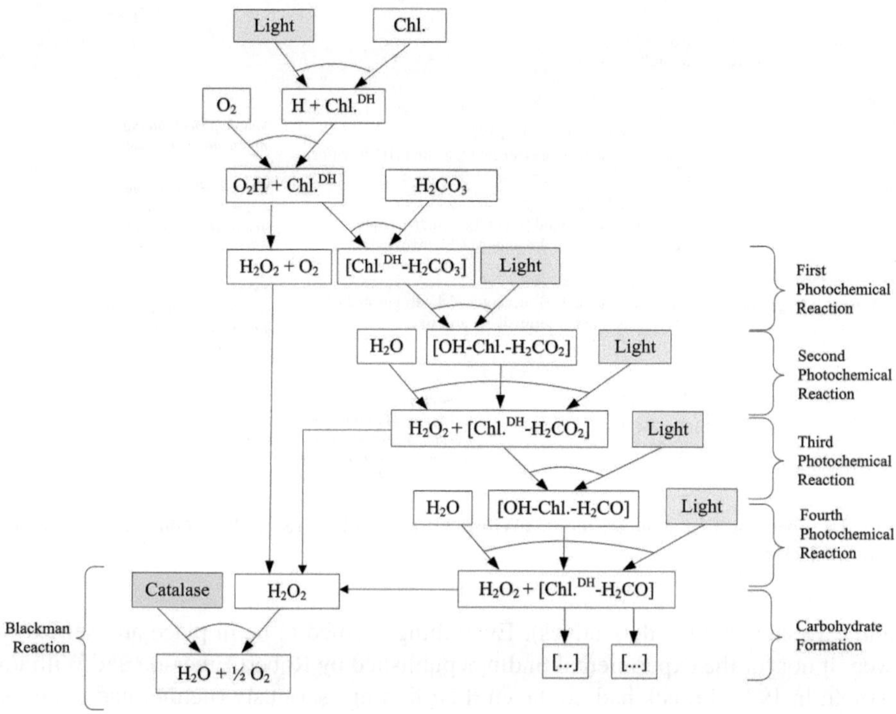

Fig. 4.3 Franck's model of photosynthesis in 1935.

step of the process (since Hans Gaffron had shown that photosynthesis also occurred in anaerobic conditions, without any oxygen present), although these supplements would be less effective: "The result of a lack of oxygen would then be that the induction period is lengthened", Franck concluded.[29]

Franck's suggestion, which is reconstructed in a graph form in Fig. 4.3, became widely accepted as the new "standard model" of photosynthesis. It was a modified and extended version of the original Willstätter–Stoll model, in which the photochemical and biochemical steps were adapted to the Warburg–Negelein value of the energy requirements of the process (4 light quanta per one molecule of oxygen). The fact that this adaptation was possible was seen to strongly support the general Willstätter–Stoll approach. Chlorophyll was still considered to be the site *and* the agent of photochemically driven carbon dioxide reduction, and oxygen was thought to be released from carbon dioxide, via the decomposition of hydrogen peroxide by catalase. The latter seemed to be the enzymatic, thermochemical Blackman reaction, while light acted in a series of reactions that took place in a complex of chlorophyll

[29] Franck (1935b, p. 437).

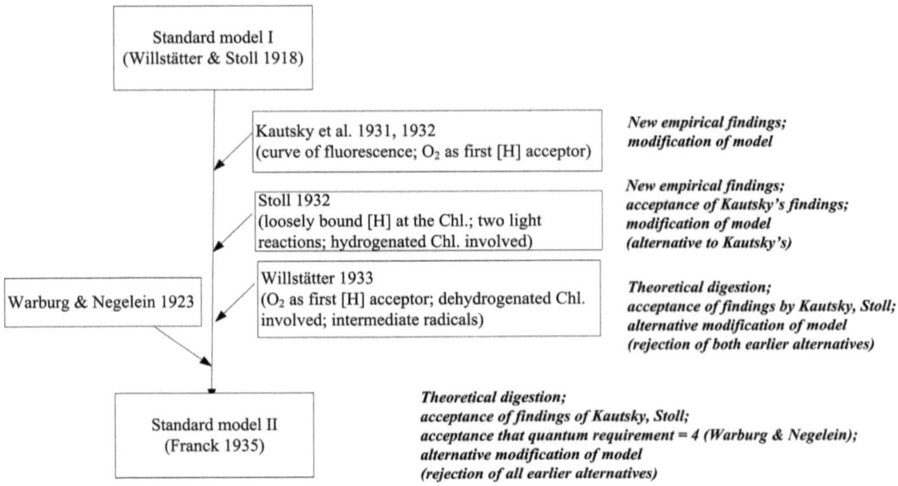

Fig. 4.4 From the first standard model (Willstätter and Stoll 1918) to the second standard model (Franck 1935).

and carbonic acid (or derivatives). Everything seemed to be in place and settled—were it not for the experimental findings published by Robert Emerson and William Arnold in 1932. Franck had not taken their findings seriously (neither had many of his colleagues); yet with hindsight it seems that Franck's model was already outdated by the time it was published. However, before I turn to these experiments, I shall briefly consider the different contributions so far from a systematic point of view.

4.1.3.3 Reflective Summary

A rough sketch of the relationship between the different proposals is given in Fig. 4.4. The first input considered in this chapter was of an *empirical* nature: based on a new methodical approach, Kautsky and his group presented their new finding, namely the curious shape of the fluorescence curve, which indicated the existence of underlying processes that had so far not been explained by the standard model. Kautsky's group thus suggested that molecular oxygen had to be integrated into the standard model of photosynthesis as the first hydrogen acceptor. Kautsky did not, however, present an extended model suggestion himself but left this task to others.

Kautsky's work was examined by Stoll, who was not convinced that oxygen was the first hydrogen acceptor, although he did admit that, in order to explain Kautsky's fluorescence curve, the standard model needed to be modified. From his own studies, Stoll reported another empirical finding, namely that chlorophyll has two loosely bound hydrogens, which (from Stoll's point of view) supported the assumption that it acted not only as a sensitiser but also took part in the actual redox reactions in photosynthesis. Thus, in his modification of the model, Stoll assumed that metastable, hydrogenated states of chlorophyll were involved and that two photochemical reactions took place (one of which would provide the energy for reducing the carbon dioxide in the chlorophyll complex; while the other, newly suggested reaction was thought to decompose water molecules, which Stoll regarded as a possible hydrogen donor). Stoll retained the earlier assumption that the Blackman reaction consisted of the decomposition of peroxides through the effect of catalase.

This suggestion was rejected, only one year later, by Stoll's former mentor and colleague Willstätter, who had no new empirical findings to add, but had thoroughly digested the earlier findings from a theoretical point of view. Willstätter accepted that the new empirical findings (of Kautsky, of Stoll and also of Warburg and Negelein) had to be accommodated by a modified photosynthesis model *and* he also regarded Kautsky's suggestion—that oxygen might be the first acceptor—as plausible. Furthermore, Willstätter assumed that the photochemical process consisted of several cycles of partial reactions, all of which involved radicals. Subsequent reactions also required the involvement of radicals; likewise, the Blackman reaction was associated with the reaction between catalase and hydrogen peroxide.

Both Stoll and Willstätter still considered their earlier approach to be "by and large" accurate; they retained the central elements such as the chlorophyll–carbon dioxide complex, the path of sugar formation via formaldehyde and the enzymatic dark reaction that yielded oxygen. The interpretation of the latter as the catalase reaction, removing hydrogen peroxides under oxygen release, had been added in the 1920s to the standard body of knowledge. Rather, their goal was to refine certain aspects of the model, while leaving other parts untouched. For example, they disagreed on where hydrogen peroxide was produced in the process, in what kind of state the chlorophyll would react and, most importantly, what exactly the photochemical reaction of photosynthesis consisted of.

Franck's contribution of 1935 marked the final digested state of the model. Franck also accepted the new empirical findings, but he took the 1923 finding of Warburg and Negelein—that photosynthesis required only 4 quanta of red light to produce one molecule of oxygen—far more seriously than the others had done. The quantum requirement, in fact, was taken by Franck to be the central parameter to which all adequate photosynthesis models had to adhere. This restriction made it highly improbable that metastable states of the chlorophyll molecule or radicals of any kind were involved in the reaction (as both Stoll and Willstätter had assumed), while Franck accepted Kautsky's suggestion that molecular oxygen was a primary hydrogen acceptor. Franck also retained the concept of the catalase reaction, and he did not even question the synthesis of sugars via formaldehyde. The resulting model hypothesis was as conservative as possible, as explanatory as possible (in view of the available empirical evidence) and as innovative as necessary (in suggesting a central sequence of four photochemical reactions in a series that did not include radicals). It was a well-balanced attempt to save the phenomena that had been established thus far and, at the same time, the generally accepted photosynthesis model.

4.2 The Crucial Experiments of 1932

With hindsight, the models of Kautsky, Stoll, Willstätter and Franck were only passing phenomena—while they were highly debated at the time and nicely illustrate how researchers struggled, around 1930, to reconcile new empirical findings, established bodies of knowledge and acknowledged theoretical requirements. By contrast, one of the most important developments of lasting impact of the early and mid-1930s was the concept of a "photosynthetic unit", which originated from the 1932 experiments carried out by Robert Emerson and William Arnold.

They found, in the course of flashing light experiments on photosynthesis, that, even under optimal conditions, only one molecule of oxygen was evolved in the alga *Chlorella* per about 2400 molecules of chlorophyll.[30] This result was quite unsettling, because up to then it had been taken for granted that every chlorophyll molecule would be as active as the other in binding carbon dioxide and, subsequently, reducing it. Emerson and Arnold had no idea how to make sense of this finding. It was only in 1936 that Hans Gaffron and Kurt Wohl would provide a theoretical interpretation and coin the actual term "photosynthetic unit"—but even then, this interpretation was not immediately well received. In the following, I shall first provide some background information on Emerson, who was the senior researcher of this project and is also a major figure in later chapters, after which I shall proceed to the 1932 experiments and examine how they were interpreted.

4.2.1 Robert Emerson: Harvard, Berlin, Caltech and Stanford

Robert Emerson dedicated his entire professional career to the study of photosynthesis with manometric methods, and his findings profoundly influenced and promoted this field of research.[31] When Emerson first went to Harvard University in 1921, he studied animal physiology, with the intention of (eventually) becoming a doctor. However, his interest soon shifted from animals to plants. Emerson himself ascribed this change of mind primarily to the influence of the botanist and plant physiologist Winthrop J. V. Osterhout, who took on Emerson as his laboratory assistant.[32] Osterhout is regarded as one of the founders of general physiology, which was the label

[30] The experiments and the many difficulties in realising the set-up have been described many times; see Myers (1994); Arnold (1991) and Govindjee (2001). Govindjee et al. (1996) is a special issue of the journal *Photosynthesis Research* dedicated to William Arnold; Govindjee (2014) provides a biography.

[31] The biographical information on Emerson has been taken from the memoir by Rabinowitch (1961), complemented by the details given in Govindjee (2004b), and by Emerson's own CV of 1936, which is held in his estate: *Curriculum vitae and bibliography of Robert Emerson*, Robert Emerson Papers, 1923–1961, Record Series 15/4/28, Box 1, University of Illinois Archives.

[32] Cf. Govindjee (2004b, p. 184). On Osterhout, see Blinks (1974).

given to an increasingly successful movement initiated by the eminent physiologist Jacques Loeb.[33] When Emerson was at Harvard, this new physiology was strongly promoted as an attractive alternative to the morphologically dominated curricula of traditional botany and zoology. Osterhout's research interests at the time were centred on the study of membrane properties, a subject to which he contributed some pioneering work; and he also worked on photosynthesis for a brief period.[34] Osterhout was one of the first professors at Harvard to integrate his own research and the recent work done by other scientists at other institutions into his lectures, which was rather unusual at the time. Furthermore, Osterhout gave a laboratory course that was for some time the only place at Harvard where students could undertake practical work in biochemistry.[35] With his engaged way of teaching, Osterhout succeeded in attracting many gifted students to his new experimental approach to studying life processes, Emerson being one of them.

Having received his first degree at Harvard in 1925, Emerson continued his graduate work in the country that was then the centre of science: Germany. He intended to study the formation of chlorophyll in plants, so he first planned to go to Munich and work with Willstätter, the leading chlorophyll expert of the time. However, since Willstätter had resigned his university position in 1924, as a public sign of protest against strong anti-semitic tendencies among faculty members, Willstätter advised Emerson to go and work with Otto Warburg in Berlin instead.[36] Emerson followed Willstätter's advice, and in 1927 he was awarded his doctorate from Berlin's Friedrich Wilhelm University.[37] It was during the course of his PhD studies and in this laboratory that Emerson became familiar with manometry and *Chlorella* as an experimental organism, both of which would play a significant role in the rest of his professional career. It was also in Berlin that Emerson first isolated the "Emerson strain" of *Chlorella pyrenoidosa*, which quickly became the standard experimental organism in photosynthesis research.[38]

---

[33] On the development of general physiology and biochemistry in the USA around 1900, see Hall (1975); Kohler (1982) and Pauly (1987b). On Jacques Loeb, see Osterhout (1928); Pauly (1987a) and Fangerau (2010).

[34] Osterhout was the first to notice the induction period of photosynthesis and to attempt to study systematically the antagonism that exists between respiration and photosynthesis. See Osterhout (1918, 1919); Osterhout and Haas (1918, 1919).

[35] Blinks (1974, p. 224).

[36] Govindjee (2004b, p. 184). See also Willstätter's autobiography for background information on his resignation. Wiesen (2000) discusses the ambiguous reception of Willstätter's memoirs after 1945.

[37] The title of the thesis was (translated into English) "On the effect of hydrocyanic acid, hydrogen sulphide and carbon monoxide on the respiration of different algae". The thesis was officially handed in by the university's botanist Hans Kniep, which at first glance implies that Kniep was Emerson's supervisor. However, this (nominal) arrangement was due to the fact that only universities were authorised to award doctoral titles, while Kaiser Wilhelm Institutes and their members were not.

[38] See French (1959, p. 437).

Emerson then returned to Harvard, while in 1930 he moved to California, to take up the post of Assistant Professor of Biophysics at the California Institute of Technology (Caltech) in Pasadena. This position was part of a newly founded programme in biochemistry and biophysics, which, in structure and approach, closely resembled the departments in general physiology being established elsewhere. Besides Emerson, the group consisted of Henry Borsook, a biochemist who had also been trained in general physiology in Toronto (Canada),[39] and the plant physiologist Kenneth V. Thimann. This group of young and talented scientists was set up to work at the forefront of experimental biology.[40] However, in 1931 Emerson complained in a letter about the attitudes of the Caltech biologists, who were all "milk-bottle-molasses and beef-hash-muscle in outlook", while he found the biochemistry section too medical and "very narrow".[41] Emerson remained at Caltech until 1946, taking a leave of absence in the years 1937–1940, which he spent at the Carnegie Institution of Washington at the campus of Stanford University. From 1946 until his untimely death (in an aircraft crash in the East River, New York, on 3 February) in 1959, Emerson was Research Professor of Botany at the University of Illinois at Urbana–Champaign as well as the Director of the Photosynthesis Project there. Emerson succeeded in recruiting Rabinowitch as a second director, which resulted in one of the most productive and fruitful centres of photosynthesis research.

Emerson became one of the leading experts in photosynthesis. His painstaking accuracy in designing experiments and constructing set-ups are legendary; his research questions were to the point and his interpretations careful and convincing. Emerson's strategy was to specialise to the point of perfection: he hardly ever used a technique other than manometry; and he rarely worked on a theme that did not involve oxygenic photosynthesis in aquatic algae. This also was the experimental context of the crucial 1932 experiments, to which I shall turn in the next section.

4.2.2 Emerson, Arnold and 2500 Molecules of Chlorophyll

4.2.2.1 Setting the Stage

The crucial results of 1932 had their roots in Emerson's course in Plant Physiology at Caltech. William Arnold, an undergraduate student of physics, had ended up in Emerson's class, because he could not fit the obligatory course in Elementary Biology into his timetable. Emerson and Arnold, the professor and the student,

[39] Borsook's recollections of this period are preserved in the interview carried out with him in 1978 by Mary Terrall, as part of the Caltech Archives Oral History Project; see Borsook (1978). On Thimann see, e.g., Stowe (1999).

[40] Kohler (1982, p. 318).

[41] Quoted in Kohler (1982, p. 322). Emerson to William J. Crozier, 24 March 1931. The original is held by the Harvard University Archives, Pusey Library: Crozier Papers. On Crozier and how he became a central figure in general physiology, see Pauly (1987b, in particular pp. 201–204).

who were only one year apart in age, took a liking to each other and engaged in scientific conversations that went far beyond the actual course work. As Arnold later recalled, Emerson was at the time very interested in the study of photosynthesis at intermittent light periods, studies that had been carried out by Brown and Escombe in 1905 and by Warburg in 1919 (see Chapter 3). The curious phenomenon was that one could omit as much as three-quarters of the light without there being a drop in the photosynthesis rate; and if conditions were optimised, one could actually increase the photosynthesis rate by using flashing light instead of continuous illumination. Emerson believed that these findings were important and considered adding a light source with rotating sectors to his own Warburg apparatus. Having heard this, Arnold suggested that Emerson might use neon lights for this purpose. Arnold was familiar with neon lights as a friend of his working in the Physics Department was involved in the development of these new light sources. Emerson agreed to Arnold's suggestion and he assured Arnold that installing the system would fulfil the laboratory work component of the Plant Physiology course. The experiment worked well, and when Arnold graduated in 1931, Emerson asked him to stay on and carry out some flashing light experiments with him. "Since I had been unable to find a place to do graduate work in astronomy, I agreed to continue as his assistant a while longer", Arnold explained later.[42] In the end, this stay would extend to another 15 months, and Arnold would never again return to either physics or astronomy.

Coming back to the 1932 experiments, the greatest difficulty the team had was building an appropriate flashing source and then implementing it into the manometric set-up. Arnold finally found that he could mount the neon tube on the water bath of the Warburg apparatus, directly underneath the vessels. This arrangement was eventually able to produce very short flashes of light. As usual, the rate of photosynthesis was measured manometrically. To ensure that the illumination was controlled only from underneath the reaction vessels, the sides and top of the vessels were silvered and then, to protect the silver, covered with copper jackets. The control vessels, which contained cultures grown in continuous light, were also illuminated by a bound neon tube mounted a few millimetres below the vessels. The only drawback was that the high precision necessary for equal illumination in all incidents only allowed them the use of a maximum set of three vessels at a time, two of which contained cell suspensions and one that was used as a zero control.

4.2.2.2 Separating the Photosynthesis Reactions

The first remarkable finding obtained using this set-up was that if the light period were sufficiently short and the dark period sufficiently long, photosynthesis rates could be increased by up to 400 %.[43] Emerson and Arnold's interpretation of this finding was in line with Warburg's earlier suggestion—that the photochemical reaction proceeded

---

[42] Arnold (1991, p. 74).

[43] Emerson and Arnold (1932b, p. 417).

rapidly until an equilibrium concentration of its product was reached, which had to be removed by the thermochemical Blackman reaction before the next cycle could start. Emerson and Arnold believed, like many others at the time, that the Blackman reaction would release the chlorophyll from its complex binding to the carbonic acid derivatives, so that it would be able to react once more with carbon dioxide molecules.[44] The important achievement of this first paper of theirs was, however, that it provided the first realistic estimation of the time scale of a full cycle of photosynthesis. Emerson and Arnold maintained that "the dark reaction requires less than 0.04 seconds for completion at 25°C, and about 0.4 seconds at 1.1°C", while the light reaction, which was not affected by temperature, could take place in about a hundred-thousandth of a second.[45] These were numerical parameters almost as fundamental as the minimum quantum requirement, which all subsequent photosynthesis models had to accommodate.

4.2.2.3 The Photochemical Reaction in Photosynthesis

Emerson and Arnold published another paper in 1932, the scope of which was to establish the ratio between the number of chlorophyll molecules present in a cell suspension and the number of molecules of carbon dioxide that are reduced:

> From the experiments of Warburg and Negelein (1923), we know that the green alga *Chlorella pyrenoidosa* can reduce one molecule of carbon dioxide for each four quanta of light absorbed, when conditions permit maximum efficiency. Chlorophyll is clearly the substance absorbing the light quanta, so we may inquire how much chlorophyll must be present for the reduction of one molecule of carbon dioxide.[46]

If the photochemical reaction were saturated with light and the dark periods were long enough for the Blackman reaction to process all the photochemical products, then the number of carbon dioxide molecules reduced per light flash would reveal how many "units" of photosynthesis were present in the sample (the "unit" was regarded as an abstract entity, that is, as "the mechanism which must undergo the photochemical reaction to reduce one molecule of carbon dioxide".[47]) The chlorophyll content of the sample divided by the number of "units" would yield the number of chlorophyll molecules per unit. The background of this experiment was provided by a long-standing assumption, originally put forward by Willstätter and Stoll, that the rate of photosynthesis was independent of the chlorophyll content of a leaf, which Emerson had already challenged in an earlier paper, although he had not been able to clarify completely the relationship between chlorophyll and the rate of photosynthesis.

Their main technical problem was how they could produce flashes of sufficient light intensity to ensure light saturation. Emerson and Arnold finally succeeded by

---

[44] See Emerson and Arnold (1932b, p. 418).

[45] Emerson and Arnold (1932b, p. 417).

[46] Emerson and Arnold (1932a, p. 191).

[47] Emerson and Arnold (1932a, p. 191).

concentrating the light incident on the cells by means of a concave mirror mounted below the neon tubes. Even then, light saturation was only approximated. They determined the chlorophyll content using a spectrophotometer, calibrated with standard samples of chlorophyll (the material for which, incidentally, had been provided by Hans Gaffron, who at the time was spending a brief period of research at Caltech). Algae cultures with varying chlorophyll content were grown by exposing them to light of different colours. However, more factors than previously suspected seemed to influence the rate of photosynthesis in the algae, as Emerson and Arnold admitted: "The chlorophyll concentration produced appears to depend on the intensity of the light and the age of the culture, as well as on the colour of the light. The neon light cultures mature faster than the incandescent light cultures, the mercury cultures much more slowly".[48] These observations proved to be typical: the complex behaviour of the cells, the performance of which was highly dependent on a plethora of environmental factors, would remain a challenge for all photosynthesis researchers using these organisms. Finding the optimal conditions for cellular growth and implementing these conditions as a standard became a central activity in all laboratories researching photosynthesis.

When the set-up was finally established, Emerson and Arnold found, to their utter surprise, a constant value of one molecule of oxygen evolved in *Chlorella* cells per about 2500 molecules of chlorophyll, which, consequently, had to be considered a "unit". Emerson and Arnold were as stunned as their audience and completely at a loss as to how to interpret this finding.[49] In the meantime, photosynthesis research was being decisively influenced by the developments taking place in a rather different field—one that, up to around 1930, nobody would have considered even remotely relevant to questions concerning photosynthesis: the discipline of microbiology. The driving force behind these developments was the Dutch microbiologist Cornelis B. van Niel; and his contributions will be examined before I turn to the conceptualisation of the photosynthetic unit.

4.3 The Generalised Equation for Photosynthesis

4.3.1 Cornelis B. Van Niel and General Microbiology

Cornelis B. van Niel (1897–1985)—or "Kees", as he was known to his friends—truly revolutionised the field of microbiology. In addition van Niel greatly advanced the field of photosynthesis research by bringing to the fore the fact that photosynthesis

---

[48] Emerson and Arnold (1932a, pp. 193–194).

[49] The findings were confirmed two years later, by Arnold and Kohn (1934), who discovered, as stated in the abstract, that "in six species of plants, representing four phyla, the minimum number of chlorophyll molecules present for each molecule of carbon dioxide reduced appears to lie between 2000 and 3000".

occurs not only in plants but also in certain bacteria, and that the study of these organisms could contribute enormously to scientists' understanding of the workings of higher plants and algae.[50]

Van Niel's first degree of 1922 was in Chemical Engineering, which he studied at the then Delft Technical College in the Netherlands (which today is known as the Delft University of Technology). However, he found that microbial fermentation made up a large share of the curriculum of this subject, so that he switched to the Microbiology Department for his graduate studies. The department was then headed by Albert J. Kluyver, who had succeeded the eminent Martinus W. Beijerinck in this position—much to the surprise of his colleagues, since up to then Kluyver had been better known for his chemical expertise than for his knowledge of microbiology.[51] However, his appointment turned out to be most fortuitous: Kluyver became the founder of comparative microbiology, which soon made the college in Delft internationally famous. This tradition enormously influenced the young van Niel. Kluyver was convinced that the study of microbiology was highly relevant to a better understanding of the biology of higher organisms. His first microbiological paper, entitled "Unity and diversity in the metabolism of micro-organisms" (1924), was a comparative study within the bacterial realm. Only two years later, in 1926, Kluyver published, together with his associate Hendrick J. L. Donker, the classic and much more ambitious paper "Unity in Biochemistry".[52] In this paper (which, unfortunately, appeared in a rather obscure German journal), the authors proposed no less than a general theory of metabolism, aimed at unifying the study of biochemistry.[53]

In their 1926 paper Kluyver and Donker endorsed Heinrich Wieland's theory of redox reactions as being hydrogen transfers. (This notion was fiercely opposed by Warburg, who defended the view that oxygen had to be involved in oxidation reactions.[54]) Kluyver and Donker believed that hydrogen transfers were at the core of all metabolic reactions. From their point of view, even the most complicated biochemical processes could be reconstructed as a series of hydrogen transfer reactions; and Kluyver and Donker were able to provide ample evidence for this assumption from the realm of bacterial metabolism. This pointed perspective attracted much attention among fellow scientists; and Kluyver used his sudden popularity to promote comparative microbiology, which, he believed, deserved to become as widespread and influential as comparative anatomy once had been.[55]

In 1923, van Niel was made Kluyver's assistant and became responsible, from 1923 to 1928, for the large Delft culture collection of bacteria, yeasts, algae and protozoa. Van Niel thus had to familiarise himself thoroughly with the handling of a tremendous range of identified microbes, which he would later describe as

[50] On van Niel see, e.g., Spath (1999); Barker and Hungate (1990) and Hungate (1986).

[51] See, on Kluyver's life and work, e.g., Woods (1957) and Kamp et al. (1959).

[52] Kluyver and Donker (1926). The original German title reads "Die Einheit in der Biochemie".

[53] On this paper's background and further implications, see Friedmann (2004).

[54] See Werner (1997) for an analysis of the controversy between Warburg and Wieland.

[55] See Spath (1999, Chapter 1, pp. 36–37).

having been a privilege. While carrying out this work, van Niel came across, for the first time, a group of purple bacteria that belonged to the family *Thiorhodaceae*, which was then the subject of great controversy. Enormous confusion prevailed as to whether the bacteria in this group could be considered, metabolically speaking, chemosynthetic, photosynthetic, neither or both.[56] By 1926, van Niel had found evidence, first, that they were actually photosynthetic (that is, they derived the energy for their metabolism from light) but that they still depended on the presence of hydrogen sulphide. Van Niel was even more excited when he found that some non-sulphur purple bacteria (*Athiorhodaceae*) "could develop in the same medium either anaerobically, but only if illuminated, or aerobically in complete darkness, so that for these organisms light and oxygen appeared to be equivalent".[57]

Despite the enthusiasm of his student, Kluyver could not be persuaded to accept this as the basis of a doctoral thesis. Instead, he encouraged van Niel to work on propionic acid bacteria, which were well known for their function in the ripening of certain types of Swiss cheese. Kluyver insisted that this would be a much better preparation for the work in industry that he anticipated for his students. Furthermore, Kluyver believed that, as purple bacteria grew so slowly, it would take van Niel too much time to get anywhere in his thesis. Van Niel reluctantly agreed to Kluvyer's suggestion, and in 1928 he received his PhD for a thesis on the biochemistry and morphology of propionic acid bacteria. Among other things, one of the questions that van Niel examined was the origin of the holes in Swiss Emmental cheese. Albeit amusing, this was definitely not the kind of fundamentally important microbiology in which van Niel had hoped to engage.

Yet by 1928, van Niel was fully convinced of the importance and justification of the line of research that Kluyver had initiated. Although microbiology was a science that had to deal with both practical and fundamental problems, van Niel believed (in agreement with Kluyver) that it was inappropriate to associate it primarily, as was usually done, with medical "bacteriology" or with the technical application of microbes in industrial fermentation, such as the brewing and dairy industries. Rather, he believed that microbiology should be conceived of as a branch of biology, and that it should be practised in a broadly encompassing and comparative way: general microbiology was what van Niel had in mind.[58] His vision had much in common with the programme pursued by general physiologists in the USA; it was also close to the gist of Otto Warburg's scientific approach as well as to the way (general) biochemistry was being practised by Frederick G. Hopkins at the University of Cambridge (UK), as I shall discuss later in this chapter. It was the search for fundamental communalities in the different realms of life that this generation of researchers shared. Van Niel

[56] See van Niel (1941, pp. 264–269), for a review of the field up to van Niel's own work. Chemosynthetic bacteria are able to reduce carbon dioxide (or methane) in order to produce organic matter; while they use the oxidation of inorganic molecules (e.g. hydrogen gas, hydrogen sulfide) or methane as a source of energy, rather than sunlight.

[57] See van Niel (1967, p. 11).

[58] How he happened to develop this broad vision is explored in Chapter 1 of Spath (1999).

realised that he would be hard put to find an academic position in the Netherlands, where he could fulfil his vision. Thus, in 1928 van Niel and his family moved to California, where he had been offered a position at the Hopkins Marine Station in Pacific Grove, affiliated to Stanford University. In the Jacques Loeb Laboratory of this station van Niel would work for the next 35 years of his life.

At the time, experimental biology was being strongly promoted at Stanford University (as at many other institutions in the USA). It was a field that had started to thrive not the least because it was being fostered by the Rockefeller Foundation.[59] The establishment of the new laboratory at the marine station was part of this general development. Emulating, perhaps, the successful profile of the Zoological Station in Naples (Italy), where Otto Warburg once had learned about the latest trends of the field, the laboratory was not organised along disciplinary boundaries and thus brought together proponents of very different branches of biology, pursuing a number of different research themes. The young biophysicist Lourens G. M. Baas Becking was head of the laboratory; and while on sabbatical leave in the Netherlands in 1928, looking around for suitable staff to complete the marine station's profile, he succeeded in recruiting van Niel as an assistant professor. Although van Niel was happy to have found work at the marine station, he was nevertheless disappointed to discover that cooperation between the different scientists in residence did not work out quite as he had expected. This brought van Niel to develop a rather pragmatic attitude towards interdisciplinary cooperation:

> This experience taught me that the attack on a problem which requires the joint efforts of diverse specialists is likely to be successful only if it develops through the gradual accretion of a group whose members have already evinced a desire to work on specific aspects of that problem.[60]

Van Niel quickly settled down to working on purple bacteria, the growth of which he found to be greatly accelerated if they were continuously illuminated at high light intensities. He also continued thinking about the unity of metabolism—for example, the unity that existed in different types of photosynthesis. It is to van Niel's work on the photosynthesis of purple bacteria that I shall now turn.

4.3.2 *Bacterial Photosynthesis and the Consequences*

In 1929, van Niel presented, for the first time, the results of his six years of work on purple bacteria. This he did in form of a talk at a gathering of the Western Society

---

[59] Opinions differ as to how to interpret this general shift to experimental biology. Kay (1993) suggests that there prevailed an agenda of "social control", exerted through a concerted campaign by the Rockefeller Foundation, a conservative American elite and some influential scientists. Spath (1999) points to the fact that, although the effects of this move were convergent, the interests and aims of individual scientists at the numerous institutions were very different.

[60] See van Niel (1967, p. 10).

of Naturalists, which that year was holding its traditional winter meeting in Pacific Grove. Van Niel's findings, as he later wrote:

> ... supported the view that photosynthesis can be considered as a light-dependent reaction in which different substances, specific for different kinds of photosynthetic organisms, serve as H-donors for the reduction of CO_2.[61]

This is an extremely dry formulation of what was then a completely revolutionary idea. In 1930 photosynthesis was still defined as a process in green plants (and algae) that produced oxygen—which *by definition* excluded the possibility that photosynthesis might occur without oxygen evolution.[62] Van Niel now set out to persuade people that the process familiar to plant scientists was only one out of a whole range of possibilities. This was going even further than the bold (and not very well received) suggestion that some years earlier had been proposed by a French plant physiologist, René Wurmser, reviving an earlier idea that had also been suggested by Georg Bredig (see Chapter 2), that the oxygen produced during photosynthesis might come from water.[63] In 1935 van Niel gave a succinct summary of his findings (given in the form of ten points):

(i) There exist bacteria which can develop in entirely inorganic media containing H_2S, in the complete absence of oxygen, but only in the light.
(ii) No development of these organisms takes place if H_2S is omitted.
(iii) In media containing a sufficient quantity of $NaHCO_3$, ammonia-N (nitrogen in the form of ammonia), K, P, and Mg the amount of development is strictly proportional to the quantity of H_2S present.
(iv) No development takes place in the absence of CO_2 (carbonate, bicarbonate).
(v) Oxygen is not produced.
(vi) During the development of these organisms H_2S becomes converted into S (green bacteria) or into H_2SO_4 (Thiorhodaceae).
(vii) The reaction of the medium becomes more and more alkaline due to the disappearance of CO_2.
(viii) Chemical analyses show that there exists a stoichiometrical relationship between the quantity of H_2S oxidised and the amount of CO_2 which has disappeared, to wit: for one molecule of H_2S oxidised to S, 0.5 molecule of CO_2 disappears (green bacteria); for 1 mol. of H_2S oxidised to H_2SO_4 almost 2 mol. of CO_2 (1.8) disappear.
(ix) The carbon of the CO_2 which has disappeared can be recovered as organic carbon in the form of bacterial substance.
(x) In the dark, in the absence of oxygen, no development takes place; H_2S is not converted into S or H_2SO_4, and there is no disappearance of CO_2.[64]

From these findings van Niel concluded that, if these bacteria really did convert carbon dioxide into an organic substance under the influence of light, one would be entirely justified to call this process "photosynthesis" and to describe the organisms

[61] See van Niel (1967, p. 19).

[62] For a timeline of research in anoxygenic photosynthesis, see Gest and Blankenship (2004).

[63] See Wurmser (1921, 1926) and, especially, Wurmser (1930). Ideas on photosynthesis being a redox process were also expressed in Thunberg (1923), although he was still looking for an acceptable pathway to formaldehyde.

[64] van Niel (1935, pp. 138–139).

that carried out this type of metabolism as "photosynthetic". The light-dependent process in the metabolic reactions of "photosynthetic bacteria", in the sense described above, was formulated by van Niel as follows:

$$2\,CO_2 + H_2S + 2\,H_2O \xrightarrow[\text{light}]{} 2\,[CH_2O] + H_2SO_4$$

This was then compared with the usual formulation of the widely accepted summary equation for photosynthesis in green plants and algae:

$$CO_2 + H_2O \xrightarrow[\text{light}]{} [CH_2O] + O_2$$

Emphasising how strikingly these equations resembled each other, van Niel proposed the following general equation for photosynthesis:

$$CO_2 + H_2A \xrightarrow[\text{light}]{} [CH_2O] + 2\,A + H_2O$$

In this form, photosynthesis was understood to be the photochemically driven reduction of carbon dioxide with a variety of hydrogen donors. Of course, this concept also bore enormous consequences for the understanding of photosynthesis in plants and algae. Van Niel himself addressed this question in his 1935 paper:

> If one tries to understand the meaning of the generalized equation for photosynthesis it becomes clear that all those mechanisms proposed for the photosynthetic reaction which imply the formation of a carbonic acid-chlorophyll complex which is subsequently transformed into a formaldehyde peroxide are not quite in accordance with the formulation of photosynthesis as an oxidation–reduction process. Such schemes fail to give a satisfactory explanation for the photosynthetic processes carried out by the green and purple bacteria. From a unified point of view, as laid down in the generalized equation, green plant photosynthesis should be considered as a reduction of CO_2 with hydrogen obtained from H_2O, and the oxygen produced during illumination as dehydrogenated H_2O.[65]

This was a severe blow to the standard model. The weak point in van Niel's argument, however, was that he was unable to propose a viable alternative—the generalised equation in itself was only a summary of the process, not a mechanism. As far as the photochemical part was concerned, van Niel was ready to follow Franck's principal line of reasoning. The absorption of 4 quanta for the reduction of one molecule of carbon dioxide in green plants, van Niel argued, strongly suggested the activation of four water molecules in the photochemical reaction; and this activation was obviously brought about by the chlorophyll. As to the thermochemical part, which van Niel considered to be the reduction of carbon dioxide (and not, as was usually assumed, the removal of hydrogen peroxide!), he was convinced that some intermediate products had to exist, since all the redox reactions known then proceeded in small steps of one, or at most two, hydrogen atoms at a time; but what these products were remained an open question.[66]

---

[65] van Niel (1935, pp. 142–143).

[66] van Niel (1935, p. 143).

As one might expect, researchers in photosynthesis were not yet ready to accept van Niel's new concept at face value. The idea that the molecular oxygen originates from water and not from carbon dioxide, still seemed outrageous to many scholars—although it had been suggested by various actors before that at least parts of the oxygen might come from the splitting of water; Bredig, Wurmser and Thunberg have already been mentioned in this book. It still would require an enormous cognitive leap (which many regarded as over the top) before scientists working in the field could drop this long-established assumption. The same held true for the other standard elements of photosynthesis models, such as the chlorophyll–carbon dioxide complex and the assumptions that the reduction of carbon dioxide was part of the light reaction, while the (dark) Blackman reaction was the removal of hydrogen peroxide. The suggestion that bacterial metabolism might be considered "photosynthetic" was equally preposterous to many of van Niel's fellow scientists. Yet, even though not many scientists were prepared to accept his suggestion in detail, van Niel's proposal provoked much discussion, and scientists around the world reacted to it—if only by attempting to provide a convincing rebuttal of this unthinkable possibility.

Far from considering his ideas to be revolutionary, van Niel himself believed, even in hindsight, that, given his background and his exposure to Kluyver's work, the suggestion that photosynthesis be investigated in the more general framework of hydrogen transfer reactions, was self-evident:

> It was a logical extension into the realm of photosynthesis of the general concept, then being developed by Kluyver and his co-workers, that fermentative as well as oxidative metabolic processes can be considered as composites of more or less elaborate series of consecutive and chemically intelligible step reactions, each one of which represented an inter- or intramolecular transfer of hydrogen atoms from a donor to an acceptor molecule or site of a molecule. The results I had obtained by 1926 had shown that the purple sulfur bacteria, or *Thiorhodaceae*, can grow in strictly mineral media but only when exposed to light. This meant that they had to be considered as photosynthetic organisms. On the other hand, the requirement for H_2S and their failure to produce O_2 could now be interpreted to mean that they use H_2S as the specific H-donor for the reduction, or assimilation, of CO_2. The gist of this idea was incorporated by Kluyver & Donker in their epoch-making treatise on "Unity in Biochemistry".[67]

This surely does not do justice to the amount of conceptual work that van Niel must have undertaken to come up with his general equation (it was, for example, by no means self-evident that an organism that requires light for growth should be considered "photosynthetic"). However, the exposure for some years to Kluyver's general theory that metabolic reactions occur by hydrogen transfer is very likely to have promoted the generation of these ideas. The parallel between van Niel's concept and Kluyver's programme was already clear at the time: comparative microbiology was the theme of van Niel's first seminar at the Hopkins Marine Station. As early as his lecture of 1929, six years before he published his concept of the general photosynthesis equation, van Niel had laid great emphasis on the fact that the study of the metabolic pathways of such inconspicuous organisms as *Thiorhodaceae* could

---

[67] van Niel (1967, p. 10).

contribute to a better understanding of the metabolism of a wide range of higher organisms. He wrote:

> And here especially lies the importance of the study of these "abnormal" photosynthetic processes, because a comparison of the factors and conditions which are required for their accomplishment will enable us to find those characteristics which are common to all. It will then be possible to derive the fundamental laws underlying all photosynthetic processes and to correlate these into a general view.[68]

Van Niel's own achievements were the most compelling evidence for this sweeping statement. He provided a new and utterly unexpected link between general microbiology, general physiology and general biochemistry. His amenable personality made this link appear even more compelling and inspired many young researchers to study bacterial photosynthesis—a research theme that did not exist before 1929. Van Niel's summer courses in General Microbiology quickly became internationally renowned, and after a few years the Hopkins Marine Station had turned into a thriving research centre for this discipline. In the next section, I shall take a look at the work of a contemporary scientist, who at first strongly opposed van Niel, but eventually became one of the latter's staunchest supporters: the aforementioned Hans Gaffron, a German chemist-turned-microbiologist, who, together with Kurt Wohl, developed in 1936 the bold hypothesis of a functional and physical "photosynthetic unit" in plants.

4.4 The Photosynthetic Unit

Hans Gaffron spent the first ten years of his life in Lima, Peru, where his father Eduard had settled as an affluent physician.[69] In 1912, after his father retired, the family returned to Germany, and in 1920 Gaffron started his studies in chemistry at the universities of Heidelberg and Berlin. His academic career began in 1925, with a doctoral thesis completed at the Chemical Institute of Berlin's Friedrich Wilhelm University under the supervision of Wilhelm Traube. In the same year, Gaffron was appointed to the post of Research Assistant in Otto Warburg's department at the Kaiser Wilhelm Institute (KWI) for Biology. He was thus working with Warburg when Robert Emerson arrived to take up his doctoral studies; and a friendship between Gaffron and Emerson developed that was to last for the rest of their lives. Like Emerson, Gaffron became thoroughly familiar with the technique of manometry, which also continued to be his preferred measuring method after leaving Berlin.

Gaffron stayed with Warburg for six years, an unusually long period for a research scholar in this laboratory, which testifies to the good working relationship they must

---

[68] van Niel (1930, p. 168). Quoted in Spath (1999, p. 117).

[69] The biographical information on Gaffron was taken from Rürup (2008, pp. 199–201). For a tribute to Gaffron and his co-workers, with special emphasis on Gaffron's work on the hydrogen metabolism in green algae, see Homann (2002).

have had. In 1931, Gaffron went to Caltech in the USA for a year as a guest, where he worked in close proximity to Emerson and Arnold, who were carrying out the crucial flashing light experiments for their 1932 paper. Gaffron then spent some time at the Zoological Station in Naples (Italy), before moving back to Berlin, in 1933, to accept a position as a Research Assistant at the KWI for Biochemistry. In 1936, when the institute's Jewish director Carl Neuberg was forced to take early retirement, all his employees, including Gaffron, were dismissed. Although Gaffron was able to find a temporary position in Friedrich von Wettstein's department at the KWI for Biology for the next year and a half, his prospects looked bleak, particularly given the fact that, according to his family, he not only opposed the Nazi Government but also sympathised with the Communists.[70] Thus, at the end of 1937, Gaffron and his wife left for the USA and took refuge in the laboratory of van Niel (see below) at the Hopkins Marine Station in California. It seems that Otto Warburg was unusually supportive in organising Gaffron's emigration: Gaffron himself believed that he would not have settled down so easily, had it not been for a letter that Warburg wrote to the officials of the Rockefeller Foundation, which contained a "magic spell" that opened up doors for Gaffron—and provided him with a Rockefeller Fellowship for the first 6 months of his stay at the marine station.[71] Later, in the autumn of 1939, Gaffron was invited by James Franck to become his research associate at the University of Chicago; Gaffron would remain affiliated to this institution for the next 20 years of his life.

Bacterial photosynthesis was to become Gaffron's main research theme, although he is perhaps best known for his discovery of hydrogen metabolism in green algae, which had decisive consequences for conceptualising photosynthesis at large. Gaffron first moved into studies on bacteria in 1933 with emphasis on the metabolism of non-sulphur purple bacteria (*Athiorhodaceae*). His achievements here were seminal in their own right, even if Gaffron later had to revise some of his interpretations. In 1933 Gaffron explained his interest in this bacterial group with the aim of adding yet another variant of photosynthesis to the alternatives that had been known up to then (chlorophyllous photosynthesis and van Niel's study of *Thiorhodaceae*).[72] The interpretation of some of his data on the metabolism of purple bacteria sparked off much contention between Gaffron and van Niel, in particular on the question as to whether or not organic substances could serve as hydrogen donors in purple sulphur bacteria. In 1931, van Niel had purported that this was the case, while Gaffron

---

[70] Warburg was allegedly instrumental in securing Gaffron this position in the KWI for Biology through his good connections with Friedrich Glum, Administrative Director of the Kaiser Wilhelm Society at the time; cf. Werner (1988, p. 246).

[71] Werner (1988, p. 246); Rürup (2008, p. 201).

[72] Gaffron (1933b, p. 2). This general aim during these years was shared by Charles Stacy French, another giant in twentieth century photosynthesis research. In French (1937, p. 71), the latter wrote: "It is with the hope of finding a new approach to green plant photosynthesis that several workers are now studying the different kinds of photoassimilation in these bacteria. Probably by defining the differences between green plants and purple bacteria CO_2 assimilation, the chemical mechanism of both will become clearer".

claimed, in 1934, to have found evidence to the contrary. This was countered by van Niel, one year later, with the conjecture that Gaffron had used cultures that were contaminated—which understandably infuriated Gaffron.[73] The argument was only settled after van Niel carried out joint experiments with Gaffron at Warburg's laboratory in Berlin. Neither of them could ever have imagined that, only a short while later, Gaffron would obtain a research position in van Niel's laboratory at the Hopkins Marine Station in California.[74]

It was also in these years that Gaffron became intrigued by the role of molecular hydrogen in bacterial metabolism. In 1934, the Dutch microbiologist Pieter Roelofsen, who was working in Utrecht, the Netherlands, claimed that molecular hydrogen was able to support the photosynthetic reduction of carbon dioxide by sulphur bacteria.[75] Gaffron checked this out with his own strains of bacteria and was able to fully confirm the finding. Hydrogenase, the enzyme responsible for the oxidation of hydrogen, and its role in photosynthesis became henceforth one of his main research themes.

4.4.1 Context and Scope of the 1936 Paper

The contribution to photosynthesis research for which Gaffron's name is most vividly remembered is the paper that he co-authored with the German physicist Kurt Wohl in 1936.[76] In it, the two young men set out to argue why the standard model of photosynthesis was no longer tenable. (It is important to note that, at the time, Gaffron had not yet accepted van Niel's generalised equation for photosynthesis). In view of the results of the 1932 Emerson–Arnold experiments described earlier, Gaffron and Wohl maintained that one had to drop the idea, once and for all, that every carbon dioxide molecule was assigned to one specific molecule of chlorophyll. Rather, the energy absorbed by a large number of chlorophyll molecules could be made available, in an unlocalised sense, to a single carbon dioxide molecule. Hence, chlorophyll did not act as a "photoferment" in the reaction but as a sensitiser.

These last two proposals had already been made three years earlier by Gaffron.[77] However, Gaffron and Wohl acknowledged in their introduction that it was only thanks to the impact of the discussions that took place at a seminar organised by the German physicist Max Delbrück in Berlin that these ideas had been elaborated and used as the basis for a new concept of photosynthesis.[78] This seminar was the very discussion circle that had produced, in 1935, the celebrated landmark paper

[73] van Niel (1931); Gaffron (1934); van Niel (1935).

[74] See Homann (2002, p. 94). Gaffron (1963) provides an account of his dispute with van Niel.

[75] See Roelofsen (1934).

[76] Gaffron and Wohl (1936).

[77] Gaffron (1933a).

[78] See Gaffron and Wohl (1936, p. 81).

"On the Nature of Gene Mutation and Gene Structure", co-authored by Delbrück, the geneticist Nikolay W. Timofeev-Ressovsky and the physicist Karl G. Zimmer.[79] The paper is also known as the "Three Man Paper" (or 3MP; originally German: *Dreimännerwerk, Dreimännerarbeit*) and is often regarded as the stimulus behind what would later become the discipline of molecular biology.[80] Thus, the whole context is worth a little further consideration.[81]

Delbrück, who was originally trained as a physicist, had gone to Berlin in 1932 to work at the KWI for Physical Chemistry as an assistant to Lise Meitner (mentioned earlier as Franck's close friend and colleague). Before that, Delbrück had spent some months in Copenhagen with Niels Bohr, who was then elaborating on the deeper meaning of quantum mechanics—in particular, the complementarity principle. Later Delbrück recalled that Bohr had vigorously spoken of the possibility that this new quantum mechanical dialectic might also be relevant to other areas of science, such as how "life" was related to physics and chemistry. It was through these discussions that Delbrück became acquainted with (and fascinated by) current problems in biology and their potential reinterpretation from the viewpoint of physics.[82] In Berlin, Delbrück decided to explore, together with like-minded scientists, the potential application of current knowledge and expertise in physics to biological phenomena. In an interview of 1978, Delbrück recalled how he had invited a group of five or six theoretical physicists to join him for informal discussions at his family home in 1934. The group met at irregular intervals, sometimes weekly, sometimes once a month, until Delbrück left Germany in the summer of 1937 (that is, shortly after Gaffron had left for the USA). This is how Delbrück described the early phase of this discussion circle:

> This little club which started out as theoretical physics, and then brought in genetics, also brought in biochemists and photosynthesis physiologists. The photosynthesis man was Hans Gaffron, and he and Kurt Wohl lived together [with their families] in the same house in Dahlem [a district of Berlin]. As a result of the talks that we had in our club on photosynthesis, they published a series of papers on the kinetics of photosynthesis. [...] There were some more sophisticated experiments on this kinetics that had been published. Wohl and Gaffron discussed these experiments, and essentially already described what is now accepted; namely, that photosynthesis is done in photosynthetic units, which consist of about 1000 molecules of chlorophyll all funneling their energy into one photosynthetic reaction center.[83]

---

[79] Timofeeff-Ressovsky et al. (1935).

[80] It is not entirely clear how this name relates to the other famous "Three Men Paper", likewise known in German as *Dreimännerarbeit*: the paper by Max Born, Werner Heisenberg and Pascual Jordan of 1926 in which they introduced the matrix mechanics formulation of quantum mechanics; see Born et al. (1926).

[81] For more detailed information on the Delbrück seminar and the development of biophysics in Berlin at the time, see Sloan (2009).

[82] See the interview with Delbrück, carried out as part of the Caltech Archives Oral History Project, for the latter's recollections of these years; Delbrück (1978, in particular p. 41).

[83] Delbrück (1978, p. 55).

Thus, even so many years later, Delbrück still vividly remembered that the discussions had centred on photosynthesis. Delbrück had been interested in the photochemical reactions in biology long before he began to dwell on the gene and its structure. Genetics was really only one of the themes that kept the group together. In addition to Gaffron, two other "photosynthesis physiologists" are reported to have participated in Delbrück's circle, at least occasionally. The American plant physiologist Charles Stacy French wrote in his autobiography that, when he went as a postdoctoral student to Otto Warburg's laboratory in Berlin in 1935, he did not see much of the famous Dahlem science institutions outside the laboratory, "except for a few seminars on photosynthesis at Max Delbrück's house with Hans Gaffron and Eugene Rabinowitch".[84] Although these may have been private discussions that were not part of the official Delbrück seminars, it may not be far-fetched to consider them within the same context. Thus, French and Rabinowitch should probably be included in the list of discussants.

Rather than a series of publications, as Delbrück recalled, the main output of these discussions was a single common paper of 1936, published in two parts. This was followed, four years later, by a comprehensive review of the same problem in English by Wohl.[85] The principal argument of the Gaffron–Wohl paper was very similar to the thrust of the Three Man Paper on the mutation and structure of the gene: Data obtained in biological systems were systematically and carefully subjected to an interpretation from a quantum physics point of view.

4.4.2 Critique of the Standard Model

Gaffron and Wohl began their paper with an outright rejection of the standard model of photosynthesis, which was described earlier in this chapter. They first turned to Franck's notion of the photochemical process, which involved four reactions, each of which was initiated by one light quantum. This, according to Franck, eventually led to the formation of formaldehyde and hydrogen peroxide, according to the following equation:

$$CO_2 + 3\,H_2O \xrightarrow[\text{light}]{} CH_2O\ (\text{formaldehyde}) + 2\,H_2O_2$$

Gaffron and Wohl pointed out that this pathway was impossible, since the reduction of carbon dioxide to formaldehyde required more energy than the 4 red light quanta could provide, which were found sufficient for photosynthesis to take place. The energetic account would look more promising, if one assumed that, instead of free formaldehyde, a carbon moiety developed at the same oxidation stage, which then

[84] French (1979, p. 7). On the life and work of French, see Govindjee and Fork (2006).

[85] Gaffron and Wohl (1936); Wohl (1940). Being of Jewish origin, also Wohl had to leave Germany in 1933 and emigrated to Oxford, England, where he was able to find a position at Balliol college; see Jost (1963).

remained bound to other components. Yet, even if this were the case, there were other difficulties: *first*, each of the photochemical steps had to run very efficiently; *second*, each of them had to have a very small activation energy threshold (since only about 40 kilocalories (kcal) were available to initiate the four reaction steps, in addition to, as was generally assumed, forming a peroxide); and *third*, the intermediary photoproducts had to have a very long lifespan (of, at least, some seconds) before the next light quantum arrived. "Unfortunately, one usually has to choose between these three desired properties: high yield, low activation energy, long lifespan", Gaffron and Wohl pointedly concluded.[86] Back reactions and unstable intermediates were only some of even more problems, which the H/OH exchange mechanisms suggested by Franck implied.

The second major problem of the standard model was that chlorophyll was thought to be an actual reactant. This was at variance with the fact that, at least in vitro, chlorophyll had been found to be rather inert, particularly in terms of photochemical reactions. And, although Stoll had suggested that in the living cell chlorophyll would be more reactive (since it was in a colloidal state and bound to specific cell proteins), Gaffron and Wohl argued that Stoll had been unable to produce any evidence to support this claim. They also found it highly improbable that magnesium should be the site where carbon dioxide was bound, as the standard model suggested (following Stoll and Willstätter). No measurable change in the chlorophyll's absorption spectrum had ever been detected during photosynthesis, although this is what one would have expected if magnesium were the binding site.

4.4.3 The Unit as Explanatory Alternative

The authors then turned to the experimental evidence that had been left unexplained by the standard model. First, the well-known finding of Warburg and Negelein (1923) that photosynthesis needed only 4 light quanta to reduce one molecule of carbon dioxide and to release one molecule of oxygen. Gaffron and Wohl set out to check whether, in these classic experiments, every single chlorophyll molecule did actually receive 4 light quanta, as set out in the standard model. Under the conditions chosen by Warburg and Negelein, the result was clearly negative: "In this experiment only 0.8 % of the chlorophyll molecules could have received the four light quanta necessary to reduce 'their' molecule of carbonic acid".[87] The key to the solution of this puzzle, Gaffron and Wohl maintained, was provided by the experiments carried out in 1932 by Emerson and Arnold, who had found that, at maximum efficiency, one molecule of carbon dioxide was reduced per (roughly) every 2500 molecules of chlorophyll, which, therefore, they called "one unit".

---

[86] Gaffron and Wohl (1936, p. 82).

[87] Gaffron and Wohl (1936, p. 86).

Gaffron and Wohl suggested that this had to be taken very seriously: "This means that a molecule of carbonic acid will be reduced as soon as four [light] quanta are absorbed by any of the chlorophyll molecules within this unit".[88] This could not possibly be reconciled with the Stoll–Franck theory, they argued, but it was able to explain the excellent quantum yield in Warburg and Negelein's experiments. Further confirmation was provided from Emerson and Arnold's estimation of the Blackman reaction's time span, which was no more than 0.02 seconds: "This means that, according to our earlier considerations, every molecule of carbonic acid that is bound in a state in which it is susceptible to [photosynthetic] assimilation [. . .] must receive in these 0.02 seconds, at the given stationary [light] intensity, four light quanta".[89] Under the given experimental conditions, this was only possible if approximately 1000 molecules of chlorophyll acted together, which Gaffron and Wohl considered the number of *active* chlorophyll molecules that formed one photosynthetic unit.

The mechanism underlying this cooperative action of the pigments was, however, far from clear. Gaffron and Wohl considered two possibilities. Either the carbonic acid and its photochemically produced derivatives were not fixed but moved in a continuous diffusion from one chlorophyll molecule to the other, picking up energy quanta on the way. Or, alternatively, the carbonic acid was bound to one determined location, where it was reduced, while the energy that had been absorbed within the assimilatory unit moved around very quickly until it passed the site of reduction and was used up. Gaffron and Wohl favoured the second option, and speculated that the carbonic acid might be bound to one of chlorophyll's nitrogen atoms (although they had to admit that there was no conclusive evidence for this hypothesis). In any event, Gaffron and Wohl emphasised that, in view of the fact that for the whole of photosynthesis only 4 light quanta were required, the involvement of high-energy intermediates, such as peroxides, was out of the question. Yet, Gaffron and Wohl, unfortunately, could offer no convincing alternative to the underlying chemical pathway.

4.5 Franck's Conservative Alternatives

4.5.1 Franck and Herzfeld's Proposal

The critical objections raised by Gaffron and Wohl against Franck's photosynthesis model of 1935 were, in principle, accepted by Franck, particularly the objection that the intermediate products would have to be too long-lived to provide a realistic option. In his subsequent papers Franck himself added a number of further points of critique, and by 1937, he had developed a new proposal, together with the physicist Karl F. Herzfeld. They contested the view that recent experiments "make the assumption of

[88] Gaffron and Wohl (1936, p. 87).

[89] Gaffron and Wohl (1936, p. 88).

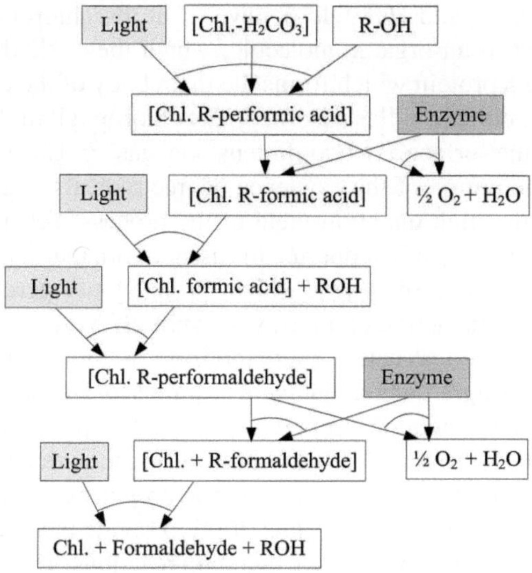

Fig. 4.5 The model of Franck and Herzfeld (1937).

a photosynthetic unit necessary, in which a large number of molecules cooperate in a way not encountered *in vitro*".[90] Their strongest argument against the existence of a photosynthetic unit (which most researchers in photosynthesis shared) was the difficulty involved in imagining a model of the underlying mechanism that would allow for the effective transfer of energy in the chlorophyll from the site of absorption to the site of carbonic acid reduction.

Franck and Herzfeld started by considering the fact that leaves emit fluorescence, which seemed to indicate that the chlorophyll in vivo was not in an aggregated but in a unimolecular state. This, the authors thought, could not be reconciled with the concept of a photosynthetic unit. In their alternative model, Franck and Herzfeld tried to accommodate this finding, as well as the assumptions that no more than 4 quanta were necessary for the process (which for most people at the time was taken to indicate the existence of four photochemical steps) and that chlorophyll formed a complex with carbonic acid. The pathway they suggested went through the stages of a peroxy acid, formic acid and a peroxy aldehyde. These, as Franck and Herzfeld did not fail to mention, "are the same intermediate compounds as in auto-oxydation processes, so that the similarity between these two inverse processes is striking". The unexpectedly low light saturation point measured by Emerson and Arnold was explained "by back chain reactions initiated by photolytical decomposition of the per-compounds".[91]

[90] Franck and Herzfeld (1937, p. 238).

[91] Franck and Herzfeld (1937, p. 237).

Furthermore, Franck and Herzfeld believed that the chlorophyll–carbonic acid complex was bound to an organic molecule, which they called ROH: "The ROH may for instance be a protein which forms the main body of the chloroplasts, on the surface of which the chlorophyll is adsorbed. The chlorophyll molecules will then be able to move along the surface as a two-dimensional gas".[92] This idea was introduced in order to avoid the notion of the existence of free radicals, which appeared to be incompatible with the high quantum yield of the process. Yet, unlike Gaffron and Wohl, Franck and Herzfeld still supported the assumption that formaldehyde was the first reduction product, in the course of which "probably two peroxide molecules are formed, which, under the action of an enzyme, split off oxygen".[93]

The mechanism is reconstructed in a graph form in Fig. 4.2. Four light reactions corresponded to the 4 light quanta that were required for the process, while the two enzyme reactions constituted the temperature-dependent part, that is, the Blackman reaction. Franck and Herzfeld defended Willstätter's assumption that the photochemical reaction steps consisted of an exchange of hydrogen versus hydroxyl groups in the carbonic acid molecule; however, they tried to divide these exchange reactions into four energetically reasonable single quantum reactions. The first photochemical reaction consisted of the formation of performic acid from carbonic acid and the ROH in a complex loosely bound to chlorophyll; the third was the formation of performaldehyde from formic acid, in which again the ROH was involved. Both these reactions were "followed by dark reactions in which the peracid or the peraldehyde is reduced under the influence of enzymes to the acid and the aldehyde", which restored the ROH.[94]

There remained, however, the uncomfortable observation made by Emerson and Arnold that the maximum assimilation rate was much lower than one would expect if every chlorophyll molecule worked as an autonomous entity. Franck and Herzfeld suggested that the photochemical products, which normally released the oxygen, would at high light intensities be frequently hit by further light quanta and, consequently, disintegrate, thereby initiating chain reactions which destroyed other photoproducts and, hence, reduced the process's overall efficiency. Both authors considered the concept of a hypothetical photosynthetic unit (in other words, a hitherto unheard-of photochemical process) to be unacceptable.

While Franck and Herzfeld's proposal was the most influential alternative to the concept of a photosynthetic unit, other explanations of the low oxygen yield at the light saturation point were also being offered. Emerson himself, for example, considered it likely that the rate of photosynthesis in flashing light of high intensity was limited by an essential molecule, which was present in an amount of only

[92] Franck and Herzfeld (1937, p. 240).

[93] Franck and Herzfeld (1937, p. 239).

[94] Franck and Herzfeld (1937, p. 240).

about 1/2500 of the chlorophyll.[95] Optical models, which tried to fill in the gaps of the "unit" concept, were also proposed, beginning with two papers by Wohl.[96] In these models it was assumed that there was a rapid transfer of absorbed energy from excited chlorophyll molecules to specific reaction sites (of unknown material nature and one per several thousand molecules of chlorophyll) to which the carbon dioxide was attached. However, none of these options was unreservedly accepted in the community. The general state of the discussion, therefore, was succinctly summarised in 1938 by the plant physiologist Winston Manning:

> The existence of a photosynthetic unit has thus far been neither proved nor disproved. Its existence would offer an explanation for several different groups of experiments, but on the other hand, various arguments largely based on physical grounds, can be offered against it.[97]

4.5.2 Franck's Further Attempts

In the years that followed, Franck used a number of further approaches to reaffirm his criticisms of the concept of a photosynthetic unit. One of these was a paper on the migration of the excitation energy in crystals, which he co-authored in 1938 with Edward Teller, a nuclear physicist of Hungarian origin, who would later find fame as the "father of the hydrogen bomb".[98] Although at first glance, the paper seems to have nothing to do with photosynthesis, on closer inspection one realises that Gaffron and Wohl's explanation of the energy transfer in a photosynthetic unit directly stimulated the work.[99] Gaffron and Wohl had considered the possibility that the chlorophyll molecules of a photosynthetic unit might be organised in the form of a one-dimensional crystal to which carbon dioxide molecules were attached, one at each end.[100] Energy absorbed at any point of this crystal would then migrate through the crystal and be channelled towards the carbon dioxide molecules. Franck and Teller maintained that this was highly improbable, because the migration of

---

[95] Emerson (1936). In his obituary of Emerson in 1961, Rabinowitch considered this to have been the most generally accepted interpretation. See Rabinowitch (1961, pp. 118–119). Emerson also used the terms "catalyst" or "photoenzyme". The concept aligns neatly with today's conception of "reaction centres".

[96] Wohl (1940, 1941).

[97] Manning (1938, p. 156).

[98] On Teller see, e.g., Rhodes (1995); Teller (2001).

[99] Franck and Teller (1938, p. 861).

[100] Again, it is striking how closely this resembles the idea, which was elaborated in the Three Man Paper, of the gene being similar to a crystal; cf. Timofeeff-Ressovsky et al. (1935).

excitation energy would be bound to trigger a much higher level of fluorescence in photosynthesising leaves than was actually the case.[101]

By 1941, Franck and Herzfeld presented yet another model—not because of inherent weaknesses of the 1937 version, the authors explained, but because of decisive new developments in the field which had made most of their earlier work obsolete.[102] Most importantly, by the end of the 1930s, a number of photosynthesis researchers working in the USA had started to doubt the validity of Warburg and Negelein's proposal that 4 light quanta were the minimum requirement for photosynthesis. A value of 10–12 light quanta seemed to be more realistic, and this boost of the energy budget (by the factor 2–3!) radically changed and, in fact, greatly alleviated, the task of modelling the process (the ensuing controversy is discussed in chapter 5). Furthermore, Samuel Ruben, Martin Kamen and their co-workers in Berkeley had found, with the help of radioactive carbon isotopes, that in photosynthesis carbon dioxide reacted with an acceptor molecule, RH, in a carboxylation process, the result of which was the formation of R–COOH (see Chapter 6 for a discussion of Ruben and Kamen's work). And, finally, Franck and Herzfeld cited work on the chlorophyll fluorescence of photosynthesis, partly carried out by Franck himself. They had found, for example, that the addition or removal of carbon dioxide produced changed the rate of chlorophyll fluorescence in proportion to the rate of photosynthesis. Franck and Herzfeld saw this as a strong indication of the fact that carbon dioxide was in direct energy exchange with the chlorophyll molecules: "Theories which assume that the photochemical part of photosynthesis results merely in a production of some reducing substance, which in turn reduces carbon dioxide in a mechanism chemically independent and spatially separated from the chlorophyll, are not in accordance with these observations".[103]

Franck and Herzfeld were still convinced that the number of light quanta required corresponded closely to the number of photochemical steps involved and, hence, the number of intermediates produced. In view of the new quantum yield value, they estimated that the number of steps had to be eight, in order to allow for some inefficient absorbance. The carboxylation reaction identified in Berkeley was cyanide sensitive, which, Franck and Herzfeld believed, demonstrated that it was promoted by a catalyst, which they called *A*. Franck and Herzfeld considered the product of this reaction, R–COOH, to be the substance that underwent further photochemical changes.

Molecules of the type R'H were thought to act as hydrogen donors, while the energy required for the transfer of hydrogen was supplied by the light energy that the chlorophyll had absorbed. The reduction of one molecule of carbon dioxide required the transfer of four hydrogen atoms, while the four remaining R' radicals regained

---

[101] On the paper's argument, see also the review by Franck and Gaffron (1941, p. 210). However, Franck and Teller assumed that there existed a one-dimensional structure (a linear chain of chlorophyll molecules); the application of two- or three-dimensional models, in, e.g., Bay and Pearlstein (1963), led to very different results. See Pearlstein (2002) for a short review.

[102] See Franck and Herzfeld (1941).

[103] Franck and Herzfeld (1941, p. 979).

their hydrogen by oxidising water molecules. Again, the energy had to be supplied by four additional instances of absorption. In view of the fluorescence experiments cited earlier, Franck and Herzfeld assumed that the carbon dioxide reduction was directly connected to the photochemical steps, so that the hydrogen transfers were produced by the chlorophyll's excitation energy. This implied, they believed, that the chlorophyll molecule took part in these reactions:

> R'H molecules then have to be members of the molecular complex containing the chlorophyll molecule itself and RCOOH or its derivatives. It simplifies the picture if one identifies the R'H molecules with the chlorophyll itself. In other words, one adopts the often-discussed idea that the chlorophyll not only acts as a sensitizer but also undergoes chemical reactions during photosynthesis. Indeed, the results of some new experiments with chlorophyll in organic solution make that hypothesis very probable.[104]

In order to explain the light saturation curve of photosynthesis as well as the flashing light experiments of Emerson and Arnold, the authors introduced what they described as a "very simple hypothesis":

> The limiting dark reaction is a process in which catalyst molecules present in a concentration several thousand times smaller than the concentration of chlorophyll operate on a photochemical product which is chemically very unstable. The catalytic reaction stabilizes the photoproduct. All the photoproducts not stabilized during their lifetime are eliminated by back reactions.[105]

The catalyst responsible for stabilising the reaction was called *B*. Franck and Herzfeld believed that each catalyst B molecule stabilised only one molecule of photoproduct, while all the others would be subject to back reactions. If the time interval between two flashes were greater than B's recovery period, on the arrival of the next wave of photoproducts all the B molecules would be available and, hence, the efficiency of the process would be at its maximum. Furthermore, Franck and Herzfeld thought that, since all the photochemical steps were so similar—they were shifts of hydrogen atoms from one bond to another—catalyst B would stabilise the products of *all* the photochemical steps. They formulated the reaction sequence as follows:

$$\text{Chl.-RCOOH} \xrightarrow[\text{light}]{} \text{Chl.'}_1$$
$$\text{Chl.'}_1 + \text{Catalyst B} \longrightarrow \text{Chl.*}_1 + \text{Catalyst B}$$

Chl.*$_1$ would then undergo the next photochemical step, the intermediate product of which (Chl.'$_2$) would be converted into Chl.*$_2$, and so forth. The chlorophyll would be replenished with hydrogen again through the formation of peroxide radicals, which had to be removed by the action of a third catalyst, C. The rest of the paper then focused on a detailed analysis of the differential equations that were supposed to demonstrate the model's validity—although one can safely assume that Franck's more biologically oriented colleagues were hardly able to appreciate them. They were far more interested in getting to know more about biochemical work that was

[104] Franck and Herzfeld (1941, p. 982).

[105] Franck and Herzfeld (1941, p. 985).

being done, at the same time, in Cambridge, UK, which will be introduced in the following section.

4.6 Isolated Chloroplasts and Water Splitting

4.6.1 Robert (Robin) Hill and the Chloroplast Reaction

The biochemical work done by Robert Hill[106]—or "Robin" as he was usually called by friends and colleagues—began at the Cambridge School of Biochemistry of the University of Cambridge (UK), which was strongly dominated by Frederick G. Hopkins's vision of general biochemistry.[107] At the time biochemistry was, for the most part, restricted to the study of animal and human metabolism, and the approach adopted by the Cambridge department (together with a few other British institutions, such as the groups headed by Sir Rudolph Peters at the University of Oxford and by David Keilin at the Molteno Institute, Cambridge) was a notable exception.[108] The general biochemistry practised at these places covered a broad range of fundamental biological topics, such as growth, development, nutrition and energy transformation, which were then studied within all forms of life: bacteria as well as animals, plants as well as invertebrates. Already in his celebrated 1913 lecture to the British Association for the Advancement of Science (BAAS), Hopkins had underlined that all forms of life were unified at the metabolic level and that this unity had to be represented in the way metabolic processes were studied.[109] Hopkins even used this as an argument for introducing the study of biochemistry as an independent discipline (which, at the time, frequently met with the objection that it was too narrow a field of study).

When Hill arrived (in 1919) as an undergraduate student at the University of Cambridge to specialise in chemistry (although he was also deeply interested in plants), Hopkins's department was on the point of entering a period of enormous expansion.[110] While in 1920 the Hopkins group counted ten workers, by 1925 this number had soared to 59.[111] Hill had already demonstrated an extremely broad range of talents, being "equally master of plant morphology, physiology, and organic and physical chemistry"; he was also described by his contemporaries as "the shy genius

[106] A special 1992 issue of *Photosynthesis Research* was dedicated to the memory of Hill; see Rich (1992). See Bendall (1994) for a biographical account and Walker (2002) for a tribute to Hill's work on chloroplasts. Hill (1965) provides an autobiographical perspective.

[107] On Hopkins and his institute, see, e.g., Needham et al. (1949) as well as Kohler (1982); Chapter 4.

[108] On the history of biochemistry in the early twentieth century, see Holmes (1986).

[109] See Hopkins (1949). He argued along similar lines in Hopkins (1926).

[110] Hill was admitted as a scholar to Emmanuel College in 1917; however, he only started seriously reading the Natural Sciences Tripos after the end of the First World War. Bendall (1994, pp. 145–146).

[111] See Kohler (1982, p. 81).

type".[112] Originally, Hill had intended to embark on a serious, biochemical study of natural dyes and plant pigments, a subject that had engrossed him for some time already.[113] However, Hopkins was less than enthusiastic about this research theme and advised Hill to work on haemoglobin instead. Hill did as directed, and from 1925 he produced a series of fine papers on this subject.

In 1924, Hopkins's Biochemical Laboratory moved into its new (and now famous) building on Tennis Court Road, thereby coming into the immediate vicinity of the Molteno Institute, where, in the years 1920–1925, Keilin carried out his seminal studies of cytochromes.[114] On the occasion of a public presentation of plant pigment solutions, for which he had prepared a number of specimens, Hill met Keilin, and was invited by the latter to join him in his work on cytochromes. Hill happily accepted and became a regular (if not daily) visitor to the Molteno Institute; he spent a full year trying, by all the means available, to isolate cytochrome *c*.[115] Hill continued to work with Keilin until the latter's death in 1963; and their collaborative effort exerted an enormous influence on the rest of Hill's career. It was while researching into cytochromes and related compounds, under the supervision of Keilin, that Hill learned the spectroscopic methods that he would later utilise to measure the activity of isolated chloroplasts. And it was Hill's thorough knowledge of the chemistry and biophysics of cytochromes, acquired in Keilin's laboratory, that led him to propose (in 1960 with Fay Bendall) the mechanism that would later become known as the "Z-scheme" of photosynthesis (see Chapter 7).

Keilin came to his research into cytochromes somehow accidentally through his investigations of peculiar haemoglobin phenomena in the larva of a horse parasite (it turned out that this bug was able to store oxyhaemoglobin for emergency use under anaerobic conditions). This noteworthy case led Keilin to study cellular oxidation; and he rediscovered a compound that had already been described in 1886 by the English physician Charles A. MacMunn. Keilin gave it the name "cytochrome" (which is the Greek for "cellular pigment"). In a celebrated paper of 1925, Keilin argued that this compound was "one of the most widely distributed respiratory pigments" in existence.[116] Keilin was able to characterise this pigment by its unique absorption spectrum of four bands, which he found uniformly present in many different forms of life. He also noted that the property of being reversibly oxidised seemed to be a characteristic of the compound. This first communication was complemented

[112] Kohler (1982, p. 83).

[113] Hill never lost this interest in plant pigments, and he became a well-known expert in the chemistry of natural dyes. Hill invariably grew the material for these and other studies in his own garden. He was also very skilled in extracting pigments and used them, among other things, for his own watercolour paintings. See Bendall (1994, p. 143).

[114] Keilin was made director of the institute in 1931. On Keilin's life and work, see Mann (1964). Keilin started his career with a strong interest in beetles and became a proficient entomologist. Even during the years of his research into cytochromes, Keilin never gave up his pursuit of questions on the morphology and physiology of insects.

[115] See the papers by Keilin et al. (1931) and Keilin and Hill (1933).

[116] Keilin (1925b, p. 315).

by a second (and equally celebrated) paper later in the same year in which Keilin made his first suggestions concerning the cytochrome's active function in cellular oxidations and reductions.[117] This short sketch of Keilin's work shows that he shared several points of common interest with Hill: both had worked on haemoglobin and its oxidised state; both became skilled in using spectroscopic methods; and both had organismic and biochemical interests. In his further studies in cellular respiration, Keilin was instrumental in conceptualising the respiratory electron transport chain; Hill, who was clearly inspired by Keilin, would later model photosynthesis along very similar lines.

In 1932, after having spent several months in the tropical surroundings of Singapore (in order to shake off a bout of depression), Hill returned to Cambridge to take up his work on haemoglobin that Hopkins had assigned him. He embarked on a study of this compound's reversible oxygenation; and for doing so he developed precise spectroscopic methods, which enabled him to monitor quantitatively the conversion of haemoglobin to oxyhaemoglobin, and vice versa. "The central problem was how haemoglobin could combine reversibly with molecular oxygen when haematin could not", is how Hill later formulated the goal of his studies.[118] Hill found, among other things, that myoglobin (muscle haemoglobin) had an even higher affinity to oxygen than the usual haemoglobin. Yet, despite these promising findings, the chemistry of related plant pigments, such as chlorophyll, remained in Hill's mind as a field into which, at some point, he still wanted to move. In 1936, Hill finally gave it a try, although he was hardly well-prepared to do so: Hill later believed that he had "crashed in" on the photosynthesis research scene. His biographer Derek Bendall described the situation as follows:

> Armed only with a reading of Spoehr's monograph [on photosynthesis, published in 1926], F. F. Blackman's analysis of limiting factors (he had attended Blackman's undergraduate lectures), and the realization that the path of his own research, where Hopkins had pointed firmly towards blood, led indirectly towards the green leaf. Others in the Biochemical Laboratory had successfully studied oxidation–reduction reactions in cell-free extracts of animal tissue; the same approach applied to leaves was to revolutionize the study of photosynthesis.[119]

The successful use of cell-free extracts by his colleagues in animal biochemistry encouraged Hill to try out the same approach in photosynthesis. If respiration, which had long been considered to be invariably bound to cell structure, could occur in certain suspensions, why not photosynthesis? In his first attempts to prepare an appropriate suspension of leaf extracts, Hill failed to observe any biochemical activity at all, in agreement with the traditional claim that the cell's structure was indispensable. Yet, there was this observation, well-known from Willstätter and Stoll's 1918 monograph, that dry leaf powder was able, for a short time, to produce oxygen, if illuminated. Hill decided that he would follow this up.

---

[117] Keilin (1925a). See Keilin (1966) on the history of research into cytochromes.

[118] Hill (1965, p. 124).

[119] Bendall (1994, p. 153).

His persistence paid off. By ingenious, albeit unconventional, means, Hill finally succeeded in preparing a satisfactory suspension of isolated chloroplasts: with a pestle and mortar he ground up leaves of, for example, the common chickweed (*Stellaria media*) and the white dead-nettle (*Lamium album*) in a buffered sucrose solution (pH 7.9) and then filtered them through glass wool. (This was before the ultracentrifuge had become a standard instrument in biological laboratories). Hill wanted to find out under which conditions these chloroplasts were able to produce oxygen. So he added a very sensitive indicator, namely myoglobin, which from his earlier work he knew would be converted into oxymyoglobin in the presence of only minute amounts of oxygen.[120]

Hill found that oxygen was, in fact, produced—yet only if an aqueous leaf extract preparation was added to the suspension. He first interpreted this finding as being due to a lack of certain enzymes, which in the chloroplast extract might no longer be present in their active forms. However, in developing these experiments further, Hill observed that a yeast extract, which certainly contained no plant-specific enzymes, could also promote the release of oxygen, and that the efficiency of the latter was proportional to its content of organic (ferric) iron compounds. Finally, it transpired that oxygen evolution could even be triggered by simply adding to the suspension inorganic iron salts, for example, in the form of ferric potassium oxalate. Catalase-inhibiting agents did not affect the production of oxygen in the system and neither did cyanide. The former was contrary to expectations, given the usual assumption that oxygen was produced in the chloroplasts by the decomposition of peroxides through the action of catalase. Hill was able to demonstrate that the participation of peroxides in this system was highly improbable. However, the most remarkable fact was that carbon dioxide was unable to act as a hydrogen acceptor. While carbon dioxide was the only known substance that could cause oxygen evolution in natural photosynthesis, ferric iron was the only reagent that was able to cause oxygen release in Hill's chloroplast suspensions. This was rather disappointing. It seemed to indicate that the reaction in Hill's extract did not, after all, represent cell-free photosynthesis; and it was completely unclear whether the reaction was related in any way to the process in living plants and algae.

In his 1937 publication, Hill carefully avoided jumping to any rash conclusions. The experiments required enormous skill and circumspection, for example, in order to ensure that the production of oxygen was not due to some property of the myoglobin, and hence were likely to produce artefacts. Another point of concern was the observation that the level of oxygen production was rather low, reaching only about one-tenth of the yield of normal photosynthesis. Nevertheless, Hill became more and more convinced of the validity of his findings, and in 1939 he gave a bold explanation of what his results might imply. First of all, Hill emphasised the fact that the chloroplasts' reaction was not specific to ferric oxalate; the latter was only a

[120] See Hill (1937, 1939) for the first publications and Bendall (1994, pp. 153–154), for an illuminating retrospective description. Hill himself never used the term "myoglobin" but always spoke of "muscle haemoglobin", which is frequently (and misleadingly) abbreviated to "haemoglobin".

means of demonstrating a general property of the chloroplast. It seemed that ferric oxalate or other hydrogen acceptors were able to oxidise a substance in the chloroplast, which was reduced in the course of the photochemical process and was vital for the release of oxygen. In the words of Hill:

> There must therefore be [in the chloroplast] some primary substance which is reduced [in the light], while at the same time giving oxygen. If this primary substance is A, and the reagent B, such as ferric oxalate, represented in terms of hydrogen transport, we have the following reactions:
> $2\,A + 2\,H_2O \longrightarrow 2\,AH_2 + O_2$
> $2\,AH_2 + B \longrightarrow BH_2 + A$
> [...] It must be concluded that the substance A is not easily removed from the chloroplast because great dilution of the suspending fluid did not diminish the rate of reaction with ferric oxalate.[121]

Hill thus suggested that the chloroplast might contain a mechanism that operated independently of the living cell, "which under illumination simultaneously evolves oxygen and reduces some unknown substance [A] which is not carbon dioxide". Hill assumed that this substance A was a kind of "respiratory catalyst".[122] And it was this substance A that transferred hydrogen to suitable acceptors, such as ferric oxalate, which could therefore be restored to its original state and be used again (while without ferric oxalate, all of this substance would be quickly reduced and the reaction would come to a standstill).

Together with Richard Scarisbrick, who became a long-standing collaborator of his, Hill elaborated and refined these studies over the next year.[123] They were able to show that the low limit of oxygen production, observed in Hill's earlier studies, was due to the reoxidation of ferrous oxalate to ferric oxalate, which consumed a large share of the oxygen that had only just been released. When the reduced compound (ferrous oxalate) was removed from the system, by the additional supply of ferricyanide, the full amount of oxygen released became apparent at high pressure: "The chloroplast then, with ferric oxalate as a hydrogen acceptor, behaves in a similar way to the whole cell as regards the production of oxygen during photosynthesis".[124] The reaction was also demonstrated to be highly sensitive to urethanes, which corresponded to what Warburg had found in his *Chlorella* experiments, and, like during the process of photosynthesis, it was influenced by varying light intensities. In view of these findings, Hill and Scarisbrick felt entitled to conclude that:

> ... the measured activity of the system in the isolated chloroplasts responsible for the production of oxygen in light represents a part of the process of normal photosynthesis. [...] The new conclusion that can be drawn from the work on isolated chloroplasts is that oxygen itself is formed in a photochemical reaction during which there is no reaction involving carbon dioxide.[125]

---

121 Hill (1939, p. 207).

122 Hill (1939, p. 209).

123 Cf. Hill and Scarisbrick (1940a, b).

124 Hill and Scarisbrick (1940a, p. 61).

125 Hill and Scarisbrick (1940b, p. 254).

This reaction—the production of oxygen by chloroplasts in suspension supplied with artificial hydrogen acceptors—later became known as the "Hill reaction", a term that was coined in 1941 by Charles Stacy French and the Cambridge-based protein chemist Mortimer Louis Anson. (Hill himself never adopted this term but always spoke of the "chloroplast reaction"). French, who at the time was working as a research assistant to James Franck in the Fels Laboratory at the University of Chicago, recalled that Anson had dropped in, on his way back from Arizona to Princeton, "to tell James Franck about Robin Hill's discovery of oxygen evolution by isolated chloroplasts".[126] Anson stayed for a month, and together with French repeated Hill's experiments in many variations, and even improved upon the technique. (They found, for example, that the efficiency of the reaction could be greatly enhanced by working at low temperatures). Franck tolerated these studies, although he believed "that all this had nothing to do with photosynthesis", a widespread attitude at the time.[127] Eventually, French and Anson prepared a paper to be presented during the physiological section of the annual meeting of the Botanical Society of America, which took place from 29 to 31 December 1941 in Dallas, Texas. However, since neither of them was able to attend the conference, their friend and colleague Jack Myers read out the paper to the audience.[128] In this paper, French and Anson explicitly looked at whether the production of oxygen in isolated chloroplasts used "the same enzymes as the oxygen production step in normal photosynthesis".[129] The paper was not exactly a sweeping success. In fact, as Myers later recalled, "it was greeted by a rather stony silence".[130] Some years were to pass before the importance and accuracy of Hill's findings would be realised.

4.6.2 *Implications of the Findings*

Hill's two main contributions to photosynthesis research of the 1930s were: *first*, he succeeded in separating the photosynthetic production of molecular oxygen from the reduction of carbon dioxide to carbohydrates. By doing so, he provided convincing evidence that these two parts of photosynthesis occurred separately. *Second*, his findings suggested that the photochemical part of photosynthesis comprised the release of oxygen, without carbon dioxide being involved as a hydrogen acceptor. Thus, Hill's experiments strongly reinforced the hypothesis (which van Niel had arrived at

[126] French (1979, p. 10). The review Franck and Gaffron (1941, p. 219), states, however, that Hill's findings only came to their notice upon publication of Hill and Scarisbrick (1940a).

[127] French (1979, p. 10).

[128] See French and Anson (1941) for the abstract of the paper. In the accounts of this episode in French (1979) as well as in Myers (1974), the name of the society was inaccurately reported.

[129] French and Anson (1941). Incidentally, in these experiments French and Anson were the first scientists to use spinach as a source of chloroplast; it remains a popular source to this day.

[130] Myers (1974, p. 422).

from a totally different starting point) that the photosynthetic oxygen originated from the light-induced hydrogen transfer from water to an appropriate acceptor. Hence, water, and not carbon dioxide, was the source of photosynthetic oxygen. These findings had a marked effect on the field, as can be taken, for example, from Gaffron's autobiographical essay of 1969:

> As late as 1936 Wohl and I were thinking about a hypothetical way to reduce a carbon dioxide compound directly à la Willstätter–Warburg. Only when Hill's chloroplast reaction [. . .] made any other than van Niel's view untenable was I ready to give in.[131]

Besides these conceptual consequences concerning the mechanism of photosynthesis, Hill's achievements opened up completely new avenues in terms of methods and materials. Hill was the first to succeed in preparing in vitro suspensions capable of photosynthetic reactions, which up to then had been considered impossible. Furthermore, Hill's findings stimulated the search for other reagents that might be used as hydrogen acceptors; this eventually led from ferric iron to TPN (i.e. NADP; (see Chapter 7).[132] Finally, Hill singled out not only a biochemical process but also a cellular component—the chloroplast—which subsequently became the subject of a broad range of other biochemical and biophysical studies.

4.7 On the Verge of New Perspectives

4.7.1 Biological Studies Generalised

The period examined in this chapter shows strikingly convergent developments in very different fields of experimental biology, such as physiology, biochemistry and microbiology: they illustrate, *first*, the firm conviction held by researchers at the time that physical and chemical tools, concepts and methods were indispensable for studying life processes (and, hence, had to be included in the curricula); *second*, they reflect the wide-spread searching for broad and comparative perspectives within the life sciences. Biochemists started to take an interest in plants; bacteria began to be used, for the first time, as experimental organisms of value in the study of the metabolism of higher organisms. Biochemical unity at the metabolic level became part of the body of generally accepted knowledge.

Far more people than ever before became interested in photosynthesis research. New research questions emerged, such as clarifying the relationship between photosynthesis in plants and the processes in bacteria; or exploring the physical nature of the energetic transitions in the light reaction stage of photosynthesis. Parallel to this process of ramification of photosynthesis research, one can observe a marked increase in the frequency and popularity of conferences and more informal meetings

---

[131] Gaffron (1969, p. 11).

[132] Cf. Myers (1974, p. 422).

on the subject. Many people strongly felt that the problem was much more complex than had previously been envisaged and that a multidimensional approach was required if all the questions on photosynthesis were to be answered. This became a strong incentive for interdisciplinary communication and cooperation—resulting, by the 1940s, in the foundation of the first interdisciplinary research groups to be exclusively dedicated to the study of photosynthesis: notably, the Photosynthesis Project at the University of Illinois at Urbana–Champaign, headed by Robert Emerson and Eugene Rabinowitch; the Fels Laboratory at the University of Chicago, led first by James Franck and later by Hans Gaffron; and, starting in 1946, the photosynthesis division of the Bio-Organic Chemistry Group at the University of California at Berkeley, headed by Melvin Calvin and Andrew A. Benson. I shall come back to these institutions in later chapters.

As was mentioned in the introductory section to this chapter, the 1930s also saw the appearance of the first "professional" researchers in photosynthesis—scientists who developed more than a passing interest in the subject. All the central actors discussed in this chapter belong to this category. It is worthwhile dwelling a little on their career paths, that is, on how they originally came to work in photosynthesis. Two related factors deserve special attention. First, all the major players in this period were educated at institutes or departments that were in the process of eroding or, at the very least, undermining, traditional disciplinary matrices. James Franck's doctoral thesis was supervised by Emil Warburg, whose interdisciplinary interests were discussed in chapter 3, and who inspired Franck to explore the physical basis of photochemistry (while contingent circumstances, such as the lack of an appropriate infrastructure for studies in nuclear physics clearly contributed as well to Franck's shift of research focus). Robert Emerson, William Arnold and Charles Stacy French were all trained in programmes with an emphasis on general physiology, the thrust of which was largely paralleled by the development of general biochemistry, which left its mark on Robin Hill, and general microbiology, which led Cornelis van Niel to study bacterial photosynthesis. These were the disciplines that, during the 1930s, greatly enhanced and fostered the application of physical and chemical methods to biological problems. It was also in this decade that Warren Weaver launched the Rockefeller Foundation's programme to support projects along these very lines (which later he would call "molecular biology").[133]

This intellectual climate was obviously a good preparation for a successful career in photosynthesis studies. However, the spread of the theme was also strengthened by the close interpersonal links between the players—the second factor to be observed among this chapter's protagonists. Gaffron, for example, came to photosynthesis by way of Otto Warburg (he worked as the latter's assistant). Gaffron's interest was

[133] Of the wealth of literature on this topic, see, in particular, Kohler (1991) and Kay (1993). Having suggested that photosynthesis, which was still considered a marginal subject, greatly profited from the advancement of "new biology"; one could even turn it the other way round and claim that photosynthesis research paved the way for the development of the "new" or molecular biology. See, e.g., the argument brought forward in Zallen (1993b).

fostered even further by the discussions he held with Max Delbrück, Kurt Wohl and other physicists in Berlin, who were all keen to solve the basic problems of the life sciences. Likewise, Emerson and French worked with Otto Warburg for an extended period; and Arnold worked with Emerson. On leaving Warburg's laboratory, Gaffron went first to work with van Niel and finally ended up working with Franck (joined shortly thereafter by French). Eugene Rabinowitch, who will enter the scene fully in the next chapter of this book, worked as an assistant to Franck in the latter's Göttingen days, before becoming Emerson's colleague at Urbana–Champaign. These links also played a role in promoting interdisciplinary discussion.

4.7.2 *The Main Lines of Thought*

As has become clear, during the 1930s, scientists approached photosynthesis from a number of very different angles and traditions. Figure 4.6 provides an overview of the resulting models and their relationships to each other. With the exception of van Niel and Hill, who came from rather different scientific backgrounds, most protagonists reacted closely to their colleagues' earlier work. At the beginning of this chapter, the relationships between the models leading up to Franck's 1935 suggestion were examined. Gaffron and Wohl harshly criticised this latter proposal, upon which Franck and Herzfeld attempted to find a better solution in 1937. When this proved untenable, Franck and Herzfeld then initiated, in 1941, a new era of models, which were constructed under the assumption that far more than 4 light quanta were available for the completion of the photochemical process—a relief that will be further explored in chapter 5. In the following sections, I shall recapitulate the general development and spell out the rationale behind each of the main lines of research.

4.7.2.1 Fluorescence Studies and a New Standard Model

The first line of thought continued the tradition that was outlined in chapter 2. The Willstätter–Stoll model, seemingly well established from the point of view of chemistry, was taken as the starting point for the analysis of photochemical details. Important new input was provided: first by the finding of Warburg and Negelein (1923) that, in order to produce one molecule of photosynthetic oxygen, no more than four to 5 light quanta were required. Second, there was the suggestion, first made by Kautsky and Hirsch (1931), that the peculiar changes of fluorescence in photosynthesising chlorophyll solutions could be used to analyse the underlying mechanism. The latter was combined with the suggestion that oxygen was, in actual fact, the first hydrogen acceptor in photosynthesis. After their 1918 monograph, Willstätter and Stoll had turned to totally different themes, each working independently: Willstätter had tried out enzyme chemistry, while Stoll had started a career in the laboratories of the Sandoz company in Basle (Switzerland), where he focused on pharmacological questions, such as the chemistry of ergot. In 1932, however, Stoll turned again to

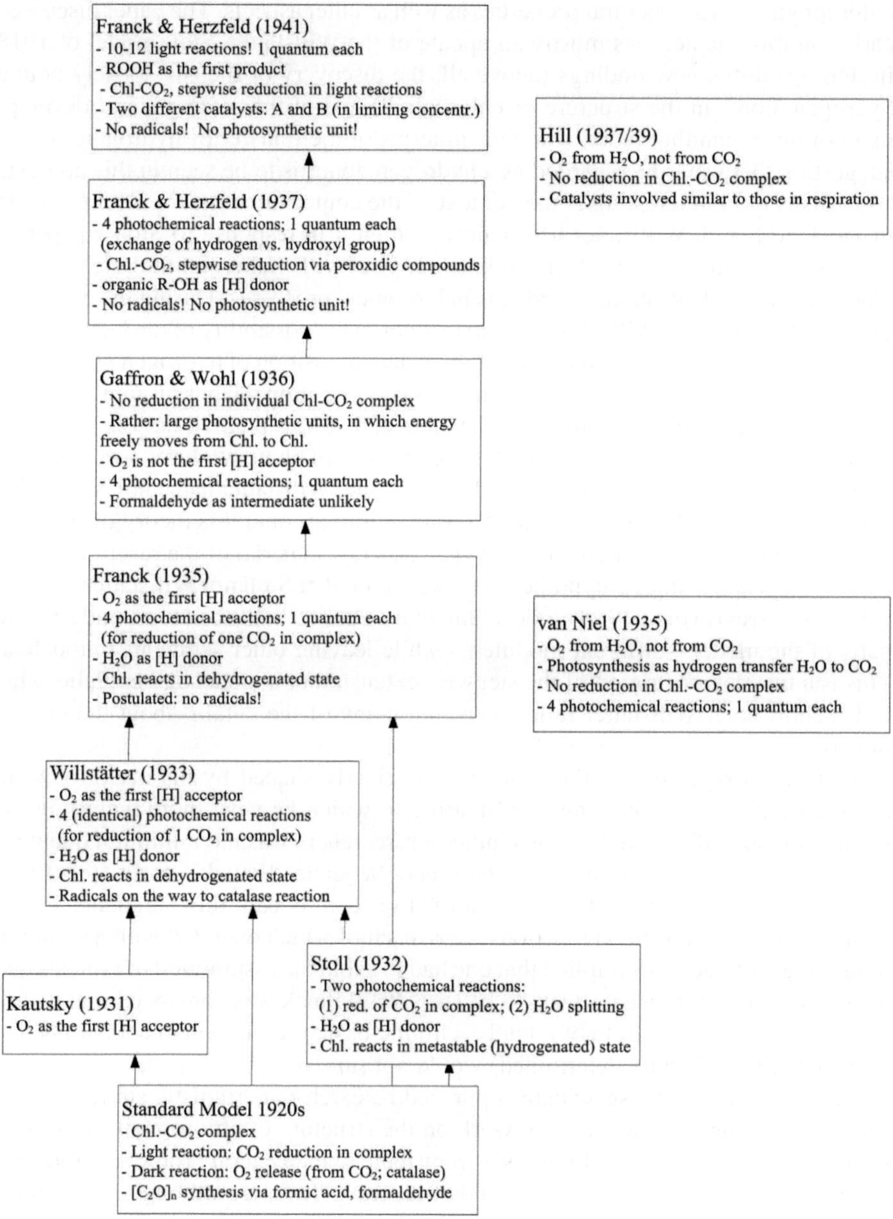

Fig. 4.6 The most important models of the photosynthetic mechanism brought forward in the 1930s and their relationships to each other. Only the main characteristics of the models that are not identical to the standard model of the 1920s have been listed.

chlorophyll—to its structural properties as well as other aspects. The paper discussed earlier in this chapter was mostly an update of the Willstätter–Stoll model of 1918 in light of Stoll's new findings (above all, the discovery of the two loosely bound hydrogen atoms in the structure of chlorophyll) and of the more general development of understanding redox reactions in terms of the transfer of hydrogen. Stoll's suggestion that water be regarded as a hydrogen donor is to be seen in this context.

Perhaps more interesting is the context of the contribution that Willstätter made in 1933. Although Willstätter had mainly written it in response to Stoll's paper, it was also a summary of his 1931 work, carried out with Haber, on the role of chain reactions initiated by chemical radicals in biological processes. During the course of their work, Haber and Willstätter had first explored the possibility of the formation of the HO_2 radical in the context of the catalytic decomposition of hydrogen peroxide in solutions.[134] This paper was cited in the 1933 contribution. Haber promptly reacted by immediately writing a letter to Willstätter, stating how pleased he was that, first, Willstätter had continued trying to solve the problem of photosynthesis, which Haber himself had been unable to sort out, and that, second, Willstätter had employed to this end their common theory of radicals.[135] Taking into account this theory of radicals, Willstätter felt that he could include oxygen as a raw material of the reaction, which was in line with Kautsky's hypothesis. However, neither Stoll nor Willstätter found it necessary to revise their 1918 model completely. Rather, both tried to extend specific parts of the model—different modules – while leaving other segments untouched. This is a fine demonstration of the stepwise extension of a model and explains why, in his short note, Willstätter failed to mention any of the details about the carbon moiety.

Franck's perspective on the problem was clearly shaped by his background in quantum physics: the one empirical finding to which he gave more weight in his work than most of the other photosynthesis researchers was the minimum quantum requirement value proposed by Warburg and Negelein. Franck's early (pre-1941) models were designed, first and foremost, to accommodate this parameter by including four photochemical reactions steps, each of which operated with a quantum requirement of one. This implied that one had to avoid the assumption of radicals and back reactions—not an easy task, to be sure. Yet, Franck was convinced that a photosynthesis model that did not comply with the basic thermodynamical parameters (which were empirically determined) would not survive.

Thus, all three of these scientists pursued research-opportunistic strategies (see Fig. 4.7): having completed some work on the structure of chlorophyll, Stoll took the opportunity to use these findings to contribute to the general problem of the photosynthesis mechanism. The same holds true for Willstätter, although his findings

[134] See Haber and Willstätter (1931), which argues that biological oxidation should be seen as a dehydration process. The elimination of hydrogen, Haber and Willstätter suggested, usually resulted in the formation of radicals (since only one of the two corresponding electrons would be removed at the same time). Cf. Willstätter (1973, p. 378); Werner and Irmscher (1995, pp. 30–31).

[135] Werner and Irmscher (1995, pp. 122–123); letter from Haber to Willstätter, 24 February 1933.

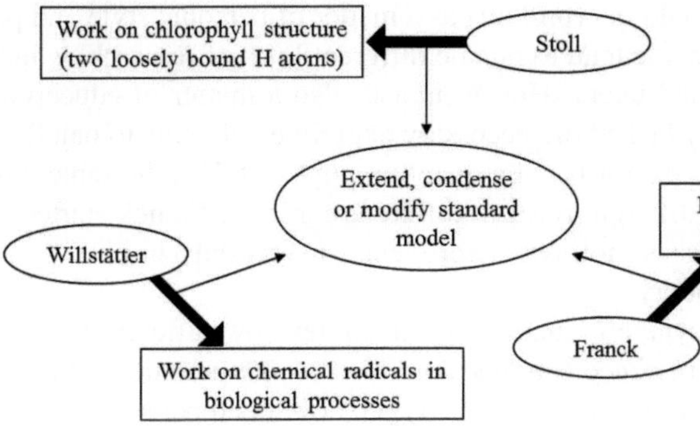

Fig. 4.7 Actors and their goals: diverging individual (superordinate) goals; extending the standard photosynthesis model as a subgoal or incidental goal. The thick arrows in bold typeface indicate the relationship "X pursues the superordinate goal Y"; thin arrows indicate that, in the course of pursuing the superordinate goal, the incidental goal of contributing to finding the photosynthesis model emerged.

were not concerned with chlorophyll, but with the formation of radicals as the central factor of the biological processes that occur in solutions. Franck used his theoretical expertise in fluorescence and energy exchange processes to try and clarify the subject. His suggested mechanism proved particularly influential. All three of them, however, left most of the elements of the standard model untouched: the formation of a chlorophyll–carbonic acid complex as the main reaction site; the reduction of carbon dioxide as part of the photochemical reactions, with chlorophyll as an actual participant; the formation of molecular oxygen as a result of the catalase-driven removal of hydrogen peroxide; and the formation of carbohydrates as a condensation process starting from the formaldehyde units.

4.7.2.2 Flashing Light Experiments and the Photosynthetic Unit

As outlined earlier, Emerson came to photosynthesis via the tradition of general physiology, as well as via Otto Warburg, the supervisor of his doctoral studies. The flashing light experiments of 1932 were intended to clarify two issues: first, a phenomenon that Warburg had noted in passing (that one could increase the rate of photosynthesis by using intermittent light and dark periods); second, the confusing observation, first noted by Willstätter and Stoll, that the rate of photosynthesis was not directly proportional to the chlorophyll content of the photosynthesising agent, as the standard model would have implied. That Emerson hit upon something that would prove to be the downfall of the standard model of photosynthesis as far as the function of chlorophyll was concerned, was, thus, hardly intentional. The resulting paper presented the surprising findings without, though, giving far-reaching interpretations. While Franck produced one conceptual model after the other and mostly played the role of theoretician in the history of photosynthesis research, Emerson was the empiricist. One could interpret this as a matter of personal style and preference; different types of people tend to pursue different lines of research, whichever are more to their liking and talent. However, it is also a matter of education and knowledge. Franck simply lacked the necessary experimental skills to handle algae and manometers; and Emerson was not a quantum physicist. Yet, the same individuals may be able to play different roles in different contexts: Franck started off as an experimentalist in physics, not as a theoretician; it was only in photosynthesis research that he kept to theory.

The case of the photosynthetic unit nicely illustrates how different background knowledge and earlier experience can lead to different interpretations. In order to account for the low ratio of one molecule of oxygen developed per several thousand molecules of chlorophyll, Emerson raised the possibility that the enzyme necessary to process the photochemical products might be present in very low concentrations. This was the factor, he thought, that was responsible for the low ratio of the end product. This was an entirely reasonable assumption, given the maxim that one should try and keep to the established knowledge of the time for as long as possible. Franck, on the other hand, shaped his theory with Herzfeld as the inverse of autooxidation processes. These he had studied intensively earlier in his career, for example, in 1931 together with Haber (shortly before the latter turned to investigating radicals with Willstätter).

Gaffron and Wohl designed their bold explanatory hypothesis against the background of the Delbrück colloquia. Their discussions were driven by the belief that in biology totally new and unexpected kinds of processes (or even laws) could be found if the insights of quantum physics were applied with sufficient competency. From this perspective, the suggestion that the photochemical reactions in photosynthesis might require the cooperative action of thousands of molecules was just what Gaffron, Wohl and their Berlin colleagues had been searching for. However, the hypothesis was not enthusiastically received by other parties. Gaffron and Wohl had summoned up convincing arguments against the standard model (it consumed too much energy; the assumption of very long-lived intermediates was unfounded; much longer induction periods would be required). But they were unable to present a mechanistic description of how the cooperation of chlorophyll molecules might work. Also Gaffron and Wohl took the Warburg–Negelein value of the quantum yield for granted; and neither did they question the fact that the oxygen had to originate from carbon dioxide via peroxidic compounds. The only part of the standard model that Gaffron and Wohl attacked was the assumption that there existed a chlorophyll–carbon dioxide complex in which the latter was reduced in a one-to-one relationship. The further pathway of the reduced carbon moiety remained largely untouched.

4.7.2.3 Microbial Photosynthesis and the Generalised Equation

Far more fundamental was the challenge that arose from microbiology, a field of study that traditionally had been far closer to medicine than to biology. It was only thanks to the Microbiology Department of the Delft Technical College in the Netherlands, where van Niel had trained, that the discipline got off the ground and that the importance of microbial investigation became recognised by other subfields of biology. From the available evidence, van Niel seemed to have pursued two goals: first, he wanted to find out more about the fascinating diversity of microorganisms; and second, and almost as importantly, he wished to use his knowledge of general microbiology to elucidate the fundamental problems of metabolism, independent of the research organism of choice. The fact that he chose the *Thiorhodaceae* as experimental organisms should not be overrated—chance clearly played a role here. When van Niel started working as Kluyver's assistant, the latter was preparing a lecture course, which, among other themes, also touched upon iron and sulphur bacteria. Van Niel's first task was, therefore, to prepare adequate cultures of these organisms for demonstration purposes. In order to do so, van Niel had to familiarise himself thoroughly with these difficult and heterogenous groups; and in the course of this work, he discovered that there were striking phenomena in the metabolism of sulphur bacteria about which a number of conflicting explanations had been claimed, none of which was entirely convincing. In addition to this spur, van Niel recalled that he "had become enamored with the aesthetically attractive purple sulfur bacteria".[136]

Notwithstanding all these contingencies, through his immersion in general microbiology van Niel was, without question, extraordinarily well prepared to conceptualise the metabolism of the purple sulphur bacteria and to compare them with green sulphur bacteria and plants. The basic assumption of metabolic and biochemical unity was not a consequence of his studies but a presupposition. The striking similarity between the summary equations of the processes in bacteria and plants or algae (they mostly differ in their use of appropriate hydrogen donors) was enough to convince van Niel of the existence of a general photosynthetic process. However, few in the scientific community were ready to accept his conclusion: after all, processes which come down to the same summary equation—i.e. the same behavioural description—can proceed by entirely different mechanisms. Van Niel's observation was merely that purple sulphur bacteria (as well as some other bacteria) were able to reduce carbon dioxide in the light, while at the same time oxidising some substances (mainly H_2S) in the medium being used. In order to call this "photosynthesis", one had to accept that the release of oxygen was not a defining feature of photosynthesis, which was quite a step, even for microbiologists. It seemed particularly audacious to assume this basic unity of process in view of the broad range of reactions observed in bacteria, which adapt so rapidly to changing environments, and the unchanging photosynthesis in plants and algae.[137]

---

136 See van Niel (1967, p. 9).

137 This discrepancy was later taken by Franck and Gaffron (1941) as an argument for the assumption that "the anaerobic type of photosynthesis is the same in all cells but that it is supplemented in green plants by the capacity of liberating gaseous oxygen. [...] Photosynthesis in plants, therefore, is the exception to the general rule" (p. 252). By contrast, van Niel (1941) assumed that in all types of photosynthesis water is reduced, while in bacteria the liberated oxygen immediately underwent

The most important implication was that one had to take very seriously the assumption that the oxygen produced during photosynthesis originated from water and not from carbon dioxide. Accepting this consequence would not only have made the standard model of photosynthesis untenable, but also implied that the neat balance between the volumes of carbon dioxide consumed and of oxygen produced suddenly was merely coincidental, which to many researchers was hard to believe.[138]

4.7.2.4 Oxygen Evolution in Chloroplasts

It was Hill's work that eventually helped dispel the reservations scientists had about van Niel's hypothesis: light-driven oxygen evolution by chloroplasts was possible without there being any need for carbon dioxide reduction. It was emphasised earlier how carefully Hill made sure that his observations were not mere artefacts but reflected the photosynthetic processes under natural conditions. Of course, evidence for this assumption was not fully conclusive. Like van Niel, Hill argued for the hypothesis that a process observed under circumstances X (isolated chloroplasts; bacterial metabolism) was the same as a similar process under circumstances Y (illuminated chloroplasts in plants), so that both could be described by the same model. Reservations were strong: even for most of the 1940s it was still being debated whether all the photosynthetic oxygen really did come from water. The absence of a convincing mechanism to achieve the decomposition of water (which required a very strong reducing agent) made people rather doubtful of the validity of this hypothesis.[139]

4.7.3 A New Conception of Photosynthesis

In addition to all these steps towards a new model of the photosynthetic mechanism there was also an important change on a more fundamental level, that slowly crystallised during the decade looked at in this chapter and deserves a moment of attention. It began to dawn on the researchers involved that the process was much more complex than previously imagined. More factors than anyone would have imagined were found to be of influence on the process—in particular, the physiological state of the experimental organism and its developmental history. The growth conditions

---

secondary reactions and the dehydrogenation of the specific hydrogen donor took place in later stages of the process.

[138] Today it is known that this "neat balance" is no longer tenable and the quantities as well as proportions can vary substantially.

[139] It was Sam Ruben and Martin Kamen's experiments in 1941 with "heavy" water, incorporating the oxygen isotope ^{18}O, that provided the strongest evidence for the hypothesis that photosynthetic oxygen came exclusively from water. See Ruben et al. (1941) for the publication, which will also be discussed in Chapter 6.

of the algae turned out to be highly relevant (such as exposure to light, pH value, atmospheric pressure, carbon dioxide concentration, and so on). Different genera and species of algae, also closely related strains, were shown to react quite differently to changes in conditions; and even when the same experimental organism, cultivated under the same standard conditions, was investigated, the data tended to vary.

The enormous flexibility of photosynthesising organisms was addressed most explicitly by Hans Gaffron in a paper of 1940.[140] Therein, he emphatically pleaded for a move away from the traditional, quasi-mechanical concept of photosynthesis, which (unlike respiration!) still was assumed to have a fixed stoichiometry. The latter was based on the long-standing assumption that the ratio of carbon dioxide consumed to oxygen released was unity, which had given rise to the hypothesis that carbon dioxide was the source of oxygen. Gaffron pointed out that this ratio became unity only under stationary conditions, while particularly at the points of transition of, for example, light to darkness, very different ratios were obtained. Further complications arose if one considered the interference of photosynthetic processes with many other reactions in the cell. The oxygen released during photosynthesis, for example, might be immediately consumed again by oxidising intermediate respiration products or by other reduced compounds of the metabolism.[141] The result would be that none or only a part of the oxygen was liberated while varying amounts of carbon dioxide were formed. Yet, how researchers were to deal with these complexities, how they ought to reorganise their work in the laboratory as well as their inferences and interpretations, Gaffron was unable to suggest.

---

[140] Gaffron (1940).

[141] This shift in the conception of photosynthesis neatly coincides with a general change in biochemistry. Up to the 1930s, a conception of metabolism as a set of linear processes prevailed. However, when ATP and various coenzymes were discovered in the 1930s, scientists slowly realised the close entanglement of the cell's metabolism, which came to be conceived of as a highly integrated system; cf. Bechtel (1986a).

Chapter 5
The Maximum Quantum Yield Controversy (1937–1955)

As has become clear in the previous chapter, the quantum yield of photosynthesis was regarded by most of the photosynthesis experts working in the 1930s as a vital piece of information. Thus, it is not surprising that, in the course of this decade, several research groups started to re-examine the standard value presented by Otto Warburg and Erwin Negelein in 1923; and this quickly resulted in serious concerns being raised about the value's validity. Warburg and Negelein's findings were increasingly questioned—most vigorously by Robert Emerson, together with several of his co-workers, who harshly criticised the methods that Warburg and Negelein had applied. In line with some other teams in the USA, Emerson argued that 8–12 light quanta were needed for photosynthesis. Papers and arguments were exchanged, but no agreement reached, which led Emerson to invite Warburg to his laboratory at the University of Illinois at Urbana–Champaign in 1948. The idea was that the two researchers would compare their experimental protocols and thereby settle the disconcerting discrepancies that had arisen—disconcerting, because both Emerson and Warburg were renowned for their mastery of manometry. Yet, Warburg's stay in Urbana proved unfruitful. Nothing was settled and the two opponents parted as enemies. Consequently, the controversy continued to grow during the 1950s, to such an extent that the biophysicist Roderick Clayton commented that "the quantum efficiency of photosynthesis became perhaps the most exhaustively measured phenomenon in the history of science".[1]

The general story of this controversy has been told many times in photosynthesis research circles.[2] However, the full complexity only comes to the fore in an analysis based on archival sources, which up to now has never been undertaken. In contrast to the chapters so far, which tried to analyse the heuristics of constructive cooperation, the focus now is on a micro-historical example, which is examined in view

---

[1] Clayton (1965, p. 40).

[2] See, e.g., the collection of historical perspectives, tributes, etc., in Govindjee et al. (2005), all of which include an array of further references.

K. Nickelsen, *Explaining Photosynthesis*, History, Philosophy and Theory of the Life Sciences 8, DOI 10.1007/978-94-017-9582-1_5

of the question how a research community reacted to a persistent—and increasingly destructive—disagreement between central players of the field.[3]

5.1 Photochemical Quantum Yield Measurements

Measuring the quantum yields and quantum requirements of photochemical reactions became standard practice in the fields of physics and photochemistry during the first decade of the twentieth century. These measurements concerned the efficiency of photochemical processes: they either showed how many light quanta were required to yield one molecule of product (this was the quantum requirement) or how many product molecules were released through the effect of one light quantum (this was the quantum yield, which is the reverse of the quantum requirement). The epistemological value of these parameters was enormous because once it was known how much energy in terms of light quanta a process required, then it became easier to reconstruct the underlying mechanism: many alternatives looked far less attractive than before when it was found that they did not fit the determined energy budget.

As mentioned in chapter 3 of this book it was the physiologist and biochemist Otto Warburg who, thanks to the work of his father Emil, had become familiar with the practice of taking quantum yield measurements and introduced it to photosynthesis research. Together with Erwin Negelein, Warburg had found that a minimum value of 4–5 quanta of light was required to produce one molecule of photosynthetic oxygen (while one molecule of carbon dioxide was consumed).[4] The important qualification here is the predicate "minimum"—for the quantum requirement of photosynthesis was found to be not a constant value but a function of a whole range of parameters. Yet, only the lowest number of quanta required was of theoretical importance: according to Einstein's Law of Photochemical Equivalence only then could one draw conclusions as to the number of photochemical steps involved and to the amount of energy necessary to make the process operate.

The 1923 Warburg–Negelein value of 4–5 quanta remained virtually unchallenged until around 1937. Not only did the experiments appear well-founded and the data conclusive, but the value of around four also nicely matched theoretical expectations: the conversion of water and carbon dioxide into molecular oxygen and a moiety of carbohydrates required a minimum calculated energy input of 112 kilocalories (kcal), while red light, which was known to be the most efficient region in the spectrum for bringing about photosynthesis, carried approximately 40 kcal per molecule of light quanta. Thus, at 100 % efficiency of the process, photosynthesis would require 2.8 light quanta per oxygen molecule; and since no process could possibly run at total

[3] This chapter was written in collaboration with Govindjee of the University of Illinois, Urbana–Champaign, who generously shared with me substantial pieces of information. Large parts of this chapter have been published as Nickelsen and Govindjee (2011). Earlier contributions to this topic include Govindjee (1999, 2004b).

[4] Warburg and Negelein (1923).

efficiency, a value slightly higher than this was to be expected. The value of four, furthermore, seemed to be highly significant to many photosynthesis researchers, because it corresponded so neatly to the four hydrogen atoms (or, alternatively, electrons) that had to change their places and bondings in the process of turning CO_2 into [CH_2O]. Yet, however well all this fitted together, a quantum requirement of four to five was still so close to the theoretical limit given by the calculation above that it was extremely difficult to devise a model of the mechanism that could convincingly explain the relevant empirical findings. These difficulties gave rise to a re-examination of the issue.

5.2 First Opponents of the Warburg–Negelein Value

The first of the several researchers, whose data on the maximum quantum yield of photosynthesis was in disagreement with that of Warburg–Negelein's value, was William Arnold. After his collaboration with Emerson in the flashing light experiments of 1932 (see Chapter 4), he entered the graduate programme in General Physiology at Harvard University. Part of his dissertation project was to measure the minimum quantum requirement of the production of photosynthetic oxygen, which he did using microcalorimetric techniques. (In microcalorimetry, the process is not monitored by registering pressure changes but by determining the resulting heat flow in a leaf or a cell suspension.) By this means, Arnold found that a minimum number of 8 light quanta were required to produce one molecule of oxygen during photosynthesis.[5] Although the number eight was two times greater than the Warburg–Negelein value of four, Arnold assumed at the time that eight was low enough to be included in the same range; this is why he had these results published only much later, in 1949.[6]

Around the same time, an interdisciplinary research team at the University of Wisconsin in Madison also started to re-evaluate the quantum yield requirement. This team included the plant physiologists Winston M. Manning, J. F. Stauffer and Benjamin M. Duggar as well as the eminent photochemist Farrington Daniels. In 1938, they presented the first published challenge to Warburg and Negelein's quantum yield value. For their experiments the group had developed a (rather cumbersome) chemical gas analysis method, which they applied to *Chlorella* cells. By this means, they had arrived at a minimum quantum requirement of 16–20 quanta per molecule

[5] Arnold (1935, pp. 35–38, Table IX; PhD thesis at Harvard University) reported, in *Vallisneria*, an efficiency of 35 % (equivalent to a minimum number of 8 photons/O_2) in red light of 420 ergs/cm^2/s, at a temperature of 22°C.

[6] Arnold's results, which he obtained using *Chlorella*, showed that the minimum number of quanta per O_2 evolved was never lower than nine per molecule of CO_2 absorbed (see Table 13.1 in Arnold 1949, p. 275). In the first sentences of the paper, Arnold stated: "This study [. . .] is being published because of the urging of Dr. [Hans] Gaffron and because Warburg has reopened the question of quantum yield in photosynthesis" (p. 273).

of oxygen evolved—diverging from the standard value by a factor of 4 or 5.[7] This publication caused quite a stir among photosynthesis researchers. The point was reinforced when a year later the group published new results (using microcalorimetric measuring techniques), showing that 12 light quanta were required to produce 1 molecule of oxygen, or, respectively, consume one molecule of carbon dioxide.[8]

Finally, given the importance that James Franck attached to the quantum yield in his papers, it is not surprising that in the late 1930s he also encouraged his assistant, the physicist Foster F. Rieke, to investigate the value again at Johns Hopkins University.[9] Rieke had decided to use Warburg's manometric techniques in order to check whether he could duplicate the original findings. He arrived at an average value of about 5 quanta for the minimum requirement: a fair confirmation of Warburg–Negelein. However, Rieke found that the values strangely varied according to the method of calculation, which led him to conclude that "either there is an obscure systematic error in one method of measurement or, under the conditions of the experiments, photosynthesis and respiration do not follow a simple course".[10] Furthermore, Rieke obtained his lowest figures only when he used a phosphate-containing medium, following the recipe of Dean Burk, a biochemist who was also working at Johns Hopkins at the time and whose help was acknowledged in Rieke's paper.[11] On only one occasion did Rieke use a carbonate–bicarbonate buffer solution, in which he measured a minimum requirement of about 8 quanta: "The quantum efficiency was reduced 40 per cent", Rieke stated in view of this aberrant run, while he surmised that some property of the buffer solution was responsible for this diverging value.[12]

When he presented his findings on a lecture trip that in 1938 brought him to visit several Californian universities, Rieke learned that yet another photosynthesis researcher had, in the meantime, turned to quantum yield measurements, and this was Robert Emerson. In 1937, Emerson had taken a leave of absence from Caltech, and spent three years at the Plant Biology Laboratory of the Carnegie Institution of Washington (on the campus of Stanford University), where he was able to benefit from working with the skilled physicist Charlton M. Lewis.[13] On 20 July 1939, Emerson turned directly to Rieke, in response to the latter's paper of April of that year, in which the value of five had been published. Emerson wrote that his group had arrived at the same low quantum requirement values as Rieke, but only when they had used a very specific medium for the cells:

---

[7] Manning et al. (1938). With hindsight one has to say that many of their experiments were not in the range where minimum quantum requirement ought to be measured, namely, at very low light intensities.

[8] Magee et al. (1939).

[9] Rieke (1939). See Guttman et al. (1970) for Rieke's obituary.

[10] Rieke (1939, p. 243).

[11] By that time, Burk had already become famous thanks to the paper Lineweaver and Burk (1934), in which the famous Lineweaver–Burk plot (or double reciprocal plot) was introduced, which became so useful in enzyme studies.

[12] Rieke (1939, p. 243). See Table IV., p. 242.

[13] See Govindjee and Krogmann (2004, p. 47), for a short obituary of Lewis.

> When we tried to repeat [the experiments] using the medium which you specified, we were unsuccessful. Thinking that you had probably followed Warburg & Negelein in using tap water, we sent to Baltimore for some tap water, and with this we were at once able to duplicate your results.

And then Emerson added the following lines:

> We are preparing some of our results for publication, and including a description of a medium in glass-distilled water in which we can regularly produce cells giving *quantum yields of about 3 quanta per* CO_2. We think it should be easy to duplicate these results in other laboratories [...]. If you care to try out our medium, I shall be glad to send you a full description of it in advance of publication.[14]

Emerson and Lewis did not believe that this value of one molecule of oxygen per 3 light quanta, which was beyond all theoretical expectations, reflected the actual efficiency of photosynthesis, but considered it an artefact of the technique.

5.3 Emerson and Lewis's Challenge

Emerson had been in touch with Warburg on the difficulties he had met with in his research into quantum yields. In a letter of November 1938, he reported to Warburg the findings that Lewis and he had arrived at so far; for example, they had encountered relatively large induction effects, which lasted no less than 5 min after the onset of illumination and would not disappear, however hard they tried.[15] They also had found that, at the change from light to dark and *vice versa*, the gas exchanges had strong oscillations, which he was not yet able to explain. One year later, in December 1939, Emerson was adamant that the values reported by Warburg and Negelein were incorrect. On a Christmas card to Warburg, he wrote:

> I shall send you a reprint in January, because I believe you will be interested in our results. The [maximum quantum] yield really is not as high as you and Negelein thought.[16]

Emerson added that "interest for Europe" was great in the United States, while he also thought "there is no point in writing anything about it". After all, this was written only a few months after Germany had invaded Poland which soon resulted in what became the Second World War.

5.3.1 *The Carbon Dioxide Burst (1939–1941)*

In the article Emerson had alluded to in his Christmas card, he and Lewis systematically explored the external factors that were relevant to the photosynthetic yield.

---

[14] Emerson to Rieke on 20 July 1939. Franck, James. Papers, [Box 7, Folder 9], Special Collections Research Center, University of Chicago Library.

[15] See the letter of Emerson to Warburg, 5 Nov. 1938. Archive of the Berlin-Brandenburg Academy of Sciences and Humanities (Archive of the BBAW), NL Warburg 262.

[16] Archive of the BBAW, NL Warburg 262. Emerson to Warburg, Dec. (Christmas) 1939.

They demonstrated that this value was strongly dependent not only on the type of water used but also on the addition of certain heavy metals, the culture's exposure to light, the age of the culture and the wavelength of light at which it had been grown. Keeping the cultures at lower temperatures also tended to increase photosynthetic efficiency. Thus, the whole issue transpired to be far more complicated than had previously been believed. As a great many factors were involved, which, in addition to their individual influence, appeared to be closely interdependent in their effects, Emerson and Lewis admitted that finding the most appropriate conditions had been extremely difficult: "We have tried as far as possible to combine the factors in such a way as to obtain the highest possible efficiency, but it has been necessary to make arbitrary choices in regard to certain factors, in order to study the influence of others".[17] (This uncertainty of whether or not one really had determined the *maximum* value of photosynthetic efficiency would remain one of the most disputed issues in the controversy.) In conditions that Emerson and Lewis suspected to be optimal, and otherwise following Warburg's experimental protocol, they arrived at the surprising yield of 0.33 molecules of carbon dioxide assimilated per absorbed light quantum—this was the value that Emerson had indicated to Rieke in his letter of 1938. Emerson and Lewis were convinced that this exceedingly high value was an artefact of the technique; and the decisive source of error was identified as being the curious gas exchange effects that appeared whenever the light source was turned on or off. In their paper, Emerson and Lewis wrote:

> [W]e found that after a change from light to dark, or *vice versa*, the rate of pressure change was subject to large deviations before coming to the new steady value. [. . .] When the light is turned on, a sharp increase of pressure occurs at once and lasts from two to five minutes. Under some circumstances the maximum rate attained may be two or three times the steady rate in the light (the respiration correction being included in each case). [. . .] When the light is turned off, the rate of pressure change returns approximately to its former (negative) value for a few minutes but then shows an increase. [See also their figure, reproduced here as Fig. 5.1][18]

Emerson and Lewis found that the sudden peak after the onset of illumination was mainly caused by the evolution of carbon dioxide. The authors concluded, rather succinctly: "This implies that for the short periods of darkness and illumination used for efficiency measurements the assumptions on which photosynthesis is computed from pressure changes become incorrect".[19] With this sentence, they effectively dismissed all previous quantum yield determinations, all of which had depended on the premise that the ratio CO_2/O_2 during photosynthesis, denoted by γ ("gamma"), was unity. Emerson and Lewis concluded that manometric measurements of the maximum quantum yield "cannot be accepted as significant until the method has been applied in such a way as to permit the simultaneous determination of both carbon dioxide and oxygen exchange".[20]

---

17 Emerson and Lewis (1939, p. 812).

18 Emerson and Lewis (1939, pp. 814–815).

19 Emerson and Lewis (1939, p. 815).

20 Emerson and Lewis (1939, p. 817).

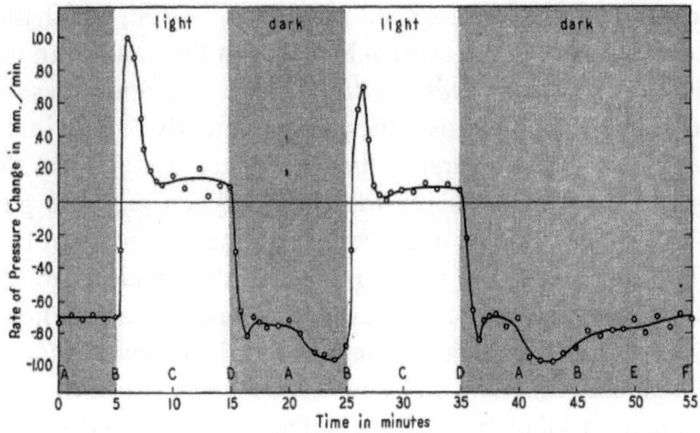

Fig. 5.1 Reproduced from Emerson and Lewis (1939, p. 815). Deviations observed in the rate of pressure change during successive periods of light and darkness. The shaded areas represent the dark periods. Light intensity: 1400 ergs/cm^2/s.

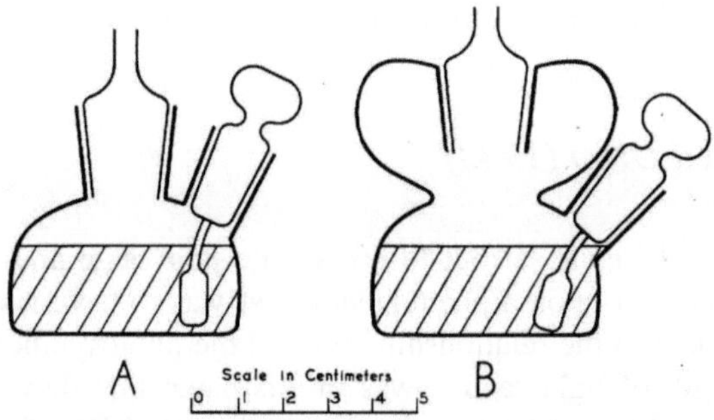

Fig. 5.2 "Diagrammatic cross sections of the two shapes of manometer vessels used for determination of oxygen and carbon dioxide exchange. The space occupied by fluid is nearly the same shape and volume in each vessel, but the gas space is much larger in vessel B. The vessels are circular in plan, diameter 5 cm". (Reproduced from Emerson and Lewis 1941a, p. 790).

This is what they set out to do, and two years later, in 1941, Emerson and Lewis presented the so-called "two-vessel method", which enabled the exchanges of oxygen and carbon dioxide to be measured simultaneously. The trick was to use two vessels containing the same quantity of identical algal suspensions, which, however, had different gas-to-liquid ratios (Fig. 5.2). The researchers thus could calculate the exchange of carbon dioxide and the exchange of oxygen independently of each other. By this means, Emerson and Lewis were now in a position to trace the development of γ in quantum yield measurements, and they found it to be extremely unstable, with the greatest variation taking place in the first 10 min of light or darkness. Emerson and Lewis concluded that this was the cause of a considerable systematic error in the Warburg–Negelein values. To measure the rate of photosynthesis, Warburg and Negelein had chosen the first 5 min of a light period, which included a sudden and significant *increase* in pressure, which was not due to a rise in photosynthetic oxygen; on the other hand, to measure the rate of respiration, which they used as their correction factor, Warburg and Negelein had chosen the first 5 min of a dark period, which comprised a significant *decrease* in pressure. Together, these ill-chosen time slots led to the efficiency of photosynthesis being greatly overestimated.

Emerson and Lewis demonstrated that the effect was mainly due to dramatic changes in carbon dioxide pressure, while the oxygen pressure gave a relatively steady course of values: "It is as if the cells contained some sort of reservoir which pours out carbon dioxide in the first minutes of illumination, and which must be filled again in the dark before the full respiration rate of carbon dioxide production can manifest itself".[21] (This phenomenon was shortly afterwards termed the "carbon dioxide burst".) Hence, Emerson and Lewis chose to calculate the rate of photosynthesis and its quantum yield from oxygen changes alone; and by this means, they arrived at values of about 0.10 molecules of carbon dioxide per absorbed quantum of light, under widely varying conditions and using no fewer than 11 different species of algae. This value, a minimum quantum requirement of ten, was in satisfactory agreement with the values reported by others, for example, the group at Madison; hence, Emerson and Lewis considered the issue to be settled.

5.3.2 *The Red Drop (1943)*

When Emerson and Lewis set out to investigate this value and related questions a little further, they hit upon a phenomenon that was to have lasting significance. In 1943, they turned to the relationship between the photosynthetic quantum yield and the wavelength of light; and it was in this paper that they published for the first time what would later become known as the "Red Drop" of photosynthetic efficiency. The starting point of their investigation was that, given Einstein's Law of Photochemical Equivalence, the efficiency of photosynthesis should be independent of the wavelength of light, at least for the range of the spectrum in which chlorophyll absorption was high. (This was due to the assumption that the primary photochemical process was proportional only to the number of absorbed quanta, irrespective of their wavelength). Warburg and Negelein's 1923 measurements had been received up to then as demonstrating this independence; yet, since Emerson and Lewis considered the work done by Warburg and Negelein as methodologically flawed, they saw a need for re-examining this question, "particularly in the red region where chlorophyll is the principal light-absorbing pigment".[22]

Emerson and Lewis found, to their utter surprise, that the quantum yield was roughly constant in the region 580–685 mμ, whereas from 685 mμ towards the infrared region of the spectrum, the yield dropped sharply (see Fig. 5.3). All attempts

[21] Emerson and Lewis (1941a, p. 794).

[22] Emerson and Lewis (1943, p. 165).

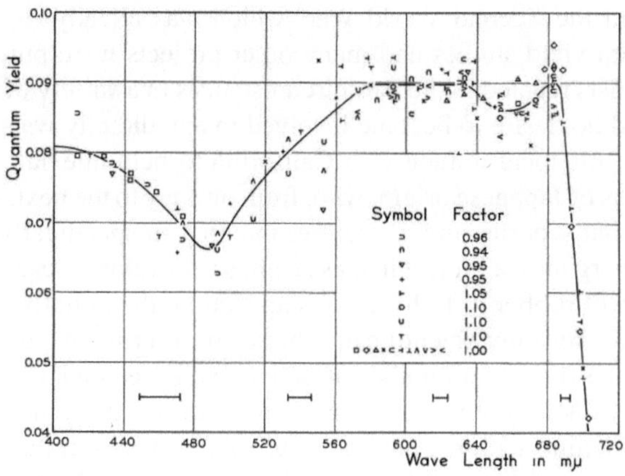

Fig. 5.3 "The quantum yield of photosynthesis as a function of wave length for *Chlorella*. The points obtained on each of nineteen separate runs are indicated by a distinct symbol". The drop of the efficiency from 685 mμ towards the infrared region of the spectrum is obvious. (Reproduced from Emerson and Lewis (1943, p. 171).

to measure photosynthetic yield in the regions beyond 700 mμ were unsuccessful.[23] The puzzling fact was that even at these higher wavelengths, that is, above 685 mμ, chlorophyll absorption was still rather high, and no pigments were known to compete with chlorophyll in this region. Emerson and Lewis were completely at a loss as to how to explain this finding. They speculated "that the light quanta of this spectral region [might] no longer provide sufficient energy for the photochemical primary process";[24] yet, it seems that even Emerson and Lewis themselves were not completely happy with this line of argument. They repeatedly emphasised the preliminary nature of the measurements and refrained from making a more comprehensive explanatory hypothesis. It was only in the second half of the 1950s (treated in Chapter 7 of this book) that scientists would finally be able to explain the phenomenon.[25]

5.4 Wartime

In the meantime, decisive events outside the laboratories happened that profoundly changed the agendas of the American photosynthesis researchers: In December 1941, after the attack on the headquarters of the US Pacific Fleet at Pearl Harbor, Hawaii,

---

[23] Today the nanometre (nm) is more commonly used than the millimicron (mμ), but the latter was used here in order to keep the main body of the text consistent with the quotes. One nm (and 1 mμ) equals 10 Å, another unit formerly used.

[24] Emerson and Lewis (1943, p. 174).

[25] See on this point also Govindjee (2001); Nickelsen (2012a).

the USA entered the Second World War, which was already in its third year in Europe. Quantum yield studies and many other projects were put aside until after 1945. The scientists reacted to the new circumstances in a variety of ways. Emerson, for example, had no desire to become involved in any directly war-related projects. Instead, he used his local influence in California to help alleviate the situation of American citizens of Japanese origin, who, from one day to the next, were considered a threat to national security and, therefore, interned in specifically erected camps. Emerson helped organise a project in these camps to develop a desert shrub, guayule, as a local source of rubber.[26] To his great satisfaction, the initiative turned out to be a major success—not only scientifically, but also in terms of giving back content and purpose to the lives of a number of deportees. Emerson's closest collaborator in this project was M. Shimpe Nishimura, who skilfully combined the expertise of professional gardening with his previous studies of physics at the Caltech (both of which had been brutally interrupted upon his internment). Nishimura would later become Emerson's assistant at the University of Illinois at Urbana–Champaign.

The situation was, of course, completely different in Europe. Franck's courageous resignation and his emigration to the United States were outlined in chapter 4. His former assistant Eugene Rabinowitch, who was also Jewish, took the same route—from Germany via Copenhagen to America—although he found it much harder to obtain a new position (not the least because he was not a Nobel Laureate). Otto Warburg's fortunes during the war were quite remarkable: Warburg not only survived the Nazi period, he even remained in his position as the Director of the Kaiser Wilhelm Institute of Cell Physiology (founded in 1931), although he was considered "half Jewish" by the Nazis. This treatment was surely exceptional, given the fact that so many other people of the same ancestry, regardless of their positions, were banished, deported or killed.[27] Particularly in the United States, many people suspected, therefore, that Warburg must have collaborated to some extent with the Nazis.[28] Yet, as far as is known today, no evidence for this assumption has come to the fore—if one does not include the lack of direct resistance as a form of collaboration. There is, on the other hand, ample evidence of the contempt Warburg felt for the new Government, which he largely tried to ignore. He repeatedly emphasised

[26] See Rabinowitch (1961, p. 115), as well as Finlay (2009, Chapter 5). See also the Oral History Interview with the plant physiologist James Bonner; cf. Bonner (1980, pp. 26–27).

[27] See, on the fate of other scientists in Warburg's discipline, Deichmann (2001b). On the history of the Kaiser Wilhelm Society in the Nazi period, see also the series of 17 volumes published in the years 2000–2007 by Wallstein Publishing, Göttingen (Germany), edited by Reinhard Rürup and Wolfgang Schieder.

[28] The plant physiologist Albert Frenkel recalled in an interview with Govindjee (on 8 Sept. 2007) that when Warburg was in the USA in 1948–1949, Gaffron had organised a party for him at Woods Hole, Massachusetts. It was at this party that a wife of one of the professors at Caltech bluntly asked Warburg why he had stayed in Germany "when the Nazis were doing such bad things". To which Warburg replied: "I wanted to protect my co-workers". He then added: "What could I have done?" Whereupon she replied: "You could have committed suicide!" Understandably, this dismayed Warburg and many of the other guests.

that he was determined not to be dispelled "by a handful of arbitrary criminals" and that he would continue to work in his institute as long as possible (which he did).[29] The special status of Warburg's institute, which was financed mostly by the Rockefeller Foundation, probably contributed to the fact that Warburg was, for a long time, exempt from the usual regulations within the Kaiser Wilhelm Society.[30] This is also how Warburg himself explained his situation, when he was interviewed, in 1945, by the Public Safety division of the American military (in the presence of the renowned chemist Roger Adams who—unsuccessfully—tried to secure Warburg a position as a scientific consultant): "Subject stated that in 1933 he was informed by the Ministry of Culture that he should have no worries about his half-Jewish ancestry as far as his work and position was concerned. He attributed this to the fact that his work was supported by the Rockefeller Institute in New York".[31] In this interview Warburg, furthermore, mentioned the fact that Hitler wanted him to continue his cancer research as the latter was highly afraid of this illness—it is hard to evaluate how much of this was founded upon facts. In the end, Warburg most probably survived because of the continuous efforts of a number of influential friends in the fields of politics, economics and science, who repeatedly managed to get him out of difficult situations.[32]

It was only after repeated bombing had damaged virtually all the laboratory's windows that the institute was evacuated, in the summer of 1943, to a manor in the environs of Berlin called Schloss Seehaus (located in the village of Liebenberg, in the district of Templin), which was completely refurbished for this purpose. One may assume that the evacuation of the institute was also done with the purpose of moving Warburg out of the direct focus of political institutions, in order to avoid further critical interest in Warburg's position.[33] Warburg himself spent most of the time in his summer residence in Nonnevitz on the Baltic Sea island of Rügen. Cut off from laboratory facilities, he mainly worked on a book, *Schwermetalle als Wirkungsgruppe von Fermenten* (Heavy Metals as the Active Group of Ferments), which summarised his earlier studies on this subject and was published in 1946.[34] The Russian front arrived

---

[29] Rüskamp (1989, p. 252); also quoted in Werner (1991, p. 285).

[30] On the relationship of the Rockefeller Foundation, which not only funded Warburg's institute, with the Kaiser Wilhelm Gesellschaft in the Nazi years, see Schüring (2006, pp. 109–119).

[31] Roger Adams Papers, 1812–1971, Record Series 15/05/023, Box 58. University of Illinois Archives (quoted p. 2). Govindjee brought this document to my awareness.

[32] See on this point Macrakis (1993, p. 64 and 226 (Footnote 53)), as well as Nickelsen (2008a). In addition to Friedrich Glum, who was, until 1937, Secretary General of the Kaiser Wilhelm Society's administrative wing, Warburg had the support of, for example, Hermann Bücher, Philipp Bouhler, Viktor Brack, Ferdinand Sauerbruch and especially Walter Schoeller. On the relationship of the Rockefeller Foundation with the Kaiser Wilhelm Gesellschaft in the Nazi years, see, e.g., Schüring (2006, pp. 109–119).

[33] See Henning (1987, p. 85).

[34] Two short notes on photosynthesis were published in 1944 in the journal *Naturwissenschaften*, in which Warburg, together with his co-worker Wilhelm Lüttgens, presented his findings that isolated (and even mechanically affected) chloroplasts were able to drive the reduction of quinone

in Liebenberg in April 1945, and Warburg's institute in Schloss Seehaus was completely cleared out: all the instruments, chemicals, benches, furniture and glasswork were taken by the Red Army.[35] Consequently, the Kaiser Wilhelm Society abandoned the building in Liebenberg, which was then taken over by the local hospital. In June 1945, the original building of Warburg's institute in Dahlem was also occupied, this time by the Allied High Command in Berlin. Thereupon, Warburg dismissed all his employees; and this was the end of his renowned Kaiser Wilhelm Institute (which, however, would later be re-established as part of the newly founded Max Planck Society[36]). Yet, from a letter that Warburg wrote to his sister Lotte on 13 January 1946, it transpires that he had already started thinking about a way out of his situation:

> I am living in my house in Gary Street again [in the Dahlem quarter of Berlin] (American Sector), and thanks to the Americans and Russians I am neither starving nor freezing. Until the end of September I stayed in Rügen, together with Jacob [Heiss], whom I saved, with much effort, from military service and the *Volkssturm*, and who is well. [...]
> Less favourable is the situation concerning my scientific work. I cannot yet say what I am going to do; of course, I have received several offers. But you know, from 1933, that I am not a friend of emigration, as this means that one's quality of life, whatever happens, deteriorates considerably. For the moment, I am staying put—with an institute, if possible; if not, without an institute—and perhaps only go and work as a guest in other countries.[37]

These then were the circumstances in which Warburg found himself at the beginning of 1946. He had no satisfactory infrastructure at his disposal and, given the state of the country, there was hardly any hope that this situation was to change in the foreseeable future. Once the war was over, however, Warburg immediately began catching up with the international scientific literature. One of the first things that he published, in 1945, was a short note on the quantum requirement of photosynthesis.[38] It was a response to the papers by Emerson and Lewis and to a review of the subject written by Franck and Gaffron in 1941, in which they had announced that the issue had been settled in favour of a minimal quantum requirement of 12.[39] Warburg strongly contested this perspective, and responded by vigorously reconfirming his earlier findings.

---

to hydroquinone, with the release of molecular oxygen. No mention was made of Robin Hill's earlier experiments—it is not entirely clear whether Warburg was aware of them at the time. See Warburg and Lüttgens (1944a, b).

[35] A diary kept by Wilhelm Lüttgens documents in rich detail the situation in Liebenberg from April to September 1945, and has been published in Werner (1991, pp. 326–334); doc. 124. See Nickelsen (2008a) for a transcript of Warburg's own diary notes in the first months of 1945.

[36] Werner (1991, p. 338).

[37] Otto Warburg to Lotte, 13 January 1946. The originally German document is quoted in Werner (1991, pp. 355–356), doc. 128.

[38] Warburg (1945).

[39] Franck and Gaffron (1941, p. 200): "We know now that the high quantum efficiency mentioned is only apparent, and that the true efficiency is only a third of it, namely, 12 quanta per CO_2 molecule reduced. The foundations on which the hypotheses concerning the amazing efficiency and the four-step mechanism rested have disappeared".

5.5 Attempts to Find a Solution

5.5.1 The Photosynthesis Project at Urbana

Soon after the end of the Second World War, Robert Emerson was approached by the University of Illinois to set up a research laboratory dedicated to photosynthesis studies on the Urbana campus. Emerson had been looking for an opportunity to leave Caltech for some time already, yet he only accepted the attractive offer from Illinois on the condition that the university also hired a physicist or physical chemist with an interest in photosynthesis.[40] The person Emerson had in mind as a potential second director was Eugene Rabinowitch, who was then at the University of Chicago studying uranium chemistry.[41] The first volume of Rabinowitch's seminal monograph on photosynthesis had just been published, which made him a particularly eligible candidate for the position.[42] From Emerson's perspective, it was no less important that Rabinowitch perfectly combined a profound knowledge in physics and chemistry with sufficient sensitivity towards the specific problems of biology. On 23 October 1946, Emerson wrote to the Dean and explained his choice by describing the intended scope of the laboratory's work, which, he argued, dearly needed Rabinowitch's skills to complement his own interests and knowledge:

> If [Rabinowitch] is appointed, it will be our plan to make a joint attack on the problem of energy absorption and conversion in the green plant. My share of the program will be the study of photosynthesis as it takes place in the intact cells of lower plants. Mr. Rabinowitch will work on artificial systems, built either from components extracted from plant parts or from non-living material, which give promise of simulating the unique energy-storing aspects of the natural process of photosynthesis.[43]

Emerson's request was granted, and thus began, in 1946, what Warburg would later mockingly describe as the "Emerson–Rabinowitch photosynthetic unit". The arrangement was to prove highly satisfactory, through the fruitful combination of their talents, characters and approaches, and the Photosynthesis Project at Urbana was to develop into one of the most active research centres on the subject.

5.5.2 Warburg Comes to the United States

On 28 November 1947, Emerson wrote his first letter to Warburg since losing touch with him after 1939. Therein, Emerson reported how he had heard, through Roger Adams, that Warburg had responded to the challenge of the Warburg–Negelein

[40] Govindjee (2004b, p. 181).

[41] On Rabinowitch, see, e.g., Rabinowitch (2005); Brody (1995) and Bannister (1972).

[42] Rabinowitch (1945); the second volume was published in two parts in 1951 and 1956.

[43] Emerson to Carmichael on 23 October 1946, Robert Emerson Papers, 1923–1961, Record Series 15/4/28, Box 1, Folder: Botany Department. University of Illinois Archives.

quantum yield values posed by Emerson and Lewis by publishing a paper. Emerson wrote: "After months, I finally obtained this manuscript from our military administrations. We translated it into English and it will be published in the *American Journal of Botany*".[44] Therein Warburg—translated by Emerson—described the papers from the Madison group as being methodically flawed and the critique by Emerson and Lewis as being insubstantial. At the same time, Warburg fully confirmed the quantum requirement values of 1923, based on new measurements that he had taken using the two-vessel method.

With hindsight one can already find in this first response some of the leitmotifs of the ensuing controversy: for example, Warburg's arrogant contempt of Emerson's work, emphasising that Emerson had learned the very techniques and approaches from him, when he was Warburg's student. Warburg wrote that he "would be astonished if Emerson had found the truth with our methods while we ourselves had fallen into error in our application of the same method". Warburg continued to explain why he considered Emerson's approach to be flawed:

> The disadvantage of this method [of Emerson's] is that the bicarbonate solutions are unphysiological from the standpoint of their chemical composition, osmotic pressure, and pH. The hydrogen ion concentration of the least alkaline bicarbonate solution is 10.0–9.4 and thus differs from the physiological hydrogen ion concentration for *Chlorella* by more than four orders of magnitude. [. . .] It would never have occurred to us to measure the most delicate of all biochemical processes, the conversion of light energy into chemical energy, in a medium in which the very survival of the cells seems remarkable. Not so Emerson. He reasons, correctly, that by determining the yield in bicarbonate solution all his concern about the assimilatory quotient is eliminated. He makes the unjustifiable assumption, however, that the effect of the unphysiological medium can be neglected for the sake of this methodological simplification.[45]

Warburg dismissed Emerson's criticism as inappropriate. He, again, had found, with "new methods" (that were barely outlined) that the assimilatory quotient γ was sufficiently close to unity, even for intervals of 5 min. Warburg underlined that he had been unable to detect anything like an outburst of carbon dioxide, while, as in 1923, he still found "a quantum requirement of 4 to 5 per molecule of evolved oxygen".[46] In these passages, the second leitmotif of the controversy emerged: Warburg's habit of not answering objections head on, but rather of presenting new data that he had obtained by altering his methods (so that the critic then needed to demonstrate, first of all, whether the earlier objections also held true for the new set-up). However, at this early stage of the controversy, Emerson chose a different strategy of response; in a letter of November 1947 he made the following suggestion:

> It is now being discussed in America how to explain the inconsistency in determining the yield of assimilation. It seems to us that it would be best if we could observe the same

[44] Archive of the BBAW, NL Warburg 262; Emerson to Warburg on 28 November 1947. The "manuscript" that Emerson refers to is the German paper Warburg (1945); it was published in English as Warburg (1948).

[45] Warburg (1948, p. 194).

[46] Warburg (1948, p. 195).

> phenomena in a laboratory together and calculate the yield in the same manner. If Germany had not been so badly damaged and if you still had your laboratory, I would suggest that I come to visit you in Berlin. But as far as I have heard, at the moment it is impossible for you to undertake any kind of scientific work. Hence, I suggest that you visit us here and carry out some comparative experiments in our laboratory. We are still far from being as well equipped as you were in Dahlem, but nevertheless our laboratory is sufficiently equipped for carrying out quantum yield measurements. You may want to bring [Fritz] Kubowitz with you and your strain of algae, a Hefner lamp and whatever other instruments need to be compared.[47]

The university's administrative department had already agreed to fund the visit, and Emerson suggested that this should be used to cover Warburg's and his laboratory assistant's travelling expenses as well as a salary for the two of them for 6 months (Warburg's assistant was assumed to be his long-standing collaborator Fritz Kubowitz). Emerson also announced that he would now apply for immigration permits from the US State Department, so that they would be ready in time. Warburg answered on 19 December. He thanked Emerson for the invitation and said that he would come with Wilhelm Lüttgens as his assistant.[48] Warburg also wanted to bring his valet and secretary Jacob Heiss with him; Warburg would pay Heiss out of his own salary, and if necessary would also cover Heiss's travelling expenses. However, Warburg still needed a personal invitation for Heiss, otherwise it would be impossible for the latter to travel. At the time, German citizens were still not free to travel abroad, and to enter the United States, and they also had to prove that they were politically unstained.

During the course of the following months, Warburg repeatedly changed his choice of assistant, his means of transport, payment and other details—all to Emerson's exasperation, since any one of these changes meant that he had to resume negotiations with both the university and the immigration officials. In the end, Warburg brought only Heiss with him, and the two of them arrived by plane. Warburg was probably never aware of all the trouble Emerson had gone to in order to organise his visit. Emerson succinctly described his feelings in a letter to Gaffron on 29 May 1948:

> Dear Hans: [...]
> [Carl] Cori [the physiologist] is right, his [Warburg's] visit is sure to lead to a lot of grief. In fact, just trying to arrange for the visit has kept me busy for a large part of the winter. After all our efforts to provide Warburg with an assistant of his own choosing, it turns out the man (Gustav Ernst Lau) cannot come because he lives in the Russian zone. Seems to me Warburg might have thought of this difficulty a few months ago, instead of now, when he is about ready to leave. Last report I had was that he and Heiss might leave by June 1st. I hear they have 400 kilos of baggage and a poodle, on all of which they expect the Univ. of Illinois to pay transportation. It will turn out that the reason Warburg wants to leave Germany is because the American administration has been unable to get any more of that good German dog-food, made of pure beef-steak, the only thing the poodle will eat. There will be Hell to

[47] Archive of the BBAW, NL Warburg 262. Emerson to Warburg on 28 November 1947. Original letter in German.

[48] Emerson, of course, did not know that in 1944 Kubowitz had denounced Warburg to the Nazi authorities; Warburg had been saved thanks to some influential friends, and would never speak to Kubowitz again. Lüttgens was the only one among his long-standing collaborators whom Warburg still fully trusted. See, for background information Nickelsen (2008a).

> pay when he finds that in America they feed horse-meat to dogs! And imagine the problem of finding housing for Warburg, Heiss, and a poodle!
> Yes, I believe Cori is right, but I hope it will be worth the trouble, to get this matter settled.
> Bob.[49]

No mention was ever made of the poodle again, so it can be assumed that it stayed behind in Germany. Warburg, though, arrived, together with Heiss and an enormous amount of luggage, on 26 June 1948. As Emerson was soon to experience, Warburg was never to make up for the bad impression he had made before.[50]

5.5.3 The Time Spent at Urbana

When Warburg arrived at Urbana, he immediately turned the laboratory upside down. As Rabinowitch wrote in his obituary of Emerson, Warburg was accustomed to working in a laboratory that completely and utterly fulfiled his wishes; and, since Emerson's laboratory was not large enough for Warburg to have been given full command of a section, he and Emerson had to tolerate each other in the laboratory's communal areas. This was bound not to work smoothly.

However, Emerson still had high hopes that the visit would pay off. He truly believed that Warburg had come to carry out experiments with him and discuss the discrepancy between their results; yet nothing of the sort happened. Warburg proceeded to work in his own usual way and was not at all interested in Emerson or his work. The only positive reaction to Emerson's challenge that Warburg showed at Urbana was the fact that he had developed a new two-vessel technique for measuring quantum yields. But instead of discussing these or other details of the technique with Emerson, Warburg spent most of his time constructing an actinometer: a device to measure radiation intensity by way of monitoring a chemical reaction (in this case, the uptake of oxygen produced during the chlorophyllide reaction by thiourea). This was surprising, given the fact that Warburg had up to then used bolometers, which were far more accurate.[51] In his paper of 1948, Warburg wrote that he had decided to abandon the experimental procedure of 1923, "because of the danger of frothing" (which he thought was the reason for the alleged carbon dioxide burst), and "because it is unnecessarily cumbersome".[52] Although it was true that a bolometer required considerable skill in handling, one still wonders why Warburg resorted to trading precision of measurement for ease of handling—if not with the purpose in mind, already then, of being able to train unexperienced persons how to reproduce his measurements.

---

[49] This letter is in private hands; thanks to the intercession of Govindjee, Peter Homann kindly made it available to the author.

[50] A more detailed description of Warburg's stay in Urbana can be found in Nickelsen and Govindjee (2011); Chapter 3.

[51] Warburg had already mentioned this alternative way of determining quantum yields as a valid possibility in his German paper for the "Fiat review", Warburg (1947, pp. 209–210), and repeated this suggestion in Warburg (1948, pp. 208–209).

[52] Warburg (1948, p. 208).

In December 1948, Emerson even organised a public workshop, to which he invited all the experts in the field, in order to force Warburg into a discussion of quantum yields with him; yet nothing came out of it.[53] Thus, the last chance to reach some sort of agreement came the day after Christmas, about 4 weeks before Warburg intended to leave Urbana. Warburg had finally agreed to carry out some experiments alongside Emerson and his co-workers, and have the results judged by "impartial observers". The two observers were the biochemist Dean Burk, who was then at the National Cancer Institute (NCI) in Bethesda (Maryland), and a young colleague of his, John Z. Hearon, at the time research fellow at the same institution.[54] It is not entirely clear, who selected these observers based on which criteria; presumably it was the botanist Oswald Tippo, at the time Head of the University of Illinois's Botany Department on the Urbana campus. Tippo might also have organised the documentation of this meeting in the local press (Fig. 5.4). However, still no agreement was reached during this 12-day period of common experimentation (26 December to 6 January). The results were widely divergent and, hence, inconclusive. Emerson later suspected that the *Chlorella* cultures had been contaminated by other microorganisms—one of the regular difficulties in dealing with these algae.

In the end both parties acknowledged that under some conditions the data purported by the opponent were, in fact, obtained; yet no agreement was reached about how significant these data were and how they should be interpreted. Burk's plan of a jointly authored note to be submitted to *Nature*, with the intention of informing the scientific public about the outcome of Warburg's visit to Urbana, was dropped: Emerson would not agree to the draft that Burk had sent him. Among other things, he felt that conditions in his laboratory were described in a more unfavourable manner than was justified by the facts; but Emerson's main point was that "the data constitute a rather insufficient basis for the establishment of conclusions so important as those we are trying to reach".[55]

Whereas Emerson later spoke of being thoroughly depressed at the outcome, Warburg announced his victory to everybody who would listen. "It was as [if] somebody put it [sic] here a drama watched by all America and the happy end was the victory of truth", Warburg wrote to Tippo, after having left Urbana.[56] Warburg also used every

[53] Many letters were exchanged in advance to that meeting; see, e.g., Emerson's invitation of French on 3 Dec. 1948, Robert Emerson Papers, 1923–1961, Record Series 15/4/28, Box 1, Folder: French, C. Stacy. University of Illinois Archives. According to Albert Frenkel (telephone interview with Govindjee, on 8 September 2007, and generously made available to me) the meeting was civilised, and without the nasty quality that the controversy would later acquire. Frenkel had been an assistant to Emerson at Caltech, where the two of them became good friends. See also Frenkel (1993) for an autobiographical account of Warburg and how the quantum controversy came to the Marine Biological Laboratory at Woods Hole (Massachusetts) in 1949.

[54] Burk and Hearon were co-authors of several papers, in particular on the oxygenation of cobalt complexes; see, e.g., Burk et al. (1946, 1947); Hearon et al. (1949). A paper by Hearon was published in August 1948 in the Journal of the NCI; see Hearon (1948).

[55] Archive of the BBAW, NL Warburg 174. Emerson to Burk, 21 Jan. 1949.

[56] Warburg, Otto to Tippo, Oswald, 17 Feb. 1949, Personal file of O. Tippo, (former) Botany Department, University of Illinois at Urbana–Champaign, now: Department of Plant Biology. I am grateful to Clint Fuller and Govindjee for pointing out this letter to me.

Fig. 5.4 A photographic record of the 12 days of conducting experiments together. From *left* to *right*: Victor Shocken (an assistant of Emerson's, who later joined Warburg's party), Shimpe Nishimura (*face obscured*), Dean Burk, Oswald Tippo (*at the rear*), Otto Warburg (*holding up two manometers*) and Robert Emerson. (The photograph appeared in the local *News Gazette* on 9 January 1949).

possible opportunity to belittle Emerson and his work. For example, on 21 January 1949 Warburg wrote to French, after the latter had inquired whether Warburg would not like to stay a little longer in the USA:

> Certainly I have not told [said] that it is impossible to work scientifically in the US. But I have told [said] that it is impossible in Emerson's laboratory. It seems to me that many scientists in this country are aware of this; but unfortunately nobody warned me. It is no crime to make mistakes in science. But it is another thing to fight established truth for years and years strewing sand into the mills of science.[57]

To the plant physiologist Frederick C. Steward of the University of Rochester, Warburg wrote on 2 January 1949, the question could only be settled if he found another place to stay for the rest of his time in the USA and inquired whether Steward would be able and willing to host him.[58] In the end, Warburg joined Burk at the NCI in Bethesda, where Burk had succeeded in securing him a 6-month position (supported by the Public Health Service). In the years to follow, Burk became in America "Warburg's bulldog", so to speak, while one has to add that Burk was far less stringent and convincing in his arguments for Warburg's cause than Thomas H. Huxley used to be, when he fought his battles for Darwin.

[57] Archives of the MPS; III. Abt., Rep. 1, Nr. 198. Warburg to French, 21 Jan. 1949.

[58] See Walker (1992a, pp. 136–137).

5.5.4 The Rediscovery of the Maximum Quantum Yield

After having spent 4 months at the NCI in Bethesda, Warburg, accompanied by Burk, moved in June 1949 to the Marine Biological Laboratory (MBL) in Woods Hole (Massachusetts), where he would spend the last month of his stay in the USA. For this purpose, the whole experimental set-up for measuring photosynthetic quantum yields, including the culture vessels, Warburg apparatus, etc., was transferred for the summer from Bethesda to Woods Hole. There, another confrontation with Emerson arose during the annual meeting of the Society for General Physiology, at which Emerson and Warburg met again to discuss quantum yields—this time the opponents became more outspoken and emotional.[59] Shortly after this last contest, Warburg returned to Berlin, where he succeeded in having his institute re-established as one of the newly founded Max Planck Institutes. He left Emerson in a rather gloomy state of mind, as can be taken from a letter that Emerson wrote to Arnold on 21 July 1949:

> Dear Bill: [...] I wish I knew what your opinion is now concerning the quantum yield of photosynthesis. Burk regards the matter as settled in Warburg's favor. I am unable to put my finger on any error in the Burk–Warburg experiments which would appear to account for the discrepancy between their results and mine, but as Franck says, there are a number of things about their experiments which are "very fishy". I felt it wasn't much use to discuss things with Burk, because to me he seemed inclined to conceal important points in a rather deceitful way. I dislike having a controversy with such people. Warburg doesn't speak to me at all any more.[60]

It becomes clear that, apart from the factual discrepancies, the controversy had, already at this stage, acquired an element of mutual lack of trust, personal defamation and suspected dishonesty. Warburg seemed to have taken Emerson's critique as a personal slight, while Emerson in turn felt exceedingly offended by Warburg's dismissive attitude towards him.

This situation was not alleviated by three papers in renowned journals (*Biochimica and Biophysica Acta* [BBA], *Science*, *Archives of Biochemistry*) that came out of Warburg's months in Bethesda, co-authored by Warburg, Burk and Sterling Handricks, together with various students.[61] The content of the papers was very similar. The value of 4–5 light quanta per molecule oxygen was again confirmed, while, as the authors stated, "a requirement of 3 quanta is open to serious consideration".[62] These data were measured, again, with a revised experimental set-up: a two-vessel method,

---

[59] This is remembered by the plant physiologist Burlyn Michel (personal communication (email) to Govindjee, 27 November 2007), who, at the time, was attending one of the Woods Hole summer courses as a student. Michel recalled how surprised he was by the "heated exchange between Emerson and Warburg following a presentation by Emerson, in which he disputed Warburg's claim of high efficiency". See also the report in Kamen (1985, p. 304), particularly the letter by Gaffron to Kamen, quoted therein and dated 25 June 1949.

[60] Emerson to Arnold, 21 July 1949, Robert Emerson Papers, 1923–1961, Record Series 15/4/28, Box 1, Folder: Arnold, William. University of Illinois Archives.

[61] Burk et al. (1949); Warburg et al. (1950).

[62] Burk et al. (1949, p. 229).

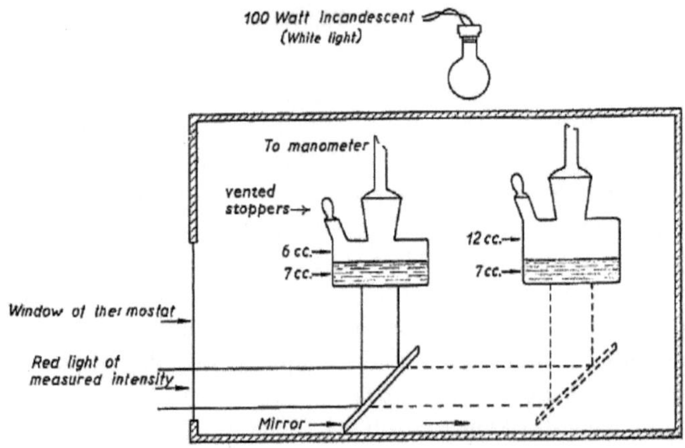

Fig. 5.5 Warburg's experimental set-up in 1950: a beam of red light (630–660 nm) of about 3–4 cm^2 in area entered the side of the thermostat and was reflected by a mirror onto the bottom of one of two vessels, alternately in the one or the other, by shifting either the mirror or the manometers. Additionally, the vessels were continuously illuminated with white light from above, which, the authors maintained, kept the process of photosynthesis at a high enough level to compensate for respiratory gas exchanges. (Reproduced from Warburget al. 1950, p. 338).

in which a slightly acidic, phosphate-containing buffer solution was used, while the light intensity was measured with the actinometer. Furthermore, Warburg and Burk had devised a procedure that, from their point of view, made all calculated corrections for respiration effects unnecessary. In their set-up (Fig. 5.5), they used white light of undetermined intensity that illuminated the vessels from above. This ensured, they argued, that the rate of photosynthesis was always much higher than the rate of respiration. An additional red beam of measured intensity caused an increase in the rate of photosynthesis, which was then manometrically recorded. Warburg and his co-authors underlined that, by this means, they no longer had to work near the compensation point, at which photosynthesis and respiration almost equalled each other, and which had rendered manometric measurements rather complicated and delicate. (It was silently assumed that the light of the red beam was completely absorbed by the cells, which the authors tried to ensure by using thick cell suspensions—without mentioning the fact that thick cell suspensions had been demonstrated to be rather difficult to control. The authors also made no effort to prove that the new way of illumination was equivalent to the technique of intermittent illumination used up to then).

Reading these papers creates the impression that Warburg and Burk were intent on taking the controversy (which was basically centred on one item of data) to another level, namely, using scientific esteem and prestige as a weapon. They richly decorated the papers with photographs that had no obvious relevance to the content but rather showed the authors at work or posing with a number of other eminent scientists, mainly Nobel Prize laureates. In addition to this visual strategy, the authors intentionally blurred the critics of their work by simply stating that the established value of the quantum yield has "sometimes been doubted by theoreticians, and it is a fact that certain investigators have raised methodological objections".[63] No mention was made of the names and arguments of Gaffron, Emerson or, who was the primary target of the "theoretician" remark, Franck; thus, nobody was in a position to defend himself.

As can be taken from the preserved correspondence, getting these pieces published was not an easy task. On 18 December 1949, Burk informed Warburg that "after some difficulty" the third paper had been accepted by the well-known journal *Archives of Biochemistry*. The editors had strongly recommended that the referees' advice for revision be followed, for example, to make the piece a bit less one-sided.[64] Yet Burk simply wrote a letter of explanation to the editor (which had the intended success) and changed nothing.[65] Notwithstanding this ostentatious self-confidence, Burk suddenly was no longer sure how to address the problem of Emerson's values, as can be taken from the same letter to Warburg. Until recently he would have firmly endorsed Warburg's notion that inadequate shaking had irreversibly harmed Emerson's cultures, he wrote, while now he believed that he had some evidence to suggest that the shaking might, after all, not be so important. Burk's solution was characteristic of Warburg and Burk's dealing with divergent opinions:

> I think we should be careful not to indicate that Emerson's 'Emerson Effect' [i.e. the carbon dioxide burst] was due only to inadequate shaking on his part, even though that might indeed be one way to produce such an apparent effect. The less said specifically about Mr. Emerson, the better I believe.[66]

Yet, on the whole Burk was enthusiastic about this piece of work, as he was quick to add: "The more I read the article, the more I like it, everything is so beautifully clear and well organized, and it is a classic in its way". He was particularly fond of the triumphant final sentence of the paper, which reads: "The fact must thus be envisaged that in a perfect nature photosynthesis is perfect too".[67]

In particular, the paper in *Science* was widely read and aroused considerable attention. Burk received a great number of requests for reprints, while he was also busy to attract public interest through popular write-ups in e.g. *Newsweek*, *Scientific Monthly*, the *Washington Evening Star* and *Time Magazine*. The rush on the theme would not die down for several months, as it was being constantly fanned by Burk's series of public speeches. These included a highly visible keynote at the annual meeting of the American Association for the Advancement of Science (AAAS), which took place at the end of December 1949. In this speech Burk framed Warburg's research in economic terms, and confronted his audience with a calculation of how

---

[63] Warburg et al. (1950, p. 335).

[64] The anonymous referee's report is preserved in the Burk–Warburg correspondence, held by the Archive of the BBAW; see the sheet dated November 9, 1949, NL Warburg 174. Publication was recommended, despite the fact that the paper was very one-sided, but summarising experimental results rather than writing them out in detail was requested.

[65] Burk's reply is likewise preserved; see Archive of the BBAW, NL Warburg 174.

[66] Archive of the BBAW, NL Warburg 174. Burk to Warburg, 18. Dec. 1949.

[67] Warburg and Burk (1950b, p. 413).

much more energy could be retrieved through sunlight if the photosynthetic process, that now, finally, was about to be understood, could be exploited on an industrial scale. The topic was also taken up by German newspapers: on 7 January 1950, for example, the *Neue Zeitung* entertained the headline, referring to Burk's paper at the AAAS meeting: "More effective use of Solar Energy can multiply food production one hundred times". On 20 January 1950, finally, Burk was able to report to Warburg that their work had made it to the front page of the *New York Times* (Burk sent off a copy of the newspaper).[68] The renowned science journalist William L. Laurence, well-known owing to his former position as the official journalist of the Manhattan Project, had featured the story. "Vital forces found in plants may increase world's food: Scientists, reporting efficiency up to 87 % in using energy of sunlight, visualize 100-fold rise in yield of algae", read the sweeping headline. (Laurence had been interested in photosynthesis before—in fact, for his presentations Burk may have used earlier writings by Laurence, in which the latter had highlighted the enormous potential of atomic energy to make the world a better place. Radioactive isotopes, for example, Laurence had announced, would pave the path to elucidate the reactions of photosynthesis and, thereby, lay the foundations for the artificial production of food and the construction of highly efficient solar power stations[69]).

The story in these newspaper articles was always the same: equipped with highly sophisticated instruments, Warburg and Burk had been able, in heroic efforts, to confirm and definitely establish the high efficiency of photosynthesis. The articles seldom failed to mention Warburg's Nobel Prize, while the existence of any criticism to this work was completely ignored. Instead, the solution of the world's energy problem as well as of the hunger crisis was announced to be imminent in the foreseeable future. Emerson was well informed on these developments, among others, from his colleague Steward at Rochester. Emerson answered:

> Dear Mr Steward,
> I appreciated receiving your account of Burk's performance in New York [at the AAAS]. One of our graduate students was there, too, and gave us his impressions, but the field is so new to him that he couldn't give us as full an account as you do. We are amused that Burk had no time for discussion of results with scientific colleagues, but had plenty of time to spill a big story for newspaper reporters. [...]
> Yes, Burk gives one this impression that he is making an intentional effort to confuse issues, rather than to clarify them. I'm inclined to agree that an ethical problem is involved, as well as a question of scientific fact. I'll appreciate advice on how to deal with the ethical issue, but I'm inclined to let it go until we have settled the facts.
> With best wishes,
> Sincerely
> Robert Emerson.[70]

[68] Burk to Warburg, 20 Jan. 1950; Archives of the BBAW, NL Warburg 174. The story already appeared in the issue of the 31 December 1949.

[69] See Laurence (1945, 1946a, pp. 167–168, 1946b).

[70] Emerson to Professor F. C. Steward January 28th, 1950; Botany Department; University of Rochester, New York. Letter kindly provided by the late David Walker.

5.5.5 *Franck's Attempt to Find a Compromise*

While Emerson struggled to find the errors in Warburg's experimental procedure and calculation method ("settle the facts", as he called it), Franck chose a different strategy. He was ready to grant Warburg that his data and methods were as sound as Emerson's, yet also Franck suggested that these data did not reflect the maximum yield of actual photosynthesis. On 14 March 1949, when Warburg was still in Bethesda, Franck sent Warburg a manuscript to look at, with the following remark:

> I would be so glad if you could subscribe to the view that the differences between the findings in the quantum yield rather indicate a difference in the observed photochemical processes than to some measurement error on the part of the observer.[71]

Two weeks later, Warburg answered rather briefly that he had studied the paper but, of course, could not agree. But he (surprisingly) concluded the letter on a warm note: "Finally, I have to say how very glad I was to see you again. Our last meeting was in Berlin, at the [Deutsche] Physikalische Gesellschaft, seventeen years ago, when you made your unforgettable speech in memory of my father".[72]

In this paper of 1949, Franck briefly reviewed the disagreement, describing how Emerson and Lewis explained Warburg's earlier measurements (namely, with the occurrence of an outburst of carbon dioxide in an acidic, phosphate-containing medium), while Warburg had rejected Emerson's criticism and reconfirmed his values under rather different experimental conditions. Franck maintained that it was "hard to accept the point of view that only Warburg's method under special conditions will permit the algae to reduce CO_2 with a quantum yield of 1/4 when all other observations systematically give $\sim 1/10$ as the highest value". Yet, he thought that the difference by a factor of two between the values of the two parties was too high to be entirely due to the confounding factor identified by Emerson and Lewis. Furthermore, Franck saw some evidence pointing to the fact that Warburg's findings might not only be quantitatively different from those of other groups but also qualitatively. Franck's suggestion, then, was intended to reconcile the results of Emerson's and Warburg's measurements, under the assumption that respiration might be interfering with the different processes of photosynthesis—in particular under conditions where the latter is not much higher than the former, that is, at very low light intensities:

> We introduce the assumption that Warburg's high quantum yield may be connected with the reduction of respiratory intermediates rather than with the reduction of CO_2. That is possible because Warburg's measurements are carried out under conditions where the photosynthetic rates are smaller than or, at best, comparable to, the respiration rates.[73]

Although other researchers had suspected that respiration interfered with photosynthesis, no systematic attempt, Franck believed, had been made to explore the

[71] Archives of the MPS [Max Planck Society], III. Abt., Rep. 1, Nr. 195. Franck to Warburg, 11 March 1949.

[72] Archives of the MPS, III. Abt., Rep. 1, Nr. 195. Warburg to Franck, 28. March 1949.

[73] Franck (1949, p. 299).

consequences of this observation. In particular, Franck suggested that, instead of using CO_2, the reducing agents of photosynthesis might be utilised, under certain conditions, to reduce half-oxidised respiratory intermediates.

However, after having developed this argument in rather sophisticated terms, Franck conceded, in a "note added in proof", that in light of the experimental conditions used by Warburg, Burk and others in their latest experiments, that is, the combination of white background illumination with a red beam of measured intensity, this theory of his had become obsolete: in these experiments, the yield remained high even under conditions where photosynthesis exceeded respiration several times. Far from being frustrated, Franck was still convinced that his observations were useful "for the reconciliation of the differences in the results of quantum yields and of the chemical nature of intermediates of photosynthesis". He still believed that there were two different photosynthetic processes, "one with the quantum yield of 1/4, the other with 8 quanta" and that the former was exceptional, taking place only under specific circumstances. Franck then suggested an alternative mechanism (which, incidentally, still in 1949 assumed the existence of a chlorophyll–carbon dioxide complex):

> [I]t might be possible that the energy stored in the phosphate bonds produced by respiration might be transferred to phosphate bonds of the CO_2 complex and of intermediate products of photosynthesis. In that way, the energy of 12 K-cal. would be available in the molecules to be reduced before each photochemical reaction and, with that additional energy photosynthesis may proceed with 4 quanta. However, this photosynthesis could, even if all other conditions are favorable to it, only proceed to a maximum rate of 1.5 times that of respiration. Any photosynthesis in algae beyond 1.5 times respiration would need 8 quanta.[74]

The idea that back reactions might be involved, which, to some extent, linked respiration and photosynthesis, or, alternatively, that the energy gained from respiratory processes was used for photosynthetic carbon dioxide reduction, would be around for the next few decades and would become very influential. Warburg himself would have shuddered at this thought, but one could claim that this idea became a major source of inspiration for Warburg and Burk's later photosynthesis model, which became known as the "one-quantum mechanism" of photosynthesis (see below).

5.5.6 *Further Exchanges of Blows*

Although Emerson was in rather low spirits by the end of 1949, he was all the more determined to find the flaw in Warburg's recent work. One aspect he set out to investigate was Warburg's claim that there was no physical time lag in their manometric measurements (which Emerson found to be strictly contrary to fact). He pursued this line of research with his assistant Nishimura and with Charles Whittingham, a postdoctoral student from the University of Cambridge (UK), who was spending a year at Urbana.[75] The results were presented at a symposium on "Carbon Dioxide Fixation

[74] Franck (1949, p. 313).

[75] See Nishimura et al. (1951) for the pertinent publication.

and Photosynthesis" in July 1950, organised by the Society for Experimental Biology and hosted by the Biochemistry Department of the University of Sheffield (UK). The conference was attended by all the major players in the field of photosynthesis, including, among others, the already well-known Emerson, Franck, French, Gaffron, Hill and, of course, Burk, as well as Daniel Arnon, Melvin Calvin and Bessel Kok, who will enter this book again in later chapters. This group was complemented by some renowned figures from the discipline of biochemistry, such as Hans Krebs and Harland G. Wood, who were working on the phenomenon of carbon dioxide fixation in heterotrophs. Warburg was also expected to attend, while he failed to show up at the last minute.

Emerson gave a detailed report of his journey to Europe in a letter of 3 November 1950 to the physicist Louis N. Ridenour, who was, at the time, Dean of the Graduate College at Urbana. According to this letter, almost a full day of the symposium was devoted to the discussion of the maximum quantum yield of photosynthesis, with centre stage being given to Emerson's and Burk's presentations. Emerson was very satisfied with his experiences there. He felt that never before had he spoken to an audience so "clearly willing to give its attention to the experimental details that have come to play such an important part in this controversy". Burk was given the first time slot to set out his position, as Emerson reported:

> [George E.] Briggs did a careful job of steering the discussion, and of giving opportunity for expression of all viewpoints. Burk was asked to give his opinion regarding the criticisms of his work which were implied by the tests of his methods reported from our laboratory. At first, he tried to avoid comment, but questioners were insistent, and he finally said there seemed to him to be nothing in our work which he would not be able to explain in a few days' time. I think it is fair to say that the consensus of opinion was that he had failed to prove his case.[76]

It is interesting to compare Emerson's letter with the letter that Burk wrote to Warburg immediately after the Sheffield symposium:

> Veni, vidi, but not quite vici. The Philistines were present in great numbers, including all our old "friends" who surpassed themselves in their attempts to muddy the waters without any new experiments. Emerson talked as in Chicago 1947, Urbana 1948, Woods Hole 1949 style, followed by a pecking at our large *Archives* article and finally by just one slide of new data, with the 2-vessel method in which he claimed that he could now get 4 quanta for certain time periods but 8 quanta if he took 30' periods and included the first 5–10 minutes.[77]

Burk's admission that he had not quite "won" strongly supports Emerson's impression that Burk had failed to convince the audience. Furthermore, it seems that Burk had failed to grasp Emerson's point, since the latter had in fact attempted to show under which circumstances one would get the (artefactually) high quantum yield—and had succeeded in doing so. In the remainder of the letter, Burk emphasised that everybody in Sheffield was greatly disappointed that Warburg had not shown up. Burk

[76] Emerson to Ridenour, 3 Nov. 1950, Robert Emerson Papers, 1923–1961, Record Series 15/4/28, Box 1, Folder: Graduate College Correspondence, University of Illinois Archives. George E. Briggs was Professor of Botany at the University of Cambridge.

[77] Archive of the BBAW, NL Warburg 174. Burk to Warburg, 10 July 1950.

had also presented photographs of the Dahlem laboratory during his talk, which, he thought, greatly impressed everybody—most of the participants had assumed that the institute was still in ruins. Burk proudly boasted: "I think it made the old-time Berliners homesick and the British and American keenly interested".

The contributions were published in a conference volume one year later. In Emerson's paper (published together with Nishimura and Whittingham), the authors clearly explained why the two-vessel method, as it was used by Warburg and his co-workers, was extremely sensitive to significant systematic errors.[78] Even tiny aberrations, which in the one-vessel method would be trifling, were likely to have enormous consequences on the final result. As Emerson and his collaborators argued in great detail, slight errors in the individual manometer readings—errors of no more than 0.3 mm, which were bound to occur all the time—could dramatically change the calculated values of γ, which, in turn, severely altered the resulting quantum yield. They identified the most dramatic source of error as being the fact that, in contrast to long established practice, Warburg and Burk had, in their measurements, failed to take into account a time interval for a physical lag in the response of the manometer to the change from light to darkness and from darkness to light. Consequently, the authors concluded that the measurements provided by Warburg and Burk failed to demonstrate the efficiencies which they reported.

What Emerson did not include in this paper, however, was his own nagging suspicion that, although he was reasonably sure that Warburg and Burk's methods were flawed, he was not yet fully in control of the situation. This can be taken from the following letter that Emerson wrote to Gaffron on 4 April 1950:

> It still seems to us that difference in physical lag [. . .] is likely to be the most important source of systematic error, but we have spells of worrying that there is something else which we have not thought of yet. One can get quantum requirements of 4 without too much difficulty, but we are not yet able to get this result regularly, combined with a γ value close to unity, as Warburg and Burk claim to have done. I suspect the trick is to have just the right combination of CO_2 burst and physical lag. The errors from these 2 factors seem to work in opposite directions, when you use Burk–Warburg vessel volumes and 10'light–10'dark cycles without allowing any interval for physical lag.[79]

5.5.7 *Controversial Themes Around 1950*

This was the state of the debate around 1950. As is obvious from the course of events, interest in the question of the maximum quantum yield was high among photosynthesis researchers. Factors "outside" scientific boundaries *sensu stricto* were looming large. Warburg was the Nobel Prize laureate, with a well-founded reputation as being an excellent experimenter, mastering manometry to the point of perfection.

[78] Nishimura et al. (1951).

[79] Emerson to Gaffron, 4 April 1950, Robert Emerson Papers, 1923–1961, Record Series 15/4/28, Box 1, Folder: Gaffron, Hans, University of Illinois Archives.

Emerson was considerably younger, far less famous, and, although he was hardly less proficient than Warburg in manometry, he had learned his technique in Warburg's laboratory, so that he always carried the stamp of a "disciple of", with the obvious connotations of inferiority. These were the circumstances upon which Warburg and Burk never failed to dwell, in order to gain maximum advantage out of them. Furthermore, they were not adverse to using other unpleasant rhetorical tricks. Warburg and Burk, for example, constantly referred to Franck as the main opponent to Warburg's work. As a quantum physicist, without any knowledge of living beings, Franck allegedly argued on theoretical grounds only, while Warburg himself let Nature speak. If Emerson was mentioned at all, he was belittled, for example, by introducing him as a "botanist" (in other words, as someone who could not possibly know much about manometry). This referred to the fact that, out of necessity, Emerson had submitted his PhD thesis to the Berlin Botany Department.[80]

However, while their influence cannot be disputed, it is not so clear how far these circumstances actually drove the course of events. It is obvious that the general public, manipulated by journalists who had been thoroughly worked on by Burk (and who were more interested in crisp headlines than in going through dull experimental minutes), tended to think that Warburg was right. And to a certain extent this also held true for the scientific public not involved in photosynthesis research—in particular those scientists who knew Warburg from other fields of his work, such as enzyme studies or cancer research. Also the "inner circle" of scientists, who were actually engaged with the subject matter themselves, and many of whom were on excellent terms with Emerson, gave serious consideration to the arguments brought forward by Warburg and Burk—but not because of Warburg's fame and authority, but because the question was so important and the answer so obscure. The experiments from which the value had to be derived were so delicate that anyone could err, Warburg as well as Emerson. It was, in particular, two factors that constantly jeopardised the reliability of the results: the physiological state of the algae cultures and the interfering effects of respiration. It was already clear then that both factors decisively influenced the outcome of manometric quantum yield measurements (although it was far less clear to what extent and in which ways) and it was known that the combined effects of these factors with other parameters, such as light intensity or temperature, might yield further unforeseeable consequences.

One can clearly see from the correspondence between the major players how deeply they were preoccupied by these problems. Cultivating the algae in a reproducible state had top priority, and was one of the most time-consuming activities in photosynthesis laboratories of the period. Indeed, Emerson employed a person specifically for that purpose (Ruth Chalmers), while he closely corresponded with the most knowledgeable algae experts of the time, such as the phycologist Ernst G. Pringsheim (who earlier had advised Otto Warburg on this question). The influence that the type of tap water (Urbana versus Baltimore tap water) used for the algae cultures had on the experimental outcome was only one among a host of

[80] See on this point Chapter 4.

intricately interwoven effects. Consequently, the strains of algae with which standard experiments—such as the Emerson and Arnold flashing light experiments—had been carried out were eagerly exchanged between photosynthesis laboratories, along with the exact recipe for this or that culture medium: this was the only way to at least approximate comparability between experimental results. The subject was so complicated that by 1950 many scientists were inclined to believe that almost any factor connected with algae culturing might influence the eventual quantum yield.

The problem of respiration multiplied these uncertainties still further. Basic manometry was unable to differentiate qualitatively between the kinds of gases that were produced or consumed. More sophisticated approaches that tried to amend this deficiency eventually became available (such as the two-vessel method). Yet, in the case of photosynthesis the matter was further complicated by the fact that not only the two gases had to be distinguished but also the rate of photosynthesis from that of respiration. The usual assumption was that the rate of respiration was the same in the light and in darkness, so that the gas exchange values in the light could be corrected by subtracting the values for respiration measured in the dark. Photosynthesis researchers, of course, worried about the validity of this method. As Rabinowitch wrote in 1945, the possibility of the effect of light on respiration was "a nightmare oppressing all who [were] concerned with the exact measurement of photosynthesis".[81] However, as no better alternative was available, people continued to work along these lines.[82] (From today's vantage point it is clear that the assumption was inaccurate: light *does* influence the rate of respiration). Furthermore, there was the additional complication that respiration, as well as many other physiological processes in the cell, possibly interfered with photosynthesis in other ways than through gas exchange, which is what Franck suspected. Alternative methods were tried to evade these problems, such as microcalorimetric techniques and polarography, but none of these were, at the time, developed to the necessary degree of precision.

The choice of medium was a third factor that strongly influenced the outcome of quantum yield measurements, although it was much easier to control than the others. It has been emphasised in this chapter that the choice of an acidic, phosphate-containing buffer solution (preferred by Warburg and Burk) versus an alkaline, carbonate–bicarbonate buffer solution (preferred by Emerson) was significant. Emerson and Lewis had demonstrated that the use of an acidic buffer solution favoured the development of a carbon dioxide burst, which tended to distort the experimental findings towards (apparently) low quantum requirements. Warburg, on the other

[81] Rabinowitch (1945, p. 569).

[82] See, e.g., a letter by Gaffron to Emerson on 4 March, 1950, in which he raised the possibility that the respiratory quotient changed during the exposure to light; to which Emerson answered, on 4 April 1950: "If the respiratory quotient changes in the dark, then there is really no sense in trying to measure the quantum requirement at low light intensity. I agree with you that this is a possibility to be considered, but up to now we have all worked on the assumption that the respiration is not much changed by illumination. We should not give this up without good reason". Both letters: Robert Emerson Papers, 1923–1961, Record Series 15/4/28, Box 1, Folder: Gaffron, Hans, University of Illinois Archives.

hand, argued that only the lowest possible value was of theoretical importance. From Warburg's point of view, the divergent values in an alkaline medium were caused by the fact that the algae were unable to photosynthesise properly at higher pH values.[83] The issue remained controversial, so that around 1950 Emerson's priority was to establish that the data obtained in an alkaline buffer solution were as valid as the data arrived at using an acidic medium.

Finally, Warburg's strategy for dealing with Emerson's critique deserves some closer attention. As Rabinowitch observed retrospectively: "[R]ather than investigating thoroughly the conditions under which the alleged high quantum yields could be obtained, [Warburg] kept publishing increasingly startling new observations, whose relation to his own earlier findings was not always clear, and which made Emerson's control experiments obsolete faster than they could be performed".[84] In his early studies, Warburg had emphasised that low light intensities had to be used in relatively short experiments, which led Franck to suggest that respiration was interfering with photosynthesis under conditions of the compensation point. However, by 1950 Warburg claimed, together with Burk, that the highest efficiencies were obtained at high light intensities over long periods of time. In the early studies, very thick suspensions, which needed to be allowed to settle down in the vessel, had been recommended, while later Warburg favoured the use of thin suspensions, which were rather vigorously shaken. Even the technique of supplementing a white background illumination with a red beam of measured intensity was dropped a few years later in favour of a "catalytic" amount of blue light (without which, it was claimed, photosynthesis was impossible). The concentration of carbon dioxide necessary for the highest yields was also slowly increased, up to 10 %. Strangely enough, the range of experimental conditions used by Warburg and his co-workers did not evoke a matching range of experimental results. These circumstances were already sarcastically commented on by Emerson in 1951 who pointedly wrote: "In no case are these specifications supported by experimental evidence. Possibly they represent only *ad hoc* assumptions".[85]

At first glance, this extended discussion on one parameter contributed nothing at all to the process of elucidating the mechanism of photosynthesis. On the contrary: financial and personal resources were consumed in abundance without the actual goal—settling the magical parameter—ever being reached. At second glance, however, these studies did result in a wealth of new knowledge being amassed, which became relevant in rather unforeseeable contexts (such as the finding of the Red

---

[83] It has been pointed out—in Govindjee (2001)—that, initially at least, Emerson and Warburg did not argue about the values measured in a carbonate-bicarbonate buffer solution. However, this situation changed in 1952, when Burk announced that the same high quantum yield had been measured in new carbonate mixtures (see p. 172).

[84] Rabinowitch (1961, p. 123). On this point, see also the summary of the (inconsistent) conditions used by Warburg when demonstrating quantum yields of four and less provided by Rabinowitch (1956, pp. 1947–1948).

[85] Nishimura et al. (1951, p. 209).

Drop of photosynthetic efficiency). Furthermore, the attempts to explain the differing results led researchers to explore aspects of photosynthesis that had so far been neglected, such as the complex relationship between photosynthesis and respiration, and the intricate details of algae cultivation. However, slowly people started to lose interest, feeling that the controversy had become stuck in a dead end. Warburg had failed to demonstrate why the experimental specifications in his set-up were relevant for high quantum yields, and Emerson had been unable to prove the same specifications conclusively irrelevant to this purpose. The shortcomings of the available methods—above all the technique of manometry—had clearly been brought to the fore, so that, on this basis, a solution hardly seemed attainable. Nevertheless, the proponents would continue to struggle (unsuccessfully) for some years yet, stimulated, among other things, by a surprising new proposal from Warburg's part: the one-quantum mechanism.

5.6 A Hardening of the Fronts

5.6.1 The One-Quantum Mechanism

In the autumn of 1950, following an extended stay of Burk's at Warburg's institute in Berlin, Warburg and Burk outmatched themselves by proposing the so-called "one-quantum mechanism" of photosynthesis. This mechanism was first published in two short notes written in German, while in the English-speaking world, it was, for a long time, only published in a semipopular version: in the form of a paper published in October 1951 in *Scientific Monthly*, co-authored by Burk together with Jerome Cornfield and Martin Schwartz.[86] It was only in Warburg's 1958 review of photosynthesis research in *Science* that the one-quantum model made its way into a high-ranking journal for original papers in English language.[87]

The 1951 paper started off with a derogative account of Warburg's critics, in which, as usual, the names of the people involved were omitted, while many general and vague accusations were raised. The debate then was featured as an argument between the open, experimental approach to science, in which Nature was allowed to speak for herself (represented by Warburg), and the anthropocentric pondering of general possibilities, prejudiced by current physical theory (the approach allegedly

[86] Warburg and Burk (1950a); Warburg (1951) and Burk et al. (1951).

[87] Warburg (1958). There was, incidentally, considerable dismay among photosynthesis researchers following the publication of this very one-sided review of Warburg's. See, e.g., the letter by Norman Good to Robin Hill, dated 2 Jan., 1959 (Cambridge University Library, Ms. Add. 9267/J.62): "What did you think of this summer's article by Warburg on photosynthesis in Science? The general reaction in America was one of considerable irritation, in no small part irritation with the Editor of Science. The article was considered, rightly it seemed to me, as a mass of willful misrepresentation and as such not worthy of a reply. However, there are many profound regrets that scientists working in other fields should be so misinformed by a journal purporting to serve all scientists".

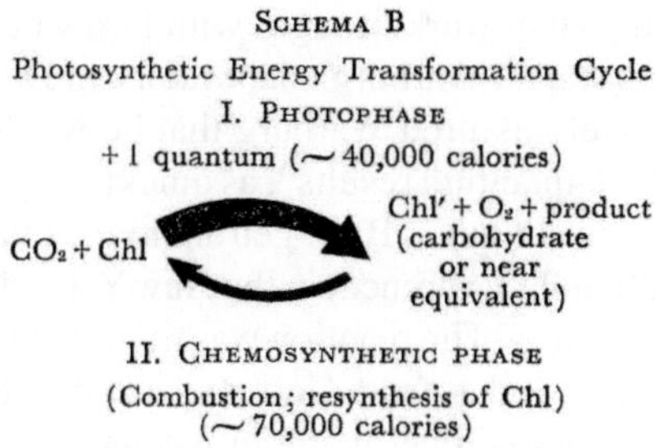

Fig. 5.6 A scheme of the photosynthetic cycle envisaged by Burk et al. (1951, p. 217). *Chl* represents the chlorophyll before illumination and *Chl'* the chlorophyll after alteration by illumination. *Chl'* is restored to *Chl* by the back reaction, at the expense of the energy derived during the chemosynthetic reaction involving the consumption of O_2 and a product.

taken by Warburg's opponents). It was the former approach, the paper argued, that eventually brought to the fore how 4 light quanta could suffice to yield high-energy carbohydrates. The important observation was, the authors reported, an effect that up to then had escaped everybody's notice: as the intervals of light and darkness were made increasingly shorter (down to 1 min each), during the dark periods a very large amount of oxygen disappeared from the system, ten times more than during normal respiration. This back reaction "prevented one from seeing the full magnitude of the forward photochemical production of oxygen", the authors claimed, while now the picture was clear:

> The over-all process of photosynthesis clearly consists of two different reactions which interlock cyclically and normally hide each other. One reaction is photochemical and proceeds in the light alone, and the other is a chemical oxidation reaction that goes on not only in the dark, but as further experimentation showed, in the light also.[88]

The main idea was that two-thirds of the photosynthetically produced oxygen would be consumed in (thermochemical) oxidation reactions. The energy thus released, amounting to no less than 70,000 calories, would then be used in the subsequent photophase, in which the complex of chlorophyll and a carbonic acid derivative would receive another 40,000 calories in the form of one absorbed quantum of red light. In summing up, then, 110,000 calories were available to produce actual carbohydrates and molecular oxygen (see also Fig. 5.6). An integrated circle of dark and light reactions would require 3 light quanta, while the photochemical process alone needed one (!) light quantum only, which brought the authors to exclaim: "What could be simpler than that nature, in harmony with Einstein's photochemical equivalence law, has one molecule of chlorophyll absorb one quantum of light to reduce one molecule of CO_2 and produce one molecule of O_2?"[89] (Simplicity, of course, was unquestionably taken to be equivalent to truth.)

---

[88] Burk et al. (1951, p. 216).

[89] Burk et al. (1951, p. 222).

These were the findings that Burk brought with him when he returned at the end of January 1951 after his stay at Warburg's laboratory in Dahlem. Having arrived in Bethesda, Burk immediately assured Warburg that he would "start the propaganda campaign in regard to the 1-quantum results", as quickly as possible.[90] This included, in addition to a whole series of talks, Burk getting in touch again with the press, including his contact (William L. Laurence) at the New York Times, who had promised to feature the story prominently. The reaction was remarkable—newspaper clippings from all over the country, as well as from Germany, reached the NCI, while the scope and style of these reports on the one-quantum mechanism clearly reflected Burk's "propaganda campaign". The *Deutsche Zeitung*, for example, featured the headline: "The chemical miracle plant: Sensational discovery is considered to banish famine". The author of the article expected that the Warburg–Burk findings would fulfil two dreams of mankind: the eradication of hunger and becoming independent of coal, petrol and wood: "The industrial production of a technological-artificial plant is about to be envisaged—an innovation which would revolutionize the world's economy". The Berlin-based *Tagesspiegel* trumpeted in a similar manner on 4 March 1951: "The solar power station of nature. Berlin scientists solve the energetic mystery of the growth of plants". On 3 May, 1951, Laurence of the New York Times finally joined in: "Plant life study yields new data. Science team reports 'cyclic' process in use of sunlight—new substances held key". In the article Laurence reported that the discovery of the one-quantum mechanism was the "crowning achievement of [Warburg's] life", and compared photosynthesis to the pilgrim step, that is, "three steps forward and two steps back". Burk's still bolder plans for the future were also quoted:

> "Now that the knowledge is available of the mechanism by which the green light captures its energy," Dr. Burk said, "the great problem is to find out what substance it is that first picks up the energy in one-quantum lots." Once this is known, it may be possible to devise ways of carrying out photosynthesis by chemical and mechanical means independently of plants. A sun energy factory might then produce a power that would easily replace fuels such as coal, oil, gas and wood. It might also be the basis for synthetic foods.[91]

As a result of all this public enthusiasm, it is not surprising that the NCI was, for the moment, appreciative of Burk and his achievements—to Burk's relief; after all, Burk had applied for a leave of absence from the NCI with the declared purpose of carrying out cancer research with Warburg, which, of course, he had not done at all. Yet, as Burk wrote to Warburg, now his superiors even defended him for not having carried out cancer research: in a strange inversion of argument, the NCI claimed that, on the strength of this extremely successful sideline of Burk's, Congress should feel further encouraged to support cancer work. Burk even was asked to prepare a report on the energetics of photosynthesis for the National Congress's meeting on the future use of natural resources. Besides Burk, only Melvin Calvin, the Berkeley-based chemist, who was working on the photosynthetic dark reactions, and the eminent photochemist Farrington Daniels, had been asked to contribute. Burk explained that this would lead

[90] Archive of the BBAW, NL Warburg 174. Burk to Warburg, 2 Feb 1951.

[91] New York Times, Thursday, 3 May 1951.

to the enormous proliferation of their findings: every congressman and senator would receive a copy, which amounted to about 10,000 people altogether.[92]

Although the response of photosynthesis researchers was not quite as enthusiastic, the papers were still eagerly received. In April 1951, Burk wrote to Heiss that Franck was most interested in the dark combustion phenomenon and had said that, "if this is definitely established something quite new and unexpected has been discovered, even if I would like to keep an open mind as to the interpretation". Another tentatively positive reaction had come from Gaffron, who, according to Burk, had conceded that "if the new observations must be interpreted as Warburg and you do, they mean of course a revolution in our general concepts".[93] In a similar vein, although very critical of the publication itself, Hill wrote to Emerson, in April 1951, that he found the idea "very stimulating", although he could not see "how phases can be sufficiently sharply defined in a 1 min alternation of relative or added light & dark".[94] It is obvious from his response to Hill that Emerson was far less charitable in his evaluation:

> I don't feel the need of further stimulation such as the new Warburg–Burk paper. I'm still confused as to the proper direction for my own further efforts. I would like to return to some of the problems raised by the Emerson–Lewis measurements at different wave lengths, particularly the sharp drop in efficiency toward the infrared, and the question whether excitation of chlorophyll with "blue" quanta can produce reactions differing in some fundamental way (higher efficiency?) than excitation with "red" quanta. To do this, I would be inclined to go back to single vessel measurements in carbonate mixture, but I feel that Warburg has put upon me a sort of curse, that I may not do this unless I can show beyond doubt that the efficiency measured in carbonate is not inferior to the efficiency in acid phosphate.[95]

Clearly, Emerson sincerely wished to detach himself from the quantum yield controversy and return to more productive work instead; yet the "curse" would remain in place for a number of years to come.

5.6.2 The Gatlinburg Conference on Photosynthesis, 1952

One obvious effect of the disagreement between Warburg–Burk and Emerson–Franck–Gaffron etc. was that it increased enormously the frequency with which letters were written and formal and informal meetings between the various photosynthesis researchers took place. One letter by Emerson may serve as an example. On 8 May 1951, Emerson wrote to Gaffron:

> Dear Hans:
> Thanks very much for the time you and Franck took yesterday to talk with us over the phone and tell us the news about Burk, Brackett, etc. We hear from [Sol] Spiegelman that Burk

[92] Archive of the BBAW, NL Warburg 174. Burk to Warburg, 23 March 1951.

[93] Quotes: Archive of the BBAW, NL Warburg 174. Burk to Heiss, 3 April 1951.

[94] Hill to Emerson, 28 April 1951, Robert Emerson Papers, 1923–1961, Record Series 15/4/28, Box 1, Folder: Hill, Robin, University of Illinois Archives.

[95] Emerson to Hill, 8 May 1951, Robert Emerson Papers, 1923–1961, Record Series 15/4/28, Box 1, Folder: Hill, Robin, University of Illinois Archives.

> made a convincing impression at Cleveland. It's interesting that his blatant self-advertising which offended people in England seems not to antagonize people in this country.
> Of course, I'm very sorry that circumstances make it so difficult for me to join you at Madison on Thursday. I would like very much to hear what Daniels picked up from Brackett. However, perhaps we can get together in Chicago, and you can tell us what Daniels found out. Eugene [Rabinowitch] says he could be in Chicago next Monday, May 14th. That would suit me very well. [. . .] Can you drop me a postal card confirming this plan? If you agree, then I will show up at your place 10:30 Monday morning.[96]

Short as it is, this letter alludes to the following events: a telephone conversation between Emerson, Gaffron, Franck and, presumably, Rabinowitch; a report by Emerson's Urbana colleague Sol Spiegelman, who had seen Burk at a conference in Cleveland; a meeting scheduled at Daniels's laboratory at Madison, at which Daniels would report about an earlier meeting with the expert in polarography and spectroscopy Frederick S. Brackett; and a future meeting in Chicago, at which Gaffron was requested to pass on to Rabinowitch and Emerson the news received from Daniels. There was, obviously, an intense flow of information going on.

The common interest was to explain, if at all possible, how Warburg and Burk obtained their implausible data. This was the main reason why researchers either wanted to go and look at Burk's experiments in Bethesda (Warburg's experiments in Dahlem were too far off for a short visit) or wished to meet independently, in order to discuss experimental methods, data and alternative interpretations. The need to exchange the results of the latest photosynthesis research also extended to other aspects of the field; so that in the summer of 1952, preparations got underway for a large conference on photosynthesis that would take place in Gatlinburg, Tennessee, at the end of October, supported by the National Science Foundation, the Office for Naval Research and the Atomic Energy Commission.[97] On 1 July, Emerson received a letter from Hendricks inviting him to participate in this conference, "for the purpose of examining those aspects of the subject which appear to be limiting further understanding". The format was intended to encourage free discussion among the participants: no formal papers were scheduled, but sessions of full half days were reserved for every theme, with an additional, uncommitted day at the end of the conference. An introductory speaker would, at the beginning of each session, briefly review the subject matter; while immediately afterwards the floor would be given to anyone in the audience: "Everyone should be prepared with slides and illustrative material on whatever is felt to be pertinent to the subject", Hendricks pointed out in his letter.[98]

Emerson was feeling pretty desperate when he received the invitation. As he wrote to Charles Whittingham on 5 September, for the greater part of the year he had been trying to free himself from Warburg's shadow, leaving the quantum yield question aside and following up some other lines of research, among other things,

[96] Emerson to Gaffron, 8 May 1951, Robert Emerson Papers, 1923–1961, Record Series 15/4/28, Box 1, Folder: Gaffron, Hans, University of Illinois Archives.

[97] Hendricks (1953) provides a short summary of the conference's discussions.

[98] Hendricks to Emerson, 1 July 1952, Robert Emerson Papers, 1923–1961, Record Series 15/4/28, Box 1, Folder: Gatlinburg Conference, University of Illinois Archives.

a thorough study of the carbon dioxide burst and its relationship to temperature and other variables. Emerson was aware that these studies would not contribute to resolving the controversy, yet, as he wrote with some exasperation: "I feel I cannot always submit to being led around by the nose by that deceitful old poker player, and must sometimes test out one or two of my own ideas". However, one full session was being reserved for the quantum requirement question at the Gatlinburg conference, and Emerson knew that, because of Warburg's latest publications, the audience would expect much of his contribution:

> Warburg's new papers, reporting high efficiencies in carbonate mixtures, etc., have excited a good deal of interest. I've tried to give them the brush-off, saying that he has not established the dark-rate on the basis of which he calculates light action, and looking always for the weak spot which I feel sure is there, concealed as cleverly as possible by the crafty old poker player. But our analysis of the errors inherent in Warburg and Burk's 2-vessel technique, adequate though it was for the refutation of their claims up to 1950 or so, is of no help in elucidating the meaning of the one-vessel measurements in carbonate buffer [solution].[99]

In this letter, Emerson alluded to the fact that, in the meantime, Warburg and Burk claimed to have measured in carbonate–bicarbonate buffer the same high quantum yields as in the usual acidic phosphate mixture. This effectively protected the new measurements against Emerson's (and his co-workers') unremitting criticism that in a phosphate buffer solution the data were distorted by the carbon dioxide burst, as no burst had ever been observed in a carbonate buffer solution.[100] Emerson had no idea how to deal with these latest experiments—but he knew that everybody at Gatlinburg would want to know what, if anything, was wrong with them.

While Emerson was tired of the whole affair, Burk, who had also been invited to Gatlinburg, was enthusiastic and motivated. He made diligent preparations, since he did not expect the "game" to be easy. "I too agree that the cold war on the quantum yield is in fact getting more intense and may soon develop into a hot one, abroad as well as here", Burk wrote to Warburg on 28 August 1952.[101] Warburg was apparently concerned about the outcome. He had declined the invitation and raised the possibility that Burk should do the same—which Burk emphatically rejected: "if none of us shows up there people will surely get, or maliciously create, the impression that we have 'lost heart' or become afraid". Warburg's alternative suggestion—to find again an impartial judge to pass an authoritative sentence after having heard the arguments—was countered by Burk with the objection that he could not think of anybody who would fit the role.[102]

In the end, there was to be not one but two sessions on the question of quantum requirements at the Gatlinburg conference, with the hope of settling the issue once and for all. How the meeting proceeded can be taken both from Burk's elaborate and highly detailed letter to Warburg, several other correspondences, and, finally, from

---

[99] Quotes: Emerson to Whittingham, 5 Sept. 1952, Robert Emerson Papers, 1923–1961, Record Series 15/4/28, Box 1, Folder: Whittingham, Charles, University of Illinois Archives.

[100] For the pertinent publications, see Warburg et al. (1952) and Warburg (1952).

[101] Archive of the BBAW, NL Warburg 174. Burk to Warburg, 28 August 1952.

[102] Archive of the BBAW, NL Warburg 174. Burk to Warburg, 12 September 1952.

a short review written by Hendricks for publication in *Science*.[103] By all accounts, it was an extremely lively meeting, with discussions lasting each day from early morning till late at night; and the sessions on quantum yields were among the liveliest. The first session, Burk reported to Warburg, was dominated by "the Old Guard", or, as Burk alternatively put it, the "Murderers' Row", which comprised "Brackett, Gaffron, Daniels, Arnold, Emerson, Brown, French, and Franck, etc.". Burk continued:

> [They got up one after the other and] beat unmercifully, and surely unscientifically, at both the 4- and the 1-quantum. The attack was far worse than at Sheffield and carefully timed and planned beforehand, to create the impression among the rest of the audience that both the 4- and the 1-quantum values were impossible, absurd and easy to explain. During all this while I didn't say anything but just sat looking unconcerned, smoking one cigar after another, while the situation seemingly got blacker and blacker.[104]

Things indeed were not looking good for Burk's camp. Brackett presented the data that he had obtained using polarographic methods, which showed that the minimum quantum requirements were six to ten per molecule of oxygen. Brown reported that mass spectroscopy gave no evidence of a substantial consumption of oxygen in the dark phases, as would be expected with the one-quantum mechanism. Gaffron demonstrated that photosynthesis could be initiated without any trace of oxygen, that is, without the probability of back reactions occurring such as those described in the one-quantum mechanism. According to Burk, Daniels presented, "with obvious facetiousness (and with some amusement)", the results obtained by students that Burk had put, as Warburg's "missionaries", in Daniels's laboratory: one of them had arrived at values of around nine, while the other had "agreed on the average with those of Warburg and Burk but individually varied from 3 to 14, in such manner that no confidence could be placed in them". Finally, Emerson reported that he had been unable to find any difference in outcome between the old and new carbonate mixtures, both of which generally gave quantum requirements of nine.[105] He also made it very clear that, in view of the wealth of data that contradicted the high Warburgian yield, it was no longer Emerson's responsibility to explain why he did not get the same results as Warburg and Burk, but that it was now up to Warburg and Burk to examine their experiments to find out why they did not get the same results as everyone else.[106]

The discussion was resumed on the last day of the conference, in a session chaired by French. On this occasion, Franck took the opportunity to talk for nearly 1 h, explaining in detail the principles underlying the energy accumulation in photosynthesis. This talk made an enormous impression upon the audience. Emerson wrote to Franck afterwards that he had never before heard Franck give so clear an exposition of the energy losses and energy requirements involved in

[103] Hendricks (1953).

[104] Archive of the BBAW, NL Warburg 174. Burk to Warburg, 13 Nov. 1952.

[105] All quotes: Archive of the BBAW, NL Warburg 174. Burk to Warburg, 13 Nov. 1952.

[106] Archive of the BBAW, NL Warburg 819. Arthur Schade to Warburg, 5 Dec. 1952.

photosynthesis.[107] In addition to that, Franck suggested again that Warburg and Burk had, in actual fact, measured the yield of a different photochemical process. This was endorsed by Kok, who was the next to speak and who pointed out that he had witnessed the low quantum requirements being reached only under conditions where respiration was likely to interfere substantially with photosynthesis. Finally, it was Burk's turn to talk, as he wrote to Warburg:

> I then got up and spoke for the rest of the session, and had the last word, so to speak, or at least the next to the last word, since there was nearly a half hour of discussion after I got through. [. . .] My general attitude was that here were the data, and our conclusions, but anybody who wished to believe otherwise, for the next five or ten years at least, could do so and see where such other beliefs might lead him.[108]

Reading these lines and Burk's further description of the event, which he consistently featured as a battle between the Good and the Evil, always on the look-out for "converts", creates the impression that he had finally lost all sense of reality. People at the conference, notably Hill, obviously started to worry and suggested that it would be wise if Burk detached himself from Warburg before it was too late. Burk also reported these details to his master:

> [V]arious "kind" people like Robin Hill and Kok and others advised me to drop any further work on the quantum yield or even photosynthesis in general before I lost my scientific reputation altogether! That you should lose yours was perhaps of lesser moment to them, than that such a nice and kind person as myself should do so!!! [. . .] Robin Hill [. . .] said to me quite frankly that in his opinion you were a "rogue" and did not mind who knew it.[109]

Not only Hill and Kok but the majority of photosynthesis researchers by then believed that Burk and Warburg were using dirty rhetorical tricks and that they were behaving in an utterly unacceptable manner when it came to dealing with factual criticism and divergent points of view. However, this did not change the fact that the value of the photosynthesis quantum yield was still unknown. Thus, even though Hill was thoroughly disgusted by Warburg and Burk's conduct, this did not prevent him from paying a visit to Burk's laboratory shortly after the conference, in order to see the disputed results with his own eyes and to discuss photosynthetic matters with Burk. Indeed, Hill had tried, together with Whittingham, to reproduce the Warburg–Burk values with his haemoglobin method; the attempt had failed, but Hill wished to try it again and therefore needed to go over some of the experimental details, as Burk proudly explained to Warburg: "He [Hill] agreed that just plain negative results, without understanding, mean nothing".[110] This was a crucial point. Already in May

[107] Emerson to Franck, 13 Nov. 1952, Robert Emerson Papers, 1923–1961, Record Series 15/4/28, Box 1, Folder: Franck, James, University of Illinois Archives.

[108] Archive of the BBAW, NL Warburg 174. Burk to Warburg, 13 Nov. 1952.

[109] Archive of the BBAW, NL Warburg 174. Burk to Warburg, 13 Nov. 1952.

[110] Archive of the BBAW, NL Warburg 174. Burk to Warburg, 13 Dec. 1952.

1952, the biophysicist Allan Brown, who would become famous for the first reliable results in mass spectroscopy,[111] had turned to Emerson with the following inquiry:

> I should like to ask you again about the extent to which you have studied the alleged "accelerated combustion" phenomena. [...] It seems that the situation is in this case different from that of the four quanta dispute. In the earlier controversy Warburg's results could be repeated if one designed the same inherent errors into the experiment, but for the "accelerated combustion" I believe no such duplication has been obtained. Is my interpretation correct that you have looked for the effect and not found it? In discussions with Burk and people of that school it is not very effective to claim that the effect is not observed. Burk counters with the argument that the conditions were not right.[112]

Emerson, indeed, had not been able to reproduce the effect and, thus, demonstrate the weak points. "How I wish I could have made some more constructive comments about the quantum yield controversy!", he wrote to Hill on 10 November 1952, shortly after the Gatlinburg meeting.[113] As Hill (and Brown) had put it, negative results, without understanding, meant nothing—particularly when the experiments required such delicate handling, as quantum yield measurements did and, hence, were bound to fail, due to methodical errors, in a high percentage of trials. Emerson, therefore, was still unable to definitely convince even his friends and colleagues that he was right; while, in the meantime, Burk received the 1952 Hillebrand Prize of the Chemical Society of Washington "for the experimental discovery of a photosynthetic energy cycle of high quantum efficiency, with demonstration of the applicability of the Einstein law of photochemical equivalence".[114] Emerson was well aware that he and his colleagues were engaged in battle with a powerful enemy, in both scientific and rhetorical terms. After the conference, Emerson wrote to Franck:

> This letter is primarily to express to you my appreciation of your presence among those of us who are working in the field of photosynthesis. You have sometimes been distressed because you felt your contribution was not as great as you would like to make it. But as I listened to you at Gatlinburg I felt, more than I ever did before, the value of the leadership which you have brought to the field. Your presence among us was an incentive to all of us to make our own contributions on the highest possible plane. There are not many people who could provide this sort of inspiration, and you are the only one in the photosynthesis group. (I must say that I think Hill may come in time to exert a similar quality of leadership, though in quite a different way, because he lacks your background in physics and photochemistry.)

---

[111] Brown, incidentally, was, at the time, writing a a review of photosynthesis research with a friend and colleague, Albert Frenkel. Therein, Brown and Frenkel strongly suggested that, in view of the clear limitations of manometry, other methods should be used to settle the debate, preferably physical techniques, such as infra-red spectrometry, polarography, mass spectrometry and others. See Brown and Frenkel (1953, p. 426).

[112] Brown to Emerson, 25 May 1952, Robert Emerson Papers, 1923–1961, Record Series 15/4/28, Box 1, Folder: Brown, Allan, University of Illinois Archives. "Accelerated combustion" refers here to the high oxygen consumption that Warburg and Burk claimed to have found, which was thought to be used for respiratory purposes (that is, for the "combustion" of carbohydrates).

[113] Robert Emerson Papers, 1923–1961, Record Series 15/4/28, Box 1, Folder: Hill, Robin, University of Illinois Archives.

[114] Archive of the BBAW, NL Warburg 174 Burk to Warburg, 17 Jan. 1953. This organisation was the local American Chemical Society Chapter for the Washington, DC area.

> I need not mention to you the names of the other men who would be dominant figures in the photosynthesis group if you were not among us, but their names and faces flitted through my mind as I listened to the talks in Gatlinburg, and I thought how glad I was that you were with us.[115]

Franck was, after all, not only known for his sharp mind and his uncompromising integrity of character—he was also the second Nobel Prize winner who was involved in the controversy; and although Warburg never ceased to refer to his suggestions as mere speculative theory, Franck still was a figure that was not easily ignored.

5.6.3 Emerson Strikes Back (1955)

After the conference, Emerson again started to work relentlessly towards obtaining results that would validate his own point of view and refute the Warburg–Burk picture of photosynthesis. In January 1954, Emerson apologised to Hill for having failed to keep in touch with him—the reason being, Emerson explained, that he had slowly started to obtain some useful experimental results:

> It was a matter of achieving a combination of very improbable states, simultaneously. Enough light energy, necessary optical parts, cathetometer telescopes, Mrs. Chalmers getting enough experience in taking readings, etc., etc. We are beginning to find out how Warburg and Burk can get *some* of the results they claim. After several years of deeply disappointing and frustrating failures, when I suddenly began to get some hopeful results, I just decided to neglect everything else. Even so, the work seems to move at a snail's pace. The cellular processes are terribly intricate, that is to say, the cells have so great a capacity for adjustment that no single experiment is ever by itself conclusive. Each day's work seems to require that 10 more days be spent to clear up the new doubts raised. But at least I am working on the cells and their photosynthesis, and not on the apparatus![116]

At that time, Emerson spent a sabbatical leave in Briggs's laboratory at the University of Cambridge, together with Ruth Chalmers, his long-standing co-worker and expert in algae culturing. The result of their efforts was a lengthy manuscript, which was completed in May 1955. Emerson submitted the paper to the journal *Plant Physiology* and, at the same time, sent out copies to a number of colleagues, whom he asked for comments. In the accompanying letter to Daniels Emerson explained why the text had grown so much in length:

> I feel apologetic about the length of the manuscript, but it is a good deal shorter than the sum total of the papers Warburg and Burk have published during the time we spent doing this work. It was my hope that I could write something which would provide readers with a basis for forming an independent opinion on the significance of the Warburg–Burk contributions, and save them the embarrassment of basing their opinions on the personal prestige of the authors. In my efforts to achieve this, I'm afraid, I let the paper become much too long![117]

---

[115] Emerson to Franck, 13 Nov. 1952, Robert Emerson Papers, 1923–1961, Record Series 15/4/28, Box 1, Folder: Franck, James, University of Illinois Archives.

[116] Cambridge University Library, Ms. Add. 9267/J.54, Emerson to Hill, 4 Jan. 1954.

[117] Emerson to Daniels, 16 June 1955, Robert Emerson Papers, 1923–1961, Record Series 15/4/28, Box 1, Folder: Daniels, Farrington, University of Illinois Archives.

Franck was the next to receive a copy of the text; and Emerson requested a very specific type of comment from him:

> I would like very much to know whether you think the standpoint from which it is written is a useful one, and also whether you think I have suppressed contentious remarks about Warburg and Burk. I wish I could make my writing as free of prickly statements as yours is. [...] I'm sorry that it is so long. Maybe I'm beating a dead horse?[118]

The manuscript succeeded in making a major impression, as one can take from a letter Emerson wrote to Whittingham shortly thereafter: "I have a long letter from Gaffron with his comments, and have spoken with Franck on the telephone about it. Gaffron tells me that for the first time Franck begins to understand my objections to the experimental work of Warburg and Burk!"[119]

On 28 July, the paper was accepted for publication and was put to print in November of the same year.[120] Its main purpose was not to prove this or that piece of data right or wrong; but consider the value of the methods with which photosynthetic efficiencies of 70 % or even higher had been found. In order to do so, Emerson and Chalmers had striven to duplicate exactly the Warburg–Burk experimental set-up, even though they found it inadequate in many respects. They diverged from this set-up in only one point, namely, in their choice of manometer. Emerson and Chalmers drew attention to the fact that, if they had used Warburg's two-vessel method, then they would have needed to take three manometer readings while the manometer was being vigorously shaken. "Even with the aid of a hand lens, a precision of $\pm$ 0.5 mm is the utmost that can be expected", Emerson and Chalmers maintained. As Warburg's findings were often taken from pressure changes of 3 mm, the resulting range of uncertainty amounted to about 30 %! "Greater precision is attainable", Emerson and Chalmers explained, "by reading the manometers with a cathetometer".[121] The latter was a horizontal telemicroscope with a scale divided into hundredths of a mm, which enabled the researcher to reduce reading errors to a precision of $\pm$ 0.03 mm. However, it required, in contrast to Warburg's practice, the use of a differential manometer.

With the help of these instruments, Emerson and Chalmers demonstrated that, under the conditions chosen by Warburg and using his methods of calculation, enormously high quantum yields could be reached, although they did not reflect the maximum quantum yield of photosynthesis. The authors painstakingly spelled out the details of their set-up as well as every possible source of error in the experiments (while at the same time they demonstrated that their errors were much smaller

[118] Emerson to Franck, 17 June 1955, Robert Emerson Papers, 1923–1961, Record Series 15/4/28, Box 1, Folder: Franck, James, University of Illinois Archives.

[119] Emerson to Whittingham, 25 June 1955, Robert Emerson Papers, 1923–1961, Record Series 15/4/28, Box 1, Folder: Whittingham, Charles, University of Illinois Archives.

[120] Emerson and Chalmers (1955).

[121] Besides the paper he wrote with Nishimura and Whittingham, Emerson also cited on this point the work carried out in the laboratory of the German plant physiologist André Pirson, namely Pirson et al. (1953). Pirson was one of the few experts working in Germany at the time who was highly critical of Warburg's claims. See Pirson (1994) for an autobiographical account.

than the ones implied in Warburg and Burk's experiments); and they arrived at the following conclusion:

> The results we have reported here support the conclusion reached earlier by a number of other investigators, that a quantum requirement of about eight per molecule of oxygen produced represents the highest efficiency that can be sustained by the evidence (equivalent to about 30% in red light). The claims put forward by Warburg and co-workers that from one to four quanta suffice per molecule of oxygen produced, appear to be founded upon experimental methods which cannot be counted upon to give results which are numerically correct, and the results, whether correct or not, cannot be regarded as an appropriate basis for calculating the efficiency of photosynthesis.[122]

It is quite impressive to see how carefully Emerson and Chalmers tried to avoid any polemic remarks; and how strictly they focused on undermining the reliability of Warburg's methods, after having been able, at last, to reproduce his data under specific circumstances. It is equally impressive to see Warburg's reaction. He had, in the meantime, turned to Samuel D. Cornell, the Executive Officer of the National Academy of Sciences (NAS) of the United States. In this letter, dated 25 July 1955, Warburg expressed his concern that several important developments in the field of photosynthesis research, achieved in his laboratory, were continuously being contested by American scientists, notwithstanding the fact that these achievements had completely changed the general understanding of photosynthesis. Warburg reminded Cornell of the fact that "in the days of Pasteur, whose discoveries were often contested in a similar way, the French Academy of Sciences settled the disputes by naming commissions to look at and to check the disputed experiments".[123] He therefore appealed to Cornell to send such a commission to Berlin-Dahlem, where Warburg had recently set up a special laboratory for demonstration purposes.

Cornell forwarded this letter to Calvin, Daniels, Emerson, Franck, Goddard and Hendricks, as the experts on this subject among the Academy's members, with a request for suggestions as to how the Academy should respond. The answers were clear and unanimous. Emerson responded that he did not see how sending a delegation to Berlin would serve any good purpose. Such a visit to Warburg's laboratory, Emerson believed, "would probably be made the basis of a new emphasis upon his prestige, a circumstance which is without direct bearing upon the problems of photosynthesis with which we are concerned".[124] Franck wrote the most elaborate reply. He expected Warburg to use a negative answer as an argument for the fact that his opponents shunned the objective testing of results—yet, he could still not agree to setting up such a committee. First, Franck underlined that it was impossible to settle the dispute this way:

---

[122] Emerson and Chalmers (1955, p. 528).

[123] A copy of Warburg's letter is preserved in: Franck, James. Papers, [Box 10, Folder 1], Special Collections Research Center, University of Chicago Library. Warburg alludes, of course, to the famous dispute between Pasteur and Pouchet; on this controversy, in which repeatedly committees of the French Académie des Sciences were involved, see, e.g., Mendelsohn (1987); Roll-Hansen (1979).

[124] Emerson to Cornell, 10 Aug. 1955. Franck, James. Papers, [Box 10, Folder 1], Special Collections Research Center, University of Chicago Library.

> Apparently Warburg supposes that it is enough to demonstrate a few examples of manometric measurements from which data may be calculated which support his views. If most of the data do not give the desired results he will explain as he has done often that it is only necessary that some of the data fit because one cannot expect that the biological material is always present in perfect conditions.

However, the main reason that Franck advised against sending a delegation was more fundamental and was based on his understanding as to how science should be made to work:

> I believe that it is not the task of our academy to sit in judgment about scientific differences of opinions. [...] [T]he decision what is right and what is wrong should be left to the normal process of the development of science which is after all, a very efficient way to weed out errors even if the processes might not be as quick.[125]

This was also Calvin's opinion: the question whether or not an individual's results and interpretations were accepted by others should be determined in the usual way, Calvin wrote to Cornell, "namely, by the willingness and interest of the scientific world in the form of the collection of individual scientists to undertake to test the results and theories proposed by Prof. Warburg".[126] Finally, Daniels informed Emerson that he had discussed the subject in informal talks at a conference in Geneva, Switzerland, with Rabinowitch, Calvin and the physiologist Detlev W. Bronk, who was then President of the Academy. All four of them had agreed that "it would be a bad precedent for the NAS to appoint a committee when scientists disagree. There would be no end of such committees". To this, Daniels added in his letter to Emerson: "Personally, I do not feel that Warburg is entitled to any more consideration than was given to him by you and your laboratory a few years ago. Warburg's letter is really quite astounding. The less attention we pay to it, the better".[127]

5.7 The Aftermath

5.7.1 The Enhancement Effect

By Emerson and Chalmer's 1955 paper, most people in the field had become convinced that the number of 8–10 quanta as a minimum requirement of photosynthesis was at least approximately the accurate one; while any further (and, perhaps, more definitive) resolution of the question had to wait for the development of new methods. A general saturation point had been reached and many participants felt that the problem had been discussed for far too many years and in far too much depth. This

[125] Franck to Cornell, 15 Aug. 1955. Franck, James. Papers, [Box 10, Folder 1], Special Collections Research Center, University of Chicago Library.

[126] Calvin to Cornell, 26 Aug. 1955. Franck, James. Papers, [Box 10, Folder 1], Special Collections Research Center, University of Chicago Library.

[127] Daniels to Emerson, 26 Aug. 1955, Robert Emerson Papers, 1923–1961, Record Series 15/4/28, Box 1, Folder: Daniels, Farrington, University of Illinois Archives.

was reflected, for example, in the fact that the theme of the maximum quantum yield of photosynthesis was deliberately excluded from the second Gatlinburg Conference on photosynthesis, held in October 1955.[128] Not that Warburg had stopped publishing ever new variations of his experimental set-up, which always gave the same high quantum yields. In reaction to Warburg's most recent papers, Gaffron wrote the following letter to Burk in March 1956:

> Dear Dean
> I delayed this note of thanks until we had sent you our recent papers in reciprocation of your kindness in forwarding us the latest reprints from Warburg's laboratory. These publications on photosynthesis have now clarified the situation rather definitely: "Too strong tobacco to smoke in my Meershaum!" The deviations from the experiments and theories of other workers in the field are wonderfully clear. Soon there will be no need to concern oneself with the matter any further.
> What I am wondering is to what extent you personally are willing to believe in and subscribe to what comes from Dahlem? For us it would simplify the situation if we were allowed to identify you entirely with the Warburg school, but I cannot help feeling that this might do you a serious injustice. You should feel young enough to dare to deviate from the party doctrine the moment you recognize how absurd the tenets are in which followers are asked to believe. Has that moment arrived?
> Very sincerely yours, Hans G. Gaffron.[129]

Gaffron's disbelief and consternation concerned, among other things, Warburg's claim (first published at the end of 1954) that, in order to compensate for respiration, no high intensity white background light was necessary, as he had argued up to then, but that blue or green light beams of rather small intensity were sufficient. The effect of these beams was then called "catalytic".[130] Despite the fact that many of the players had had enough of quantum yields, this paper was broadly discussed. Thus, the Stanford-based plant physiologist Lawrence Blinks wrote to Emerson on 28 September 1955: "What do you think of Warburg and Krippahl's strange findings on blue light (after red)? [...] Off hand, I don't believe it, but strange things do happen".[131] Emerson replied to him on 10 October:

> As for Warburg's blue light experiments, the wave length is about the same as that in which Tony Lewis and I found evidence of strong effects of light on respiration [in 1943]. I've tried to think of an interpretation that would account for both our observations and Warburg's, but they don't seem to agree. Warburg's manometry has become almost mystical, and I'm generally pretty skeptical of his interpretations, but I do believe there are some special effects of certain wave lengths of blue light.[132]

---

128 The papers of this conference were published in the volume Gaffron et al. (1957).

129 Archive of the BBAW, NL Warburg 174. Gaffron to Burk, 7 March 1956.

130 Warburg et al. (1954). See also Warburg et al. (1955) for extensions of this idea, and the general review Warburg (1958).

131 Blinks to Emerson, 28 Sept. 1955, Robert Emerson Papers, 1923–1961, Record Series 15/4/28, Box 1, Folder: Blinks, Haxo, University of Illinois Archives. *Ulva* and *Monostroma* are genera of (multicellular) green algae.

132 Emerson to Blinks, 10 Oct. 1955, Robert Emerson Papers, 1923–1961, Record Series 15/4/28, Box 1, Folder: Blinks, Haxo, University of Illinois Archives.

The experiments on the effect of combining lights of different colours foreshadowed a fruitful line of research that Emerson pursued in the second half of the 1950s. In December 1955, Emerson wrote to Hill to inform him that he was now working again on photosynthesis efficiencies at different wavelengths, "a subject that warms my enthusiasm quite a bit more than the problems of two-vessel manometry. Maybe we shall come to trying some mixtures of blue and red light, though when one looks closely at Warburg's data, it seems that there is no clear evidence for the special effect of blue".[133] One can take from his correspondence that Emerson did indeed take up this idea, which led him to make an unexpected discovery. In a letter to Arnold on 9 April 1956, Emerson wrote: "The significance of the long-wave limit is becoming very interesting. Accessory illumination with shorter wave lengths makes the increment of photosynthesis attainable with very long wave lengths higher than it is without accessory illumination".[134] A week later, Emerson wrote to Franck, informing him of the same observation as well as adding the following lines:

> The amounts of accessory light required are considerable—larger than the amounts of the red beams being used. "Catalytic" amounts of accessory light are not sufficient. [...] This letter should reach you by Wednesday. I plan to telephone you Thursday morning at about 10 o'clock. Please excuse the brevity of this rather hasty letter. I should be setting up an experiment.[135]

Emerson was clearly excited about his findings and urgently requested Franck's opinion and advice. At the same time, he was anxious, already then, to differentiate between his findings and Warburg's catalytic effect of blue and green light.[136] Under no circumstances did Emerson want to be cited as confirming Warburg's results, although he did acknowledge that he had hit upon the phenomenon in question while double-checking the catalytic light claim. In July 1957, Emerson wrote to Hill: "I've had quite an exciting time with the experiments on mixing long-wave light with shorter wave lengths. The effects do not match Warburg's claims at all, but of course we were stimulated to do the experiments because of Warburg's claims".[137]

Emerson presented a preliminary account of his work to the annual meeting of the NAS, between 23 and 25 April 1956, in which he reported: "If the low-intensity

[133] Cambridge University Library, Ms. Add. 9267/J.54. Emerson to Hill, Dec. 26, 1955.

[134] Emerson to Arnold, 9 April 1956, Robert Emerson Papers, 1923–1961, Record Series 15/4/28, Box 1, Folder: Arnold, William, University of Illinois Archives.

[135] Emerson to Franck, 16 April 1956, Robert Emerson Papers, 1923–1961, Record Series 15/4/28, Box 1, Folder: Franck, James, University of Illinois Archives.

[136] This is a recurrent theme in Emerson's papers on the subject. The ensuing public attention of Warburg's work was critically remarked upon by Gaffron, to which Emerson replied: "As for Warburg, I agree with you that our reference to his work with supplementary light will lead to controversy, but I do not feel it would be right for us to report work with supplementary light, without at least a reference to his work. This is the sort of thing he does to us all the time, and the least I can do is set him a good example". Emerson to Gaffron, 14 Dec. 1956, Robert Emerson Papers, 1923–1961, Record Series 15/4/28, Box 1, Folder: Gaffron, Hans, University of Illinois Archives.

[137] Cambridge University Library, Ms. Add. 9267/J.54. Emerson to Hill, 6 July 1957.

light beam of measured energy is supplemented by a more intense (unmeasured) beam, then the efficiency of the small increment of measured light remains nearly constant out to 685 mμ".[138] By November, Emerson had put together a substantial manuscript, which he mailed off to the *Proceedings of the National Academy of Sciences* (PNAS). In a letter to Blinks, Emerson described the interpretation of the phenomenon that he was presenting in the paper as the nearest thing to an "idea" that he had ever produced in his life and admitted that Rabinowitch was very doubtful of its value.[139] The "idea" was the following suggestion:

> [T]he significance of the supplementary light may be that it adds excitation of other pigments besides chlorophyll *a*. The maintenance of maximum efficiency may require the excitation of some pigment with an absorption band corresponding to an energy level higher than the first excited state of chlorophyll *a*.[140]

In the green algae, the authors surmised, this pigment might be chlorophyll *b*. There was, however, a problem, as they openly acknowledged: "[This interpretation] is in conflict with the widely accepted view that transfer of excitation energy to chlorophyll *a* from other pigments takes place with practically 100 per cent efficiency".[141] This referred to the work done by the Dutch biophysicist Louis N. M. Duysens who, in his PhD thesis of 1952, had demonstrated exactly this.[142] Emerson undoubtedly realised that his contradiction to Duysens's findings was no minor point but still thought he could find a way to resolve it. In the following months he and his co-workers focused exclusively on this theme, while the project came to a tragic end before it was finished.[143]

5.7.2 Emerson's Death and Beyond

On 4 February 1959, Emerson died in an aeroplane crash. He had always distrusted aviation as a means of transport, preferring to travel around the country by train. It was only because the train service from Indianapolis to New York had been discontinued in the late 1950s that he had grudgingly turned to flying between Chicago and New York.

---

[138] See the abstract of the paper, published in *Science* 123 (1956), p. 673, co-authored by Emerson, Chalmers, Carl Cederstrand and Marcia Brody.

[139] See: Emerson to Blinks, 30 Nov. 1956, Robert Emerson Papers, 1923–1961, Record Series 15/4/28, Box 1, Folder: Blinks, Haxo, University of Illinois Archives.

[140] Emerson et al. (1957, p. 142).

[141] Emerson et al. (1957, p. 142).

[142] Duysens (1952); Duysens's work will be discussed in Chapter 7.

[143] See the abstract of the paper, published in *Science* 125 (1957, p. 746), with Robert Emerson as the sole author, although his collaborative work with Chalmers and Cederstrand was acknowledged. The failure to replicate Warburg's findings was repeated again in Emerson's presentation at the 1958 NAS meeting; see the abstract in *Science* 127 (1958, pp. 1059–1060). The suggestion that accessory pigments were responsible for the Enhancement Effect was also repeated in Emerson and Chalmers (1958).

This particular time, Emerson had wanted to attend a conference at Harvard University. Even more tragic was the fact that Emerson was originally booked on another flight. Yet when he arrived at Chicago, a flight that had been delayed was still waiting to depart for New York; at the last minute Emerson transferred to it, hoping that he would arrive at his destination a little earlier. This turned out to be a fatal decision.

Emerson left behind a great deal of experimental material at the Urbana laboratory, accumulated during the years that he had been working on the long-wave limit of photosynthesis; only some of this work was published posthumously, by Emerson's friend and Urbana colleague Rabinowitch.[144] In this paper, Rabinowitch duly presented Emerson's recent data, although he argued against the assumption that the phenomenon indicated a direct contribution of chlorophyll *b* in photosynthesis. He suggested, as an alternative, that two types of chlorophyll *a* were present in the living cell, one of which resembled the chlorophyll *b* more closely than the other, while both were necessary to sensitise photosynthesis to its maximum quantum yield. The existence of these two forms of chlorophyll *a*, which had distinct functions in the photosynthetic process in *Chlorella* cells, was established in a project pursued by Govindjee, who had been one of Emerson's last doctoral students.[145]

The curious enhancement phenomenon and the sudden emergence of different forms and functions of chlorophyll triggered a wealth of further investigations; at the same time, the question of determining maximum quantum yields gradually faded from the scene. While Warburg seemed to believe that Emerson's death had decided the matter in his favour—he was overheard stating this in public[146]—one feels more justified to conclude that the question had lost its attraction and importance. Far more exciting new developments needed to be clarified, which were more likely to lead to more immediate advances being made in understanding photosynthesis than the continued pursuit of quantum yield numbers.

5.8 How Controversies End

In Germany, the whole controversy was received rather differently. In his country, Warburg was still held in high esteem and of great influence; furthermore, the majority of German researchers was informed on the dispute through Warburg's own papers, which obviously were not impartial.[147] However, Warburg received a strong

---

[144] Emerson and Rabinowitch (1960).

[145] See Govindjee and Rabinowitch (1960) and Rabinowitch and Govindjee (1961). On Govindjee, see also, e.g., Eaton-Rye (2013). French (1961) (presented in March 1960) had independently come to the same conclusion. See also Chapter 7 of this book.

[146] Govindjee (personal communication) was recounted this episode by Rabinowitch, on his return from a conference held at Gif-sur-Yvette (south-west of Paris) in 1963.

[147] Pirson (1994, p. 215), attributes the lack of involvement of the Germans in this controversy to the general inadequacies of the universities' experimental equipment, which simply did not allow German researchers to contribute usefully to this difficult question. An additional factor may be that,

blow to his authority when in 1961 it was not him who received the Nobel Prize in Chemistry for his work in photosynthesis, but rather the American chemist Melvin Calvin, that is, one of Warburg's opponents. The weekly magazine *Spiegel* gave a sarcastic report of the situation. It described, how in 1957, Warburg had proudly declared in a public lecture that, thanks to his work, Germany had been able to maintain its international leadership in photosynthesis research, despite the Second World War and the country's collapse. The decision of the Nobel Prize Committee to award the Chemistry Prize to Calvin and not Warburg only four years later was felt to be in stark—and disillusioning—contrast to the pompous self-confidence of Warburg's. According to the article, Germany now slowly began to realise that Warburg, through ignoring the work done in other laboratories, had become more and more isolated within the international scientific community and that, as a consequence, his contributions were increasingly off the mark.[148]

Although Warburg seemed undisturbed by these developments, and would still claim in a publication of 1969 (one year before his death) that he had solved the problem of photosynthesis,[149] the late biochemist Birgit Vennesland passed on to the public the following remarkable quotation. When Vennesland asked him whether he had made any mistakes in his life, Warburg replied:

> Of course, I have made mistakes—many of them. The only way to avoid making any mistakes is never to do anything at all. My biggest mistake was to get much too much involved in controversy. [. . .] It isn't that controversy itself is wrong. No, it can be even stimulating. But controversy takes too much time and energy. That's what's wrong about it. I have wasted my time and energy in controversy, when I should have been going on doing new experiments.[150]

Since Emerson also felt that he had spent too much time and energy on the quantum yield controversy,[151] as well as almost everyone else working around them, it is surprising that the controversy had not ended much earlier. Three points need to be taken into consideration. *First*, the importance of the subject. The maximum quantum yield was, at the time, not simply an arbitrary number that was being questioned, but a key parameter on which the modelling of the photosynthesis mechanism would be based. Thus, finding the true value was of great significance. *Second*, the technical complexity of the experiments. The practical difficulties that the researchers faced were enormous. It started with the problem of finding adequate light sources, but the most vexing question was the impossibility of differentiating clearly between the gas exchanges caused by photosynthesis and respiration. This was a serious problem, and although all the protagonists acknowledged its existence, they were

---

in the spirit of self-pity prevailing in the postwar years, many Germans (particularly nonexperts) were inclined to frame the controversy in political terms, as another aspect of the unjustified repression of Germans by the US "invaders". On the atmosphere among scientists in postwar Germany, see Deichmann (2001a), which focuses on biochemists, or Hentschel (2005), on physicists.

148 See Anonymous (1961).

149 See Warburg et al. (1969).

150 Quoted in Govindjee (2004b, p. 185).

151 Cf. Walker (1997, p. 8).

unable to solve it. *Third*, the strong dependence of the efficiency of photosynthesis on a vast number of interrelated physiological factors. It was found that the algae were extremely adaptable, and reacted in totally unforeseeable ways to the slightest changes in parameters. This meant that all sorts of quantum yields could be accurately measured—a statement with which Emerson and Warburg would have happily concurred—although none of these values might be the maximum quantum yield or the minimum quantum requirement. On the other hand, one had to make sure that the highest yields that were measured (in this case by Warburg and Burk) were yields that reflected photosynthetic oxygen production rather than secondary processes or methodical artefacts. This all led to the fact that, as late as 1960, Bessel Kok stated, in a comprehensive paper on the problem, that preponderate evidence seemed to support the assumption that at least 8 quanta were required per molecule of oxygen evolved; while he immediately added: "It is rather dissatisfying that 25 years after Warburg and Negelein's first estimations we cannot justify more firmly stated conclusions".[152]

In principle, this situation was not that exceptional, since, at any given time, scientists tend to disagree with each other on uncertain issues. But not all disagreements develop into controversies. Marcelo Dascal suggested to differentiate between the following types of disagreement: (1) a "discussion", which he defines as "a polemic whose object is a well-circumscribed topic or problem". However, the contenders tend to acknowledge that "the root of the problem is a mistake relating to some concept, result or procedure" that allows for solutions. (2) A "dispute", on the other hand, is "a polemic that also seems to have as its object a well-defined divergence", while the contenders do not accept this divergence as being grounded in some mistake. It can temporarily be terminated or "dissolved" by some procedure, but it cannot be "solved". (3) A "controversy", finally, "can begin with a specific problem, but it spreads quickly to other problems and reveals profound divergences". Controversies, in Dascal's view, can neither be "solved" nor "dissolved", they are "resolved".[153]

Taxonomies like this can help to characterise an item under investigation—although, as in this case, it may resist a clear assignment to any of these categories. The disagreement on the maximum quantum yield of photosynthesis started off, through the contributions by Emerson and Lewis, as a "discussion" on a very specific issue: they discovered some flaws in Warburg and Negelein's work, suggested an improved method and, hence, a new value for the parameter in question. Some fellow scientists, such as Franck and Gaffron, immediately declared the question as being settled in favour of the new value. When Warburg continued to disagree, Emerson tried to "solve" the discussion to mutual satisfaction by carrying out experiments together in Urbana. This seemed to be rather straightforward. Warburg, on the other hand, never accepted Emerson's critique but reproached his adversary for using inadequate methods himself—thus, taking Dascal's categories again, from Warburg's point of view, the situation rather should be called a "dispute". The fact that Warburg tried to terminate, or "dissolve", the disagreement through an authoritarian sentence

[152] Kok (1960, p. 623).

[153] Cf. Dascal (1998).

by "impartial observers", that is, without actually seeking "solution", also points in this direction. Thus, one and the same disagreement can obviously be categorised, at one and the same time, as a constructive discussion as well as a (potentially) destructive dispute. However, considering the further development, one may, after all, feel justified in calling this disagreement a "controversy", as it certainly revealed profound divergences between the contenders: both, on the question of how seriously certain principles of science ought to be taken, such as thermodynamical limitations of efficiency (which made a minimum requirement of photosynthesis of 4 light quanta highly improbable, while it excluded the possibility of a value of less than four); and on the question of how to deal with fellow scientists and divergent standpoints. It certainly was a "public and persistently maintained dispute", taken seriously by the community, which is how Ernan McMullin characterised a controversy.[154] And it may even satisfy the more demanding definition by Gideon Freudenthal, who uses the term "scientific controversy" to refer "specifically to a persistent antagonistic discussion over a disagreement concerning a substantial scientific issue that is not resolvable by standard means of the discipline involved".[155] Given the fact that the maximum quantum yield of photosynthesis could not be determined with the required degree of precision by any technique available at the time, I shall continue, in the following, to refer to the subject of this chapter as the "maximum quantum yield controversy", which is also how it is called in the accounts of the actors.

What is striking is that disagreements, like the one under study here, are not only ambiguous in their character at any point of time, depending on whose perspective one adopts, but they are dynamic entities that may pass through all of Dascal's (otherwise very useful) categories in the course of their prolonged existence. The fact that the contenders may not even agree on what they disagree implies a further complication in analysing controversies. At first glance, the adversaries in this case were debating a mere factual piece of scientific information: the value of an empirical parameter; and the participants first tried to "solve" the issue by reproducing one value or the other. However, this line of action did not bring the conflict to an end. Warburg's values, for a long time, simply could not be reproduced by others, which led him to accuse his adversaries of being experimentally incompetent. This moved the item of the controversy to other levels of disagreement: *first*, from the factual to the methodological, as the parties disagreed on the appropriate methods for determining the yield; and, *second*, to the level of fundamental assumptions and theoretical priorities—when Warburg's values, finally, were reproduced, the disagreement was then on the question how to interpret the reproduced numbers.[156] Yet, the controversy also started to be strongly influenced by non-epistemic factors: rhetorical effects,

[154] McMullin (1987, pp. 51–52). The public dimension of controversies, extending beyond the dialogue of two actors, is also emphasised by others, such as Giere (2000).

[155] Freudenthal (2000, p. 126).

[156] This threefold taxonomy of potentially controversial issues—facts, methods, theoretical assumptions—is widely shared in the literature on controversies in science; a classic is, e.g., Laudan (1984).

questions of authority and reputation, personal insults and defamation as well as the burning desire to prove the other person wrong (rather than the mere desire to prove one's own point of view right). The technical issue of finding the maximum quantum yield of photosynthesis in order to elucidate the photosynthetic mechanism developed into a battle between the Good and the Evil.

The complex mixture of epistemic uncertainties and non-epistemic influences decisively contributed, so it seems, to the prolonged existence of the controversy. Warburg and Burk were playing against unwritten rules—both by belittling their adversaries and by presenting one set-up after another without arguing for the change. Yet, they still could have measured the accurate value, so that the community of photosynthesis researchers, for a long time, was not in a position to easily dismiss their results. While the style of Warburg and Burk's behaviour and their rhetoric in publication did arouse criticism, it was only after some years of intense debate and after Emerson and Chalmers had shown the misgivings of Warburg's set-up that participants started to seriously mistrust the latter's work. Gaffron, for example, was genuinely trying to keep Warburg in the United States in 1949, although he knew very well about the latter's difficult character and intolerable conduct towards Emerson at Urbana. The gain in scientific terms seemed to have outweighed these drawbacks (and, of course, Gaffron was indebted to Warburg for the help that the latter had given him in finding a new place to work in the 1930s). However, by the mid-1950s the situation had changed. Warburg and his co-workers had consistently failed to demonstrate the actual relevance of the factors that they declared were essential for observing high quantum yields in photosynthesis; while his opponents had succeeded in demonstrating why these very factors would produce high yields as an artefact. At the beginning of 1955 Gaffron wrote to Franck that he had received a letter from Emerson stating that although Emerson had again faithfully reproduced the latest experimental set-up specified by Warburg, he had obtained no change in resulting quantum yields. Gaffron commented:

> [Emerson] can't understand Warburg's results—and neither can I, of course. Unless one leaves the realm of science and says: since [Warburg] lies anyway—he lies consciously when citing other people—why not also here? If his set-up has been worked on for so long, that it yields inaccurate results automatically, who will be able to detect this, without taking everything apart and building it up again, piece by piece. [Carl] Neuberg, who always lied, and may therefore be considered an expert in this matter, said, if one fails to prove that he [Warburg] has been deceiving us, then he will have won.[157]

It is, in a way, remarkable that it was only after 1955, that is, after the paper by Emerson and Chalmers, that people started to seriously mistrust Warburg and Burk. In order to reach this point, it seems, notwithstanding the strong non-epistemic component of the controversy, the community required arguments relating to the

[157] Gaffron to Franck, dated 2 Feb. Franck, James. Papers, [Box 3, Folder 7], Special Collections Research Center, University of Chicago Library. Original letter in German. The year 1955 was reconstructed from the description of Warburg's setup, namely catalytic use of blue or green light, which Warburg had first announced in Warburg et al. (1954).

epistemic issue.[158] Emerson himself had postponed the ethical side of the debate until he had "settled the facts" (as he wrote to Steward, see p. 160). But also his attitude had changed by the mid-1950s, as can be taken from a letter of his to Daniels, responding to a remark on Warburg's latest work:

> I feel as you do that maybe someone ought to check on Warburg's electrometric results, but I think there is a limit to what we can accomplish by checking each fantastic claim as it comes along. I'm afraid I have built up a prejudice against Warburg's experimental work, because of his abuse of the manometric technique, and I tend to feel that if I took the time I would find a joker in his electrometric measurements as well. One cannot print this sort of thing, of course, nor say it for the record.[159]

These sentiments were increasingly shared by a number of other people, and Warburg's reputation suffered accordingly and with lasting effect. For example, Warburg's finding that very small amounts of carbon dioxide were necessary for photosynthesis to function was unjustifiedly swept aside.[160] The maximum quantum yield controversy, thus, found its end much in the way that Philip Kitcher found characteristic. He claimed, in a Tolstoyan manner of speaking, that "each resolution of a scientific controversy proceeds in much the same way; all unresolved scientific controversies are unresolved in their own way".[161] The typical resolution, Kitcher purports, is accomplished "because those deemed by members of the community to be authoritative with respect to the range of topics in questions concur in recommending [it]". This point of consensus among those who were the authorities in photosynthesis research was reached after Emerson's major paper, even though Warburg continued to publish his own results.[162]

This "consensus" was substantially strengthened by the fact that, by 1960, it had firmly been established that the photosynthetic oxygen came from water and that photosynthesis involved two light reactions in series which was impossible to bring about with the number of light quanta that Warburg and Burk had proposed. However, at the same time the figure had lost its importance, as, in the meantime, other methods had become available to validate possible models of the photosynthetic mechanism, above all, the different types of spectroscopy that were developed through the 1950s. This will be the theme of chapter 7 of this book. The next chapter 6, following the chronology of the events, turns to the "dark" side of photosynthesis: the reduction of carbon dioxide to carbohydrates.

---

[158] This point that was also underlined by McMullin (1987, p. 78).

[159] Emerson to Daniels, 20 May 1954, Robert Emerson Papers, 1923–1961, Record Series 15/4/28, Box 1, Folder: Daniels, Farrington, University of Illinois Archives.

[160] See Stemler (2002) for a short review of the so-called bicarbonate effect, discovered originally in Warburg and Krippahl (1958), and the detrimental effect Warburg's reputation had on the reception of this phenomenon. The bicarbonate effect later became an important theme of research in Govindjee's laboratory at Urbana; see Shevela et al. (2012) for a recent survey.

[161] Kitcher (2000).

[162] An ultimate attempt to replicate Warburg's values under the conditions he himself had specified was made in Govindjee et al. (1968). The results reconfirmed the figure of 8 to 12 light quanta as a minimum requirement for photosynthesis.

Chapter 6
The Path of Carbon in Photosynthesis (1937–1954)

While the quantum yield controversy was unsettling a large part of the photosynthesis community, others were, at the same time, deeply involved in studying the so-called dark reactions of photosynthesis. The turning point of the latter project came with the advent of radioactive isotopes (or radioisotopes), as it was found that they could be used to trace the metabolic processes of plants and animals. In particular, it was the discovery of the long-lived isotope carbon-14, made by Martin Kamen and Sam Ruben in 1940, that lay the groundwork for the subsequent elucidation of many metabolic pathways.

This chapter starts with a discussion of the path-breaking work done by Kamen and Ruben at the University of California, Berkeley, and then continues with an analysis of the research team headed by Melvin Calvin and Andrew A. Benson at the same institution. It was this group that succeeded in elucidating the reaction cycle of photosynthetic carbon reduction. Through sheer manpower and financial support (mostly provided by the United States Atomic Energy Commission [AEC]), this group rapidly outmatched all its competitors. The main part of the chapter reconstructs the work of the Berkeley Group, which can be traced, among other sources, through a series of more than 20 publications from Calvin's laboratory, all of which were entitled "The Path of Carbon in Photosynthesis". The group's work on the reaction cycle was (more or less) completed by 1954, and in view of these achievements Calvin was awarded the 1961 Nobel Prize in chemistry. Unfortunately, no acknowledgement was given to the equally central contributions made by Calvin's collaborator Benson and by Kamen, the discoverer, with Ruben (who had by then died in a laboratory accident) of carbon-14.[1]

[1] The story of how the path of carbon in photosynthesis was elucidated has been told before, mostly in the form of autobiographical reports. See, e.g., Bassham (2003); Benson (2002a, b); Calvin (1964, 1989, 1992); Florkin (1979) (Chapter 56, pp. 81–108), Kamen (1974, 1985, 1989); Lehmann (1968) and Morton (2007). The book by Rabinowitch (1956) also has a section on the "Evolution of the CO_2 Reduction Mechanism", in which the different stages of the work carried out at Berkeley are summarised (pp. 1688–1698).

K. Nickelsen, *Explaining Photosynthesis,* History, Philosophy and Theory of the Life Sciences 8, DOI 10.1007/978-94-017-9582-1_6

6.1 Early Radiotracer Experiments

6.1.1 Kamen Meets the Cyclotron

It may come as a surprise that it was the cyclotron that actually paved the path for the elucidation of the fate of carbon during photosynthesis, and, in particular, the first cyclotron in the laboratory of Ernest O. Lawrence. It has been told many times before how Lawrence, at the age of 27, arrived at the University of California's Berkeley campus in 1928, and how he set up a unique, interdisciplinary laboratory that became well known for being the first example of so-called Big Science.[2] The instrument in question was a type of particle accelerator, which was referred to by Lawrence as the "proton merry-go-round" but which publicly became known as the "cyclotron". However, home-made and clumsy the first versions of this instrument appeared (the prototype was constructed of glass, sealing wax and bronze), they still operated perfectly well. In these instruments, charged particles were spun around in a vacuum chamber by means of a high-frequency, alternating voltage combined with a steady magnetic field. This resulted in an enormous increase in the energy of the particles, which made them go round in spirals, until, finally, they were thrown upon a target at the perimeter of the chamber. These deliberately induced collisions created secondary particles, which could then be extracted for analysis (Fig. 6.1).

The first of these instruments contained an accelerating chamber that was no more than 5 inches (12.7 cm) in diameter; it was followed by a series of ever growing variants, which finally exceeded the capacity of Lawrence's laboratory. Lawrence managed to persuade university officials that he needed more space, and in 1931 an empty building with a sufficiently sturdy floor was turned over to Lawrence into which the new 27-inches (68.58 cm) cyclotron (which included a 70-ton magnet) was moved: this became the famous Berkeley Radiation Laboratory.[3] From the outside, no one would have suspected that the old clapboard building housed a laboratory with cutting-edge technology; however, it had the enormous advantage of having a solid concrete floor which was strong enough to support the enormous weight of the new cyclotron. It was in this building, affectionately called the (Old) Rad Lab, or ORL, that the route that carbon travels through a plant during photosynthesis was later mapped. In 1935, Lawrence invited his brother John, a physician, to start a biomedical research unit at Berkeley in order to explore how the radioactive isotopes could be used, which were being produced in large quantities by the cyclotron. Early on in these studies, radioactive phosphorus-32 seemed to have cured mice suffering

[2] On Lawrence, the Berkeley Radiation Laboratory and the early history of the cyclotron, see Heilbron and Seidel (1989); Heilbron et al. (1981) and Herken (2002). For a more general view that enlarges on the links with the Manhattan Project, see also Boorse et al. (1989) and Rhodes (1986). See Creager (2013) for an illuminating account of how profoundly the availability of radioisotopes through the cyclotron (and later the nuclear power stations) changed the life sciences after 1945.

[3] This is not to be confused with the equally famous MIT Radiation Laboratory, although both were also abbreviated to the same nickname "Rad Lab".

Fig. 6.1 The original clapboard Radiation Laboratory (Rad Lab), which later became known as the Old Rad Lab or the ORL for short, of the University of California, Berkeley (ca. 1925). (Courtesy of The Bancroft Library, Berkeley).

from leukaemia, and in 1937 John Lawrence had initiated clinical trials using the same means for the therapy of human patients.[4] This newly discovered therapeutic use gave Ernest Lawrence more than enough justification to raise money for an even bigger instrument: the 60-inch (1.52 m) cyclotron, with a 200-ton magnet, for which, again, a new building was required—the Crocker Laboratory. The first run was in 1939, the very year in which Lawrence was awarded the Nobel Prize in physics.[5]

During the second half of the 1930s, which was a particularly turbulent period from the point of view of nuclear physics, photosynthesis researchers also started to use the cyclotron. It all began with a young chemist named Martin D. Kamen, who started working in Ernest Lawrence's laboratory in 1936. Born in Toronto, Canada, but brought up in Chicago, Kamen was the son of Eastern European immigrants.[6] He received his doctorate in physical chemistry from the University of Chicago in 1936, after which he went to California to ask Lawrence for a job. Kamen seems to have been the right person in the right place at the right time. After he had worked in the laboratory for six months, with enthusiasm and skill but without a salary, Lawrence realised that a chemist was clearly needed to oversee the preparation of the radioisotopes. In the first place, John Lawrence's biomedical group had to be supplied

[4] Cf. Creager (2013, pp. 24–41) and Creager (2006).

[5] He was honoured "for the invention and development of the cyclotron and for results obtained with it, especially with regard to artificial radioactive elements". See www.nobelprize.org.

[6] His father was born in Belarus, his mother in "Lithuania or Latvia", as he wrote in his autobiography, Kamen (1985, p. 2). Kamen (1989) provides a shorter account of his memories. See also Gest (2005), who was Kamen's first doctoral student at Washington University, St. Louis.

with radiophosphorus; but the laboratory was also receiving an increasing number of requests for radioisotopes from elsewhere.[7] In view of these developments, Kamen became the only person in the laboratory to be employed on a permanent contract; and although his working hours were largely filled with time-consuming technical tasks, such as maintaining the cyclotron and distributing the isotopes, Kamen still managed to find time to carry out his own research.

6.1.2 *Kamen Meets Ruben*

In 1937, when he had still only been at Berkeley for a few months, Kamen bumped into another young chemist, Samuel Ruben, who was finishing his doctoral work in Berkeley's Department of Chemistry and had his desk in the so-called Rat House. The latter, located near the Rad Lab, was officially known as the Chemistry Annex. Built in 1915, it had been hardly modernised in the intervening years.[8] The two men, both at the time 24 years of age, decided to collaborate on the chemically interesting aspects of the cyclotron's output, which they believed were being unjustly neglected by Lawrence's physics-dominated group. Their collaborative work very soon took a decisive turn. Kamen described in his autobiography how 1 day Lawrence dashed into the laboratory, in a highly excited state of mind, bringing with him the assistant professor in physiology, Israel L. Chaikoff:

> E.O.L. [= Ernest O. Lawrence] informed me that Chaikoff had a proposal to use short-lived carbon-11 to study carbohydrate metabolism. [. . .] When I inquired how this was to be done, particularly how the ^{11}C could be incorporated quickly into the starting material, such as D-glucose, E.O.L. said the isotope would be given to green plants as CO_2. By photosynthesis the plants would in short order synthesize the radioactive glucose, which could then be fed to the rats used in Chaikoff's studies.[9]

Kamen immediately agreed to join the project, and enthusiastically promised to provide as much carbon-11 as was required. He was soon to learn that it had in fact been his friend Ruben who had originally come up with this suggestion; the matter was promptly cleared up, and work on the project started. Yet, what from today's vantage point might appear as a self-evident method of research—two chemists working cooperatively with a colleague from the School of Medicine on biological problems—was highly exceptional at the time. Berkeley chemists usually considered biology, not to speak of medicine, to be a second-rate discipline. Kamen was in a relatively safe position in the Rad Lab, but Ruben was only a young lecturer, employed on a limited, fixed-term contract by Berkeley's extremely competitive Department of Chemistry. Considering his situation, it was rather bold of Ruben

---

[7] Cf. Creager (2013, pp. 41–49).

[8] On the life and work of Ruben, see, e.g., Gest (2004) and Johnston (2003), Chapter 3.

[9] Kamen (1985, pp. 81–82).

(some would have said foolish) to pursue a line of research that could potentially harm his reputation as a chemist.[10]

Elucidating the pathways of intermediary metabolism had been the dream of biochemists and physiologists since the nineteenth century (see Chapter 2); yet, the insurmountable problem was the lack of direct access. Whenever the usual analytical techniques of chemistry were applied, the processes under study stopped operating. A crucial step forward was only made in the mid-1930s when (stable) isotopes were introduced to metabolic studies.[11] In 1932, the chemist Harold Urey, working at Columbia University in New York, had discovered the heavy hydrogen isotope "deuterium" (^2H or D) and succeeded in preparing "heavy water", that is, water incorporating deuterium instead of the usual hydrogen.[12] In 1934, Rudolph L. Schoenheimer, a biochemist of Jewish ancestry who had been dismissed from his academic post at the University of Freiburg (Germany), obtained a position at Columbia. Together with his young co-worker David Rittenberg, he started investigating the biological implications of Urey's discovery. After a short while, Schoenheimer devised the isotopic tracer technique, which enabled researchers to start studying metabolic processes in detail. The trick was that compounds, into which heavy isotopes of either hydrogen (D) or nitrogen (^{15}N) had been incorporated, were then rendered identifiable, for example, by spectroscopic methods. It was for this reason that the technique came to be called "tracer" methodology: it enabled scientists to *trace* the compounds with the incorporated isotopes and, hence, discover the intermediate steps of biochemical pathways. Schoenheimer and his group were extremely successful, particularly after they had found out how to combine labels (by using two different stable isotopes). This proved to be an ingenious method for investigating potential group transfers in a pathway. The findings opened up a vast range of new experimental approaches and eventually led to a completely new concept of the cell's metabolism, which involved the idea that the cell's constituents were not static but in a dynamic state of continuous turnover.[13]

Ruben was well aware of these crucial developments, as was everybody else working in the field of isotope chemistry at the time.[14] It took only a short leap from the successful use of *stable* isotopes to the idea that *radioactive* isotopes could also be useful in metabolic studies. Making use of the recently found radioactive isotope carbon-11 seemed particularly promising, not only in view of the central

[10] In his autobiography, Kamen stressed that Ruben's reputation was boosted when, after the latter had achieved some visibility in the department through the isotope studies, some graduate students chose to work with him on strictly physico-chemical problems, so that it became apparent that Ruben was not exclusively committed to biochemical work. See Kamen (1985, p. 114).

[11] See also Creager (2013, pp. 224–227).

[12] Urey was awarded the 1934 Nobel Prize in chemistry for these achievements.

[13] These studies' results were later summarised in Schoenheimer (1940). See also Simoni et al. (2002) for a review of the work by Schoenheimer and Rittenberg. The classic papers are Schoenheimer and Rittenberg (1935, 1937). Kohler (1977) provides a thorough historical study of the episode.

[14] See Kamen (1985, p. 82).

role of carbon in metabolism but also because one could (apparently) easily produce "tagged" glucose in photosynthesis: if radioactive carbon dioxide was fed to plants in the light, this radioactive compound would incorporate the radioactive label in all six carbons of the glucose. These in turn would then remain identifiable through all the other intermediate stages of the metabolic processing of glucose.

The only obvious drawback of this approach was the short half-life of the available isotope carbon-11 (which was just under 21 minutes). But Ruben thought that it was still possible to use this compound to gain fundamental insight into the complex fate of glucose in the rat's digestive system—if only Ernest Lawrence (as the master of the cyclotron) could be persuaded to become interested in this plan. They were, after all, located in one of the few places in the world where, at the time, a project of this type was feasible, as there was a ready source of radioactive isotopes at hand and Ruben and Kamen were, by then, sufficiently trained to be able to handle the isotopes. Their lack of expertise in animal and plant physiology was to be compensated for by the addition of two other members to the team: Chaikoff, who has already been mentioned, would bring with him his expertise in animal physiology; and plant physiology was soon to be covered by Zev Hassid, a friend of Kamen's who was then working as instructor in plant nutrition at the Berkeley Agricultural Experiment Station, with a particular interest in the biochemistry of carbohydrates.[15]

One can see the principle of research opportunism at work again here: if it seems that by investing a limited amount of time, resources and energy one can make a rewarding contribution to a field of study, why not go for it? It might not be your field of expertise, it might previously not have been on your agenda and you might not even have heard about the problem until very recently; but it might just be that the time is right for you to contribute your methodical knowledge or other skills to a particular theme (or temporarily join a group that is working on the subject). The fact that this strategy pays off so well explains why scientists usually do not change their techniques and methods once they have reached a certain (advanced) stage in their career.[16] The investment needed to acquire a comparable proficiency in a new method—say spectroscopy instead of manometry, or vice versa—would be disproportionately high. Rather, scientists tend to be continuously on the look out for problems that they might be able to solve using the methods and techniques with which they are familiar. This is not only how Ruben and Kamen came to photosynthesis research, but also how, for example, Otto Warburg and James Franck stumbled into the field (as was described in earlier chapters).[17]

[15] On Hassid, see Ballou and Barker (1979).

[16] This observation is supported by the case studies described in Holmes (2004).

[17] Kamen later was brought, by the same line of reasoning (although under completely different circumstances), to the study of bacterial metabolism—a topic to which he had never dreamed to contribute, but one that would greatly benefit from his specific skills and interests.

6.1.3 *The First Metabolism Experiments*

Ruben, Kamen, Chaikoff and Hassid immediately set to work.[18] Boron-10 was chosen as the appropriate target material in the cyclotron, from which, after heavy bombardment, the radioactive carbon-11 was obtained and immediately combusted to $^{11}CO_2$. This whole procedure was taken care of by Kamen and Ruben, who also caught the gas in a U-tube immersed in liquid air. They would then dash off with the radioactive material to Ruben's laboratory, which was in the Department of Chemistry's Rat House, where Hassid was waiting, so that he could administer the gas to his plants. About 10 min after the $^{11}CO_2$ had been prepared, the leaves were cut off, chopped into pieces and immersed in boiling ethanol, in order to stop any further metabolic reactions from occurring. The radioactively labelled glucose (or related carbohydrates) was extracted from this liquid, and fed to Chaikoff's rats. This extraction, however, turned out more complicated than they had expected, because only tiny amounts of radioactively labelled product were produced—far too few to be isolated using the standard methods in chemistry. Thus, the group had to make use of "carrier" molecules. In this particular case, they added ordinary, unlabelled glucose, so that a reasonable amount of precipitate could be obtained, in which the labelled compounds then would be identifiable. The need to add these carrier molecules brought the group into further difficulties as the scientists had to predict in which compounds the labelled carbon would appear. The team turned to plant physiology textbooks; but to their surprise they learned that the series of intermediate steps in photosynthesis from carbon dioxide to glucose was far from clear.

It proved extremely hard to transform this procedure into a stable protocol. Not only were the yields of $^{11}CO_2$ unforeseeable; the next steps turned out to be even more erratic in outcome. Sometimes the plants absorbed sufficient amounts of the labelled carbon dioxide, sometimes not; and worst of all was that hardly any of the labelled carbon showed up in the glucose.[19] "In the meanwhile, Chaikoff and his expectant students were becoming restive", Kamen recalled. Then, there was an unexpected turn of events:

> During a recital of these troubles Sam [Ruben] suddenly stopped, his eyes widened, and he blurted, "Why are we bothering with the rats at all? Hell, with you and me together we could solve photosynthesis in no time!" From that moment we were out of everything but the photosynthesis business.[20]

[18] See Kamen (1985, p. 83), for a description of the early stages of the project.

[19] In retrospect, one can understand the nature of these two stumbling blocks: first, as was later demonstrated, plants prefer "normal" carbon to its isotopes, so that radioactively labelled carbon dioxide tends to be absorbed at lower rates; second in contrast to standard (even very recently published) textbooks, Walker (2007, p. 182), has underlined the fact that "like sucrose, free glucose is not a major product of carbon assimilation by photosynthesising chloroplasts if, indeed, it is formed in the light at all".

[20] Kamen (1985, p. 84).

This remarkable moment, which Kamen remembered so vividly, is one of those rare instances in which a change in a research goal can be precisely nailed down. Before this moment, the whole business of having plants fix the labelled carbon dioxide and turn it into glucose was only the means to another end, namely, to find out what happened to glucose in rats. Facing the puzzling results, above all the fact that the labelled carbon disappeared somewhere in the plant, but not in the form of glucose, Ruben and Kamen realised that the means could be turned to an end in itself. What up to then had been thought of as being only a preparatory procedure to provide a measuring device became the actual focus of investigation.[21] As a consequence, a new experimental protocol was developed:

> Our strategy for the solution of the Big Problem [that is, identifying the first product of CO_2 fixation in plants] was simple in prospect, if complex in implementation: feed the plants $^{11}CO_2$, wait for predetermined lengths of time, from a few minutes up to an hour, add carrier (measurable amounts of whatever compound we guessed might be labeled in the $^{11}CO_2$ exposure period), extract, isolate, and see if any or all of the radioactivity appeared in the carrier. If we had guessed right, most, if not all, of the radioactivity would be localized in the compound defined by the carrier.[22]

Simple in prospect, if complex in implementation: this formulation appropriately catches the group's struggle during the next weeks and months. Straightforward as the principal idea sounded on paper, in practice it proved extremely challenging. Getting access to cyclotron runs was the first difficulty: the cyclotron was only occasionally available for the preparation of carbon-11 or other isotopes of biological interest. The next problem was to find a procedure that would provide enough labelled $^{11}CO_2$, without exposing the person in charge to enormous amounts of radioactivity. Even after improvements had been made to the procedure, Kamen was still so heavily contaminated with radioactive material after having prepared the isotopes, that Ruben banned him from entering the laboratory in the Rat House while samples were being counted. Finally, the short half-life of carbon-11 was a serious problem. After the labelled material had been retrieved, there was barely enough time to carry out all the necessary operations (since everything had to be done within the period of about two-and-a-half hours). Kamen gives an excellent description of the manic speed at which they needed to conduct the experiments in his autobiography:

> At the Rat House, Sam [Ruben] and Zev [Hassid] would be waiting for me like sprinters at the starting gate. Beakers would be filled with boiling water or other solvents and pipettes ready to suck up measured volumes of radioactive solutions onto absorbent blotters, which would be held by tongs over hot plates and dried. All the necessary reagents and apparatus would be in place. The [Geiger] counter would be ticking away establishing the background

[21] In the terminology suggested in Graßhoff et al. (2000), this implies turning the *Herstellungsprozess* into the actual *Untersuchungsprozess*. In the framework proposed by Hans-Jörg Rheinberger one may want to describe it as turning a "technical object" into an "epistemic object"; cf. Rheinberger (1997).

[22] Kamen (1985, p. 84). Additional background information on these early experiments can be found in Benson (1982) and Barker (1982).

activity. Each experiment had to be planned ahead in every detail so that no time was lost in confusion or delay in deciding what procedure to follow.

Anyone looking in on the Rat House when an experiment was in progress would have had the impression of three madmen hopping about in an insane asylum, what with the frenzied activity punctuated by loud classical music from the radio monitor [to register the running of spark testers in other laboratories, which would have confounded the measuring], and Sam's yells to get on with it and hand him samples while he sat at the counter table, feverishly taking background and sample counts. We had no idea of what had happened until hours later when, with all samples assayed, we sat in exhausted consultation, calculating and evaluating the results.[23]

In addition to the frenzy taking place in the laboratory, the researchers spent many hours in the library learning about the almost infinite number of substances that could be potential intermediates in photosynthesis, in which, accordingly, the labelled carbon might appear. The next step was to look up the analytical procedures for isolating each of them—formaldehyde, incidentally, was one of the first substances that Ruben, Kamen and Hassid tried especially hard to find. The demands of the project, on time and physical strength, which came on top of everybody's routine work, finally caused Hassid to withdraw.[24] His role had become dispensable by the fact that, after a short initial period during which they had worked with barley, the group had abandoned using higher plants and turned to algae and bacteria. By then, Kamen and Ruben were able to handle these, mainly thanks to a crash course delivered to them by the chemist-turned-microbiologist Horace A. Barker, a colleague of Hassid's of the Agricultural Experiment Station.[25]

Although Ruben and Kamen became increasingly proficient in their search for the reduction products in photosynthesis, they were keenly aware of their limited understanding of the biology behind it, and tried to learn as much as they could about the process. Kamen emphasised in his autobiography how eager Ruben and he were to communicate with other photosynthesis researchers in the country, in order to exchange (and critically discuss) the most recent findings. One of the centres of photosynthesis research was only a few hours' drive away from Berkeley: the Biological Division of the Carnegie Institution at Stanford, where in 1937 Robert Emerson and Charlton M. Lewis were working on quantum yields. Occasional visitors during these years, such as James Franck, complemented the range of discussion partners at this institution.[26] Hardly less important was the work going on at the Hopkins Marine Station of Stanford University at Pacific Grove, where Cornelis B. van Niel and, at

[23] Kamen (1985, p. 86).

[24] Kamen (1985, p. 87).

[25] Kamen (1989, p. 140); Barker (1982, p. 68). After receiving his PhD in chemistry in 1933, Barker had been the first postdoctoral student of Cornelis B. van Niel at the Hopkins Marine Station. This was exactly the time when van Niel was elaborating his concept of a general equation of photosynthesis; cf. Chapter 4. Again, the closely interrelated network of actors and institutions in photosynthesis research during the 1930s is obvious.

[26] It was at the Carnegie Institution that Emerson and Lewis, during the former's leave of absence from Caltech, started their work on the maximum quantum yield of photosynthesis; see Chapter 5. For a broader account of this institution's history, see Craig (2005).

the time, William A. Arnold were based (see Chapter 4). Kamen and Ruben were warmly welcomed into this Californian discussion circle. Far from showing contempt for their lack of biological training, the chemists' ideas were met with keen interest. Kamen remembered that Emerson in particular reacted enthusiastically and declared "that all work should be held in abeyance until we had the chance to fully exploit our labelling techniques".[27] As has already been mentioned in earlier chapters, these informal meetings—of as many experts as could be gathered in one place—were highly characteristic of photosynthesis research in this decade. In addition to the meetings themselves, much of the correspondence of the time included reports, to keep those informed who had been unable to participate in person. Far from keeping preliminary results to themselves, which one might expect in a competitive, new field of science, they rather took the chance of exchanging and discussing data and their interpretation freely with their colleagues.[28]

The next few years were particularly rewarding for Ruben and Kamen, who became the established leaders in radiotracer methodology and popular speakers within this discussion circle. And in spite of the severe methodical limitations, by the end of the 1930s they had made some substantial advances:

> Our experiments clearly established the existence of two systems, one a complex of dark reactions for CO_2 uptake with production of reduced cell material, and the other a light dependent process for the simultaneous evolution of molecular oxygen. These two systems in the plant had to be closely coupled so that one did not get ahead of the other, but the means for accomplishing this still remained unknown.[29]

The importance of establishing that carbon dioxide was reduced in the dark should not be underestimated. In the late 1930s it was still widely assumed that carbon dioxide reduction was achieved by the action of light on a complex with chlorophyll; the data produced from the radioactively labelled carbon decisively contributed to abandoning this hypothesis. The second point that Ruben and Kamen concluded from their data was that the first fixation product in photosynthesis was, most probably, a charged molecule containing carboxyl and hydroxyl groups.[30] One year later Kamen and Ruben tried to find out this compound's molecular weight. This was no easy task, given that the usual techniques for doing so required about a million times more of the substance than Ruben and Kamen had at hand. The two feasible techniques they resorted to resulted in widely divergent figures: while one indicated a molecular weight of 100–400, the other yielded a figure between 500 and 1000, and the prospect of acquiring more precise figures in the foreseeable future was highly unlikely. It is interesting to note how Ruben and Kamen (mistakenly) settled on the higher figure:

---

[27] Kamen (1985, p. 104).

[28] On the entanglement of cooperation and competition in the sciences, see Nickelsen (2014).

[29] Kamen (1985, p. 107).

[30] This was one of the findings that prompted the model developed by Franck and Herzfeld (1941); see Chapter 4. Ruben and Kamen's research findings were summarised by Kamen in 1949 in a contribution to the volume edited by Franck and Loomis; see Kamen (1949). For the original papers, see: Ruben et al. (1939a, b, 1940a, b) and Ruben and Kamen (1940a).

> We knew from reading the literature that the reactions of carbon dioxide to form a carboxyl product were not favored unless the molecules reacting were complex [i.e. large], such as in certain polyphenols. In these cases, the reactions to form carboxyls might occur significantly at high temperatures, such as 200°C. We assumed that in the algae special conditions existed that made possible such reactions at room temperature.[31]

Ruben and Kamen decided that the required size and complexity of the reactants (which the literature claimed was a precondition for the reaction to occur) meant that a molecular weight of between 500 and 1000 rather than between 100 and 400 was more likely. The assumption that "special conditions" in the living system would make up for other factors necessary in experimental set-ups in vitro exactly parallels the reasoning that had dominated nineteenth-century chemistry: then it had also been felt that formaldehyde was formed in plants from carbon dioxide and water by the same mechanism (or a very similar one) that was operating in the test tube, even if the latter required extreme conditions. "Cell constituents" were then assumed to make up for the lack of high pressure and temperature (see Chapter 2).

However, the vast difference that exists between physicochemical and biochemical systems became clear to Ruben, Kamen and other interested chemists shortly thereafter, when the process of phosphorylation as the main source of metabolically usable energy (materialised mostly in the form of adenosine triphosphate ATP) became widely known through the seminal papers of Fritz Lipmann and Herman Kalckar.[32] Ruben immediately tried to apply these insights to the explanation of photosynthesis. From his work with Kamen and Hassid, he accepted as established knowledge that the first step in the reduction of carbon dioxide was the carboxylation of an organic residue. This process, Ruben argued, was not favoured energetically unless it was coupled to reactions that prompted an appropriate energy release. Drawing on the work done by Lipmann and Kalckar, Ruben suggested that the organic residue might first be phosphorylated by an energy-rich phosphorous donor, which would explain the rise in its energy level; and that in the second step the phosphorylated organic residue might react with carbon dioxide, yielding a carbonic acid, an inorganic phosphate and the necessary activation energy. Ruben further speculated that the unknown organic residue might be an aldehyde, and the energy rich phosphorous donor ATP:

> Thus by a sequence of coupled equilibria such as suggested above, the dark fixation of carbon dioxide may be accomplished (also suggested by Lipmann) at the expense of the hydrolysis of an energy rich phosphorylated compound.[33]

Although the details of Ruben's suggestion did not stand the test of time, the proposal that a charged, phosphorylated intermediate might be the carbon dioxide acceptor closely resembles what is known today about this reaction sequence. However, even

[31] Kamen (1985, p. 109).

[32] Lipmann (1941) and Kalckar (1941). See also Chapter 7.

[33] Ruben (1943, p. 280).

Kamen, Ruben's closest collaborator, very much doubted that this scheme was correct. At the time, nobody was able to see how light energy could possibly be used to produce energy-rich phosphates.[34]

Ruben developed this approach not only to explain photosynthesis but also "carbon dioxide fixation and reduction by the many different chemosynthetic and heterotrophic organisms".[35] This referred to the discovery that the fixation and at the same time the reduction of carbon dioxide were not confined to plants but were generally present in a great many heterotrophic organisms, that is, organisms that cannot synthesise their own food, ranging from bacteria to mammals. A first inkling of this fundamental insight was provided by the microbiologists Harland G. Wood and Chester H. Werkman, who found, towards the end of the 1930s, that propionic bacteria were capable of carbon dioxide fixation through the reduction of oxaloacetic acid to succinic acid.[36] In subsequent years, the generality of this phenomenon gradually became understood, and the work that Kamen and Ruben undertook with radioactive tracer carbon contributed considerably to the acceptance of this surprising insight.[37] The implications for photosynthesis research were enormous. If the fixation of carbon dioxide by plants was not exclusively connected to the light reactions of photosynthesis, then the compounds that were found to be radioactively labelled and, hence, were interpreted as being derived from the labelled carbon dioxide, were not necessarily produced in the course of photosynthetic reactions.[38] Furthermore, it opened up the obvious possibility, as we can take from Ruben's work, that the reaction mechanism for carbon dioxide reduction might be the same in bacteria, animals and plants. This discredited even further the still popular idea of the chlorophyll-carbonic acid complex as the site and catalyst of carbon dioxide reduction (since there is no chlorophyll in heterotrophic organisms). The idea to model carbon dioxide reduction in the process of photosynthesis along the same lines as carbon dioxide reduction in heterotrophs was to play a central role in the debate on the path of carbon in photosynthesis, as will become clear later in this chapter.

6.1.4 Carbon-14 and the End of a Collaboration

The story of how Ruben and Kamen found, in 1940, the long-lived isotope carbon-14, that so fundamentally changed not only the study of metabolism but also innumerable other areas of science, has been told before, so that a sketch of the central events

---

[34] Kamen (1985, p. 162). However, Ruben's reaction sequence did make Kamen think intensely about the general relationship between phosphate metabolism and biochemical energy storage, which later became a central theme of Kamen's research work.

[35] Ruben (1943, p. 281).

[36] The seminal papers based on the study of propionic bacteria were Wood and Werkman (1935, 1936, 1938). See also Singleton (1997) and Krebs (1974) for historical accounts of this discovery.

[37] Ruben and Kamen (1940b).

[38] Cf. Kamen (1985, pp. 110–112).

should suffice.[39] In view of the difficulties of working with carbon-11, there was a huge incentive to find a radioactive carbon isotope with a longer half-life. The existence of such a carbon isotope had long been predicted, and there were indications that one did in fact exist, although no one had been able to isolate it. Ernest Lawrence was adamant that this discovery would be made in his laboratory, and he gave Kamen all the support and cyclotron time the latter needed to find this substance. Kamen enthusiastically seized the opportunity, and, together with Ruben, succeeded in discovering carbon-14 on 27 February 1940.[40] It took them some time, however, before they became thoroughly convinced of the accuracy of their discovery. Both Ruben and Kamen were terribly worried that they had made a mistake, even after the news had broken (Ruben refused to make an appearance on the occasion of the discovery's public announcement at Berkeley), but their anguish was unfounded. Carbon-14 turned out to have a half-life of 5700 years: more than enough time to carry out any biochemical experiment one might have in mind.[41]

Unfortunately, Kamen and Ruben were able to produce only a very small amount of carbon-14, so that no more than one single project with this new carbon isotope as a tracer was completed before December 1941.[42] In the mean time, Kamen and Ruben started a collaborative project with the Berkeley-based physical chemist Merle Randall and the latter's graduate student James L. Hyde. Randall and Hyde had constructed a large distillation column, which allowed them to isolate heavy water in which the (stable) oxygen isotope oxygen-18 had been incorporated.[43] Among other aspects, Ruben and Kamen contributed to the project through their acquired expertise in handling the algae, which were then grown in a medium with a relatively high concentration of heavy water. In these experiments, the group found a strong indication that the photosynthetic oxygen did originate from the water, and not from the bicarbonate in the solution (although the data, of course, did not provide any information on the mechanism of oxygen release from water). The crucial finding was that the isotopic composition of the oxygen produced during photosynthesis was similar to the one in water but unlike the isotopic composition of oxygen in carbon dioxide and atmospheric oxygen.[44] The obvious conclusion was that, despite all skepticism, photosynthetic oxygen did not originate from carbon dioxide, but from water. It is interesting to see that this very conclusion (which, incidentally,

---

[39] Kamen (1985), Chapter 7 is just one example; other references are given therein.

[40] The discovery was published in Ruben and Kamen (1941).

[41] The disadvantage of long half-lives is, of course, that the number of disintegrations per minute becomes very small—sometimes too small to be detected. Fortunately, though, carbon-14 turned out to be still within the biochemically useful range.

[42] This was a study in the metabolism of propionic acid bacteria, published as Carson et al. (1941).

[43] Kamen (1985, p. 140).

[44] See Ruben et al. (1941); confirmation was provided by Dole and Jenks (1944) and Holt and French (1948). Ruben et al. (1941) did not refer to the important work carried out by Robin Hill at the University of Cambridge (UK), which was described in Chapter 4; nor did Hill refer to Ruben et al. (1941), although they had arrived at similar conclusions at roughly the same time.

Otto Warburg never accepted) was arrived at independently by a scientific team around Alexander P. Vinogradov in the Soviet Union at roughly the same time.[45] This indicates, first, that researchers from all over the world jumped at the opportunities provided by the new tracer methodology; and, second, that the pertinent problem was considered pressing and relevant.

The attack on Pearl Harbor in December 1941 terminated any further pursuit of these lines of research. Kamen became involved with the uranium enrichment process of the Manhattan Project, while Ruben worked on a meteorological war project studying the biological effects of poison gas. Unjustly, however, once the war was over it was not Ruben and Kamen who were able to resume their highly promising work with carbon-14. In 1943, Ruben had died in a laboratory accident while experimenting with the highly toxic phosgene gas, and Kamen had become a victim of the McCarthy "witch hunts". Due to his alleged involvement with Communists, Kamen was fired from Lawrence's laboratory in 1944, and even though, once the war was over, he continued to make outstanding contributions to science at Washington University in St. Louis, Illinois, it took him about a decade to establish his innocence in court.[46]

6.2 The Bio-Organic Chemistry Group

Thus, rather than Kamen and Ruben, it was the chemist Melvin Calvin who, in 1946, was made head of the newly founded Bio-Organic Chemistry Group at Berkeley. Although Calvin had been at Berkeley since 1937, he only became properly acquainted with Ernest Lawrence when they worked together on the Manhattan Project; and in late 1945, Lawrence was able to win Calvin over for the setting up of the Bio-Organic Chemistry Group. The task of this group was to investigate the use of radioactive isotopes in chemical and biochemical studies, including the development of radioactive compounds to be used in the treatment of cancer. Financial support would be secured from the Atomic Energy Commission, with which Lawrence had excellent connections.[47] Already in 1945, only two months after the bombing of Hiroshima and Nagasaki, this project had been advertised for in the public by the science journalist William L. Laurence. Trying to underline the peaceful potential of atomic energy (the reputation of which had, understandably, suffered), he wrote that, with the new types of "tagged atoms" that had become available, "a new approach can be made toward solving one of the major mysteries of nature, the process whereby plants are able, by the use of the green colouring substance named chlorophyll, to harness the energy of the sun." This was reiterated in 1946, when Laurence envisioned that the

[45] See Vinogradov and Teiss (1941, 1947).

[46] For more background information, see Kamen (1985, 1989).

[47] See Seidel (1983), as well as Calvin (1992, p. 52); Seaborg and Benson (1997, p. 8).

project of elucidating the pathway of photosynthesis would enable both the artificial production of food and the construction of solar power plants of unprecedented efficiency.[48]

Calvin had received his doctorate in chemistry in 1935 at the University of Minnesota with a thesis on halogen electron affinity.[49] He had then gone, with a Rockefeller Fellowship, for two years to the University of Manchester (UK), where he spent most of his time in the laboratory of the renowned physical chemist Michael Polanyi. During these years, Calvin became interested in the electronic basis of the photochemical properties of porphyrins, such as haem and chlorophyll as well as their analogues. He went back to the USA in 1937, to Berkeley, where, in his first years, Calvin collaborated closely with the organic chemist Gilbert N. Lewis on the photochemistry of coloured porphyrin analogues and other themes. However, the war brought this work to a halt and Calvin became involved in the separation and purification of uranium and plutonium. By then Calvin had already acquired a reputation as an extremely intelligent and able scientist, with a broad range of skills and interests—some of them very close to the foundational problems of photosynthesis. He is reported to have been an exceedingly fast thinker who produced new models and theories (of any subject) at an almost terrifying pace; and he was not at all disturbed by the fact that, inevitably, a good many of his ideas turned out to be wrong.

In addition to his intellectual qualities, Calvin also proved to be a deft laboratory manager, and adept at selecting the right people for his group. In line with Ernest Lawrence's preferences, and with those of the AEC, the first projects of the newly founded Bio-Organic Chemistry Group focused on the medical applications of carbon-14, as well as on the synthesis of labelled amino acids and other metabolites, which were used in John Lawrence's laboratory. However, already from the start a subdivision was set up, headed by the chemist Andrew A. Benson, to specialise in photosynthesis research.[50] The medically oriented part of Calvin's group settled in the newly erected Donner Laboratory, where John Lawrence's group was located, while the photosynthesis division inherited the old clapboard building of Ernest Lawrence's original Rad Lab (which consequently was hitherto referred to as the *Old* Rad Lab, or simply the ORL). Together with the building, the photosynthesis division also inherited access to the integrated glass shop, the carpenter's shop and the machine shop, including a number of skilled artisans who were eager to assist the scientists to carry out their more extravagant ideas.[51]

---

[48] Laurence (1945, p. 6, 1946a, pp. 267–268, b, p. 41).

[49] On Calvin's life and work, see, e.g., Loach (1997), Bassham (1997) and Seaborg and Benson (1997). See also Calvin (1989, 1992) for autobiographical accounts.

[50] On Benson's life and work, see, e.g., Buchanan et al. (2007) and the autobiographical perspectives Benson (2002a, b, 2010).

[51] It was James A. Bassham, in particular, who emphasised in retrospect the importance of collaborating with such highly skilled glassblowers, machinists and carpenters; see Bassham (2003, p. 38).

Inviting Benson, who was then at the Caltech in Pasadena, to become the head of the photosynthesis division was an obvious way to provide the necessary continuity to the studies undertaken by Ruben and Kamen before and during the war. As an conscientious objector, Benson had not been drafted but had worked in 1942/1943 as a lecturer in the Department of Chemistry at Berkeley. In this capacity Benson had already collaborated with Ruben in the latter's carbon-14 studies (and also in the poison gas work). Benson's expertise was even more obviously needed when it became apparent that, first, Kamen had taken all his laboratory notes with him when he left and, second, that it was Benson who had received from Ruben a small vial containing carbon-14. It was only in the magnitude of millimicrocuries, as Benson recalled, but it was at the time the only supply of carbon-14 that existed.[52] (This situation, of course, changed rapidly after 1945, when radioactive isotopes became widely available by virtue of the nuclear reactors.)

Benson became the leading scientist in the new photosynthesis laboratory. He designed the chemical hoods, laboratory benches and other facilities and instruments after the excellent laboratories at Caltech. Benson also designed the legendary white table on which the chromatograms were later spread out for discussion and which very soon became the social and intellectual centre of the laboratory. One of Benson's most ingenious inventions, which became particularly famous, was a special vessel in which the algae could most appropriately be illuminated and exposed to the radioactively labelled $^{14}CO_2$. Owing to its shape, this vessel became known as the "lollipop" (see Fig. 6.2). This is how Vivian Moses, one of Calvin's collaborators at Berkeley, described this vessel in his recollections:

> A 'lollipop' was simply a glass vessel about 4 or 5 inches [10–13 cm] in diameter, circular in view, flattened so that the space between the two sides of the lollipop was relatively narrow—I would say something like 5 mm with an opening at the top for pouring liquid in and a large stopcock at the bottom. The idea was that you put the algal culture of *Chlorella* [...] in the lollipop, shone lights from both sides so that the algae were very highly illuminated, squirted in whatever radioactive material you wished to study the algal conversion of and, when you were ready to take a sample, you opened the stopcock [...] and the liquid suddenly fell out straight into boiling alcohol and killed the plants very quickly, and, as it were, "froze" everything for later investigation. It was called a lollipop simply because it looked like the top end of a lollipop on a stick.[53]

The third most important figure involved in reconstructing the path of carbon in photosynthesis was James Alan (Al) Bassham, who also joined the project at an early stage. Like Benson, he had been drawn to study photosynthesis through the

---

[52] See, e.g., Benson (2002a, p. 34). There is a discrepancy between Benson's and Calvin's accounts of who was given the vial. Calvin claims to have inherited this vial from Ruben himself, see Calvin (1992, p. 53). However, this version seems highly unlikely in view of the fact that Benson, not Calvin, was, at the time, collaborating with Ruben.

[53] In a large-scale Oral History Project, carried out by Vivian Moses, in collaboration with his wife Sheila, a substantial number of scientists were asked about their recollections of working in Calvin's group. The interviews were published online as Moses and Moses (2000). This quote is taken from the interview with Moses himself, at page 17/6 (i.e. from page 6 of the interview number 17 of the collection; all the interviews were numbered according to this pattern).

Fig. 6.2 Benson's famous "lollipop" vessel. (Photo courtesy of Lawrence Berkeley National Laboratory).

influence of Ruben, who had been one of his lecturers at Berkeley. On his return after the war, Bassham was taken on by the Chemistry Department as a doctoral student straight away and given a list of potential supervisors. Bassham recalled the following:

> I got a list of professors to go to and Calvin was [. . .] the first one on the list. So, he was the first one I went to see and he provided me with a list of his research projects. The first one he mentioned was the work with carbon-14, both with photosynthesis and also with some organic reaction mechanisms. And, of course, being with the big professor I listened politely to all of his research proposals. But I had already made up my mind as soon as I heard about carbon-14 in photosynthesis, because of my experience with Ruben. And so that's the one I told him I wanted to work on and that's how I got started in that project.[54]

Before this encounter, Bassham had neither read anything by Calvin nor had he even heard about him; Calvin was still far from being the major figure that he would later become.[55] Noteworthy is the fact that Bassham was not assigned a theme for his thesis on which to work independently, as would have been the case in almost every other laboratory at the time. Rather, Bassham immediately joined the laboratory's general agenda (he was to focus on degradation studies), and his subsequent publications,

[54] Moses and Moses (2000), interview with Bassham, p. 7/2. On this episode, see also Bassham (2003, pp. 37–38).

[55] It is clear from the interviews carried out in the course of a Oral History Project by Vivian Moses, together with his wife Sheila, that only a few of the people who came to work with Calvin were aware of his status in science. The younger scientists mostly came across his laboratory accidentally.

coauthored with many others, counted as the equivalent to the traditional single-authored thesis that usually qualified one for a doctoral degree.[56]

At this point of time, in 1947, the photosynthesis group was still very small. Indeed, it consisted only of Benson and Samuel Aronoff, another former doctoral student of Ruben's, who had spent two years at Chicago with Franck and Gaffron.[57] Aronoff's role at Berkeley was to take care of the algae and to develop appropriate culturing methods so that the state of the algae at least approximated a constant standard (no easy task, as was explained in earlier chapters). Furthermore, because of his experience with plant material, Aronoff was in charge of ensuring that the biological side of photosynthesis did not go unheeded, since none of the chemists involved in the project knew very much about plants.[58] Shortly after Bassham arrived, another graduate student, John Weigl, and two technicians, Tom Goodale and Gordon Hall, joined the project.

The one thing on which all group members agreed when recalling this period was the fact that the atmosphere in the clapboard building of the Old Rad Lab was special and intense in the early years of the project. The open-plan structure of the laboratory, which had only a few doors and compartments, strongly encouraged the continuous exchange of ideas and information among this group of young and talented scientists. (When the laboratory was founded, Calvin was not yet 35 and by far the oldest of the group!) This is how Vivian Moses, who had come as a postdoctoral student from England, remembered these years:

> It soon became clear to me in the context of the lab that collaborations between people were highly encouraged. This was not something that I felt had happened in London [. . .] but was very much the name of the game in Berkeley. The pattern was that people would talk to one another—I was going to say, continuously [. . .]. They would say "Why don't we do so and so?". Faced with something that needed to be resolved, someone would say "Why don't we do this?" or "Why don't we do it that way?" and somebody else would join in and say "We could modify . . . " and so forth. And before very long you would find a new collaboration had been started. In addition to whatever it might have been that those people had been doing before, they added a new thing. This was continuously going on: people were constantly forming and re-forming collaborative associations. And, of course, it happened to me just as it happened to everybody else.[59]

Moses found that everyone was keenly interested in what the others were doing. The members of the laboratory were aware that the whole group was not only working in the same area but also working towards the same goal, the achievement of which would be to everybody's advantage. Of course, there was a need to stay informed, since otherwise one might miss important clues for one's own work; but the prevailing spirit was that collaboration not only paid off but was also valuable in itself. The habit of openly discussing almost anything across the benches stayed vividly in the memory of many group members. "That was probably one of the most exciting of the

56 Moses and Moses (2000), interview with Bassham, p. 7/4.

57 On Aronoff's life and work, see Govindjee (2010).

58 See Moses and Moses (2000), interview with Bassham, p. 7/3.

59 Moses and Moses (2000), interview with Moses, p. 17/8.

learning experiences", the chemist Murray Goodman recalled forty years later. "You couldn't miss if you wanted to hear what was the latest, what was the conjecture, what was the abandoned hypothesis. It was all sort of put out there in a very open way for everybody to consider."[60]

The group's weekly seminars became legendary. They were held every Friday at 8 o'clock in the morning, thus, at a time at which many of the visiting scientists were not used to showing up at the laboratory, let alone ready to talk about science. In the earliest days, the speaker was chosen spontaneously at the meeting itself. Calvin believed that everyone should be able to speak about his or her research at any time, so that he would just pick out a person from whom he had not heard any news recently. However, in view of the panic and distress that this practice caused among members of the group, after a year or so Benson persuaded Calvin that it would be more profitable if at least one day's notice was given (which usually resulted in the speaker staying up all night to prepare for the seminar). Rough treatment had to be expected; in fact, some participants were interrupted after just a few sentences and given no chance to continue, since Calvin would insist on exact information on every single point. As Rodney Quayle remembered: "Once you started with a seminar, Calvin could tear you to pieces. He got lost in the science, totally divorced from any personal feelings, and he would shred you."[61] It was only later, when the laboratory's group had grown larger in number that a formal schedule for the seminar was finally put together, and members would get a week's notice; in principle, though, the atmosphere remained the same.[62]

The size and diversity of the group were the salient features of Calvin's laboratory. Far from being solely concerned with photosynthesis, Calvin also launched a project on chemical evolution and other themes that were only remotely related to the laboratory's core activities. While elsewhere rigid departmental structures still dominated (even within chemistry departments, the different subfields, such as organic chemistry or physical chemistry, operated largely independently of each other, with hardly any contact, not to speak of collaboration, between them), in Calvin's laboratory all the boundaries were blurred. Quayle nicely recalled the contrast between the Calvin–Benson group and the other laboratories in which he had worked: "[S]uddenly to come into a lab where you were a scientist. [. . .] You happened to be a chemist, but the chap next to you was a botanist and the other chap next to you was a physicist, and if you hadn't ever seen a Warburg manometer in your life before, if it was necessary to use it, you learned."[63] The strong feeling of community also included the administrative and technical staff, as Calvin's long-time secretary Marilyn Taylor vividly remembered: "We felt like a family. Coming to work was like coming home. I think most people performed [well] in that kind of a situation.

---

[60] Moses and Moses (2000), interview with Goodman, p. 14/4.

[61] Moses and Moses (2000), interview with Quayle p. 3/9.

[62] On the seminar's policy, see, e.g., Moses and Moses (2000), interview with Calvin, pp. 1/22–23.

[63] Moses and Moses (2000), interview with Quayle, p. 3/13.

When necessary, you did 150 or 200 % or whatever [...]. We all worked together as a group."[64]

This was the environment in which, through concerted effort of the group working around Calvin and Benson, the photosynthetic carbon reduction cycle eventually was elucidated, as will be reconstructed in the following sections.[65]

6.3 The First Cyclic Model

The line of research that Calvin and Benson decided to pursue once the laboratory had been set up was a direct continuation of the earlier experiments initiated by Ruben and Kamen: cell suspensions of the green algae *Chlorella* and *Scenedesmus* were exposed for defined intervals to $^{14}CO_2$ (mostly administered in the form of a solution of sodium bicarbonate). Then the course of photosynthesis was stopped by the algae being killed (boiling alcohol was poured onto them), so that the sequence of intermediate products could be discerned. However, as Kamen and Ruben had found, this procedure, although easy to outline, was extremely complicated to realise in practice.

Not only at Berkeley but also elsewhere, the search for the mechanism for carbon reduction was guided by the assumption that photosynthesis was, chemically speaking, in many respects the reverse of respiration. This was not only suggested by the fact that the products of the one process were the raw materials of the other, but additionally justified by the growing awareness that many biochemical reactions ran in both directions. Furthermore, comparative biochemistry was increasingly bringing to the fore the fact that many reaction mechanisms operated in exactly the same manner in plants, animals and bacteria. In a short note submitted to *Science* in 1938, the plant physiologist Kenneth V. Thimann, who had been Emerson's colleague at Caltech, succinctly summarised the starting point for any further study of the carbon reduction in photosynthesis:

> It is often stated that enzymatic processes are reversible. It is also a commonplace that photosynthesis is in many respects the reverse of respiration. Now there is every reason to believe that the CO_2 formed in oxidations arises from organic acids, probably by the same reactions as in fermentation [...]. What could be more natural than to suppose that in photosynthesis the absorption of carbon dioxide takes place in the reverse way, by combination with an aldehyde, or, more probably, with an organic acid to produce a new carboxyl group? Specifically, a probable reaction is the combination of CO_2 with pyruvic acid to produce oxaloacetic or perhaps with lactic acid to produce malic. The light reaction would then be the reduction, not of CO_2 as such, but of the carboxyl group.[66]

---

[64] Moses and Moses (2000), interview with Taylor, p. 1/60.

[65] See for shorter treatments of the following episode, although with differing foci, Nickelsen (2012b) and Schüring (2006).

[66] Thimann (1938, p. 506).

Thimann argued that, although it had been assumed since the time of Willstätter and Stoll that in photosynthesis carbon dioxide formed a complex with chlorophyll and was reduced in this complex by a light-driven reaction, no convincing evidence had ever been provided to support this assumption. Even less evidence had been produced for the assumption that formaldehyde was the first reduction product; in fact, Thimann believed that "the persistence of these unsupported theories must be ascribed to the absence of any plausible substitute".[67] He found it far more convincing to think of photosynthesis as a process involving a cycle of the "combination of carbon dioxide with an organic acid, the photo-reduction of the carboxyl group and the consequent intramolecular changes leading finally to the setting free of the organic acid again"—Thimann still believed that the light reaction was connected to the reduction of carbon dioxide! As in other cases, Thimann thought, some of the organic acids then would be withdrawn from the cycle and reduced to sugar.[68] Surprisingly, Thimann seems not to have been aware of the pathbreaking work by the German biochemist Hans Krebs, published in 1937, in which the organic acids mentioned by Thimann were knitted together into the metabolic cycle, which later became known as the tricarboxylic acid cycle or "Krebs cycle".[69] Thimann did cite, however, Krebs's earlier work on the urea cycle, which was the first metabolic cycle to be completely illuminated. Since then, cyclic pathways had become an integral part of biochemists' conceptual tool kit.[70]

Thimann's appeal that one should pursue this promising approach did not go unheeded; and the general idea—that there was a close parallel between the respiratory process and the process of photosynthesis—was the prevailing concept when the Calvin–Benson group started its work. By then, the biochemical pathway of cell respiration had been largely uncovered, including: (1) the process of glycolysis, that is, the degradation of glucose to pyruvic acid via the formation of phosphoglyceric acid (PGA) and triose phosphates (that is, phosphoglyceraldehyde and dihydroxyacetone in equilibrium); (2) the tricarboxylic acid cycle, that is, the stepwise oxidation of the carbon residue, whereby reducing equivalents are formed; and (3) the first ideas about how to formulate the final oxidation process, including the fact that cytochromes might be involved (see also chapter 7). Additional support to the idea that photosynthesis should be modelled as the reverse of respiration was provided by the finding that carbon dioxide fixation in heterotrophic tissues was a very general phenomenon, and that it usually involved a partial reversal of the tricarboxylic acid cycle. It was an obvious assumption that photosynthetic carbon dioxide reduction should proceed in a similar way (As background information for later comparisons, the biochemical steps of glycolysis and the path of heterotrophic carbon dioxide fixation have been reconstructed in graph form in Fig. 6.3.)

---

[67] Thimann (1938, p. 506).

[68] Thimann (1938, p. 507).

[69] Krebs and Johnson (1937).

[70] Krebs and Henseleit (1932). On the discovery of the urea cycle, see Nickelsen and Graßhoff (2008); Holmes (1991).

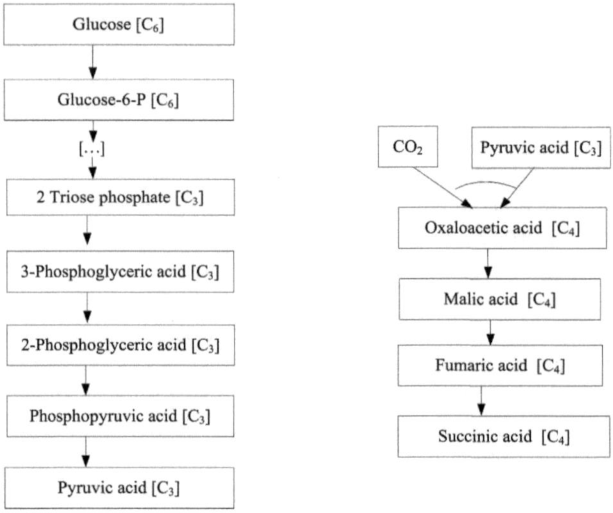

Fig. 6.3 *Left*: The sequence of carbon compounds in glycolysis (degradation of glucose). *Right*: What was held to be the process of carbon dioxide fixation in heterotrophs in 1940. The sequence from pyruvic acid to succinic acid is a partial reversal of the tricarboxylic acid cycle in cell oxidation processes.

Thus, when the Berkeley Group started working on photosynthesis in 1946, they had a well-defined goal and a convincing working hypothesis of the mechanism in question. In their first attempts to clarify the whereabouts of the radioactively labelled carbon dioxide, to which the algae had been exposed, the group used classical chemical procedures to try and identify the intermediate compounds from carbon dioxide to carbohydrates. However, they soon realised that these methods would not lead them anywhere: they were far too slow and required far too large amounts of plant material.[71] So the group changed over to using ion exchange columns—which was an improvement in procedure but far from perfect.[72]

The second major difficulty was ensuring that the labelled products (if they could be identified at all) were, in fact, products of photosynthesis and not the result of a non-photosynthetic carbon dioxide fixation. The Calvin–Benson group tried to avoid this difficulty by using periods of "pre-illumination": the cells were strongly illuminated in the absence of carbon dioxide for a period of at least ten minutes and sometimes for up to more than an hour, after which the light source was switched off and the algae were supplied with labelled carbon dioxide. The Berkeley team found that large amounts of carbon dioxide were almost immediately fixed, exceeding by

[71] See Calvin (1989, p. 9).

[72] According to Benson, this came about through Calvin's involvement as a consultant to Dow Chemicals, where new resins were being developed at the time. See Moses and Moses (2000), interview with Benson, p. 12/18.

far the usual dark fixation rate. They concluded from this that the increase in fixation was, in fact, due to the foregoing illumination period, during which the necessary amount of "reducing power" had been produced. Compared with the high rate of photosynthetic carbon dioxide fixation under these circumstances, the Calvin–Benson group thought that the risk of confounding photosynthetic with non-photosynthetic fixation products was negligible. (However, this procedure was soon criticised, in particular by members of the Chicago group around Hans Gaffron, who found it less than reliable and were not at all convinced by the Berkeley team's conclusions. It was, in fact, a problem comparable to the difficulties in differentiating manometrically between the gas exchanges caused by photosynthesis and those caused by respiration; and a satisfactory solution was equally unattainable with the methods at hand.)[73]

The third complication concerned the inherent limits of tracer studies, which in the initial excitement had been overlooked (or downplayed) by many researchers, yet in the course of time became more and more obvious. It transpired, for example, that since carboxylation reactions were reversible, all kinds of exchange reactions were occurring in the cell, so that not all the compounds, from which radioactively labelled carbon emerged, could be counted as being part of a carbon dioxide fixation chain, photosynthetic or not. Furthermore, it became apparent that in many metabolic pathways a clear preference existed for the use of ordinary carbon (^{12}C), so that far less radioactively labelled carbon was being incorporated into the plants than had theoretically been expected. How one should deal with these difficulties was far from clear.

Despite all these challenges in experimental procedure, in 1947 the Berkeley team published a tentative report: after a period of illumination of 5 min, about 70 % of the labelled carbon was found in the succinic acid [C_4] and 3 % in the fumaric acid [C_4], both of which were known to be components of the tricarboxylic acid cycle (The abbreviation [C_4] denotes that this compound contains four carbon atoms, usually as a backbone chain. Within a compound the carbons are given numbers according to their position in the chain). 15 % percent was present in a cationic fraction, presumably amino acids, while another 9 % of the products had anionic properties.[74]

One year later, in 1948, the list of labelled compounds after a period of illumination of 30 s, drastically differed from the first account: it was now found that most of the labelled carbon dioxide was incorporated into the malic acid [C_4], alanine [C_3], triose phosphate [C_3], PGA [C_3], glucose [C_6] and fructose [C_6] as well as into the latter two compounds' phosphate esters. Remarkably, succinic acid or fumaric acid among the early photosynthesis products, which had been so prominently presented in the earlier paper, were no longer detected.[75] The 1948 list of products was complemented by the group's first impressions of the labelling patterns (i.e. the temporal sequence

[73] See Benson and Calvin (1947) and Calvin and Benson (1948) for the first publications from Berkeley. The approach was criticised in, e.g., Brown et al. (1949); Fager (1949); Gaffron and Fager (1951).

[74] Benson and Calvin (1947, p. 648), Table 1.

[75] Calvin and Benson (1948).

in which the carbon atoms of a compound were labelled): in most of the acidic compounds, the labelled carbon appeared first in the terminal carboxyl groups. This was different in the two hexoses, that is, glucose and fructose. Here, the central carbons, at positions 3 and 4 in the chain, were the first to be labelled, while the labelling subsequently spread to the ends of the chain.

The identification of large amounts of 3-PGA [C_3] and triose phosphate [C_3] with this particular labelling pattern was highly suggestive, because it was well-known that these compounds were among the first intermediates of glucose degradation in the course of glycolysis (see, for the latter pathway, Fig. 6.3).[76] In fact, one might speculate that this was just what the Berkeley Group had been hoping and looking for, bearing in mind that the standard working hypothesis of the photosynthesis model was that it was the reverse of respiration. From their data, Calvin and Benson were pretty much convinced that the synthesis of glucose was the reverse of the well-known sequence of glycolysis:

> In view of the presence of such large amounts of radioactive triose phosphate and phosphoglyceric acid in the very short photosynthetic experiments (30 sec) as well as in the dark fixation, it can be taken as fairly certain that the hexose synthesis proceeds by a reversal of the usual glycolytic split of fructose diphosphate, and therefore some path must be found by which the radioactivity appears first in the number 1 carbon atoms of the 3-carbon compounds and gradually spreads into the number 2 and then the number 3 carbon atom of the triose phosphate and the phosphoglyceric acid.[77]

Calvin and Benson suggested that this reaction was accompanied by a cyclic path, which would regenerate the carbon dioxide acceptor and in which the various [C_4] dicarboxylic acids were involved. They had found that a number of compounds were being rapidly saturated with radioactivity (i.e. in a very short time span all their carbon atoms were being radioactively labelled), and their explanation was that these compounds were part of a cycle, which regenerated the original carbon dioxide acceptor and in which all the existing carbon atoms were quickly replaced.

The pathway, as formulated by Calvin and Benson in 1948, is reconstructed in the form of a graph in Fig. 6.4.[78] This graph also highlights the fact that the group around Benson and Calvin had (implicitly) modularised the problem in three partial processes: (1) carbon dioxide had to be incorporated into an acceptor molecule ("carboxylation"); (2) there had to be a pathway from the acceptor molecule to six-carbon sugars, eventually glucose ("glucose formation"); (3) the primary carbon dioxide acceptor molecule had to be regenerated so that further carbon dioxide molecules could be incorporated ("regeneration of the primary acceptor"). These three modules can be traced through later stages of their work, although they were continuously revised according to the latest evidence. This "functional decomposition" of a mechanism in several component operations that contributed to the overall functioning of the system was identified by William Bechtel in collaboration with Robert C.

---

[76] Calvin (1962, pp. 880–881, 1989, p. 9).

[77] Calvin and Benson (1948, pp. 478–479).

[78] The same model hypothesis was still published in Benson et al. (1949, p. 399).

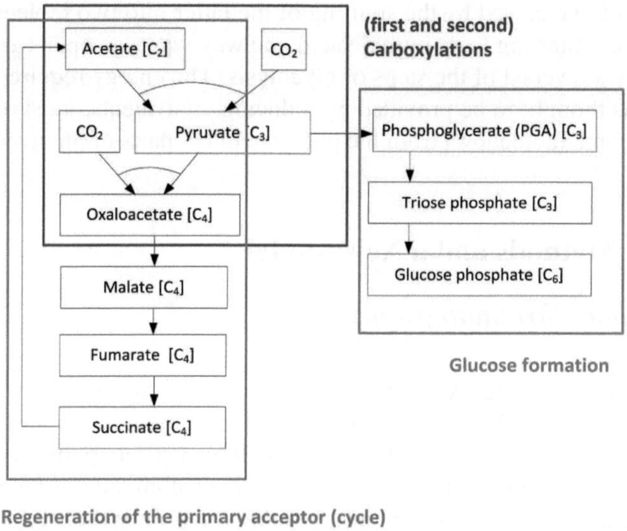

Fig. 6.4 The photosynthetic carbon cycle, as formulated by the Berkeley Group in 1948. The formation of glucose was modelled as the reversal of glycolysis. The carboxylation of the acceptor molecule as well as its regeneration in a cyclic sequence were—following the template of heterotrophic carbon dioxide fixation—conceived of as the reversal of portions of the tricarboxylic acid cycle. The latter two modules were substantially revised in the course of the following years.

Richardson and, later, Adele Abrahamsen as a typical move in the elucidation of mechanisms.[79] While it was by no means certain that the hypothesised modules corresponded to actual modules in the system, it was a valuable guiding assumption and greatly facilitated the task of the group.

The resulting model had the definite advantage that it was exclusively based on reaction mechanisms, which were part of the standard biochemical body of knowledge.[80] It was, in fact, nothing but the standard hypothesis fleshed out in detail, and up to 1949 the data fitted this model rather neatly. The fixation of carbon dioxide during photosynthesis was assumed to function in exactly the same way as the non-photosynthetic fixation of carbon dioxide, that is, by a reverse of the well-known decarboxylation reactions in the tricarboxylic acid cycle. Acetic acid [C_2] was thought to act as the first carbon dioxide acceptor, the carboxylation of which would lead to the formation of pyruvic acid [C_3]. The latter would be subject to a second carboxylation, which resulted in the formation of oxaloacetic acid [C_4]. This would be transformed into malic acid [C_4], fumaric acid [C_4] and succinic acid [C_4]; and

[79] See Bechtel and Richardson (1993); Bechtel and Abrahamsen (2005). The concepts of "functional" and "structural" decomposition also looms large in Bechtel (2006).

[80] "Using some of the reactions already established in animal tissue and bacteria, it is possible to account for the above results as well as the observed distribution of radiocarbon in short photosynthesis." Benson and Calvin (1947, pp. 648–649).

the cycle would be closed by the splitting of the latter into two molecules of acetic acid [C_2] again. Starting from pyruvic acid, the way was also open to the formation of glucose by a reversal of the steps of glycolysis. The energy required for all these reactions was thought to be provided by reducing equivalents, most of which were assumed to come (somehow) from the photochemical parts of photosynthesis.

6.4 New Methods and a New Model

6.4.1 Paper Chromatography

While the findings of the Calvin–Benson group reported so far were based on their work with ion-exchange columns, a decisive element of the group's sweeping success, in addition to using carbon-14, was their use of (partition) paper chromatography, to which they turned after 1948. This technique, originally developed in 1944 by a group of British chemists, became the principal analytical tool of the Berkeley team for identifying radioactively labelled products.[81] It all started with the arrival at Berkeley of the biology graduate student William Stepka in 1947. He had become interested in using paper chromatography to separate amino acids while being a student at Rochester University, New York, and he brought the technique with him when he moved to Berkeley to begin his doctoral studies. In the beginning, Stepka had to talk Benson and Calvin round to trying out paper chromatography to identify labelled intermediates (as the technique was thoroughly despised by most biochemists), but soon they grasped its enormous potential: paper chromatography was not only much faster than ion-exchange columns but also more precise.[82] Paper chromatography became the method of choice for the group and it was Benson, in particular, who developed the technique further to the point of near perfection.[83]

The underlying principle was simple: from the concentrated extract of an algal suspension, containing all the potentially labelled compounds, a small drop was spotted onto a piece of filter paper. The overall radioactivity of this sample was quantitatively determined, and the paper was then hung in a closed chamber, with the edge below

[81] See Consden et al. (1944) for the publication of the method; their work was based on the seminal suggestion published in Martin and Synge (1941).

[82] The plant biochemist Albert Frenkel emphasised that it was Charles Dent who had brought paper chromatography from the UK to the USA in the first place and who was instrumental in first attempting to identify ^{14}C-labelled amino acids; see Frenkel (1993, p. 106). Among the early relevant publications were Dent et al. (1947a, b) and Fink and Fink (1948). On the advent of paper chromatography in Berkeley see also Kamen (1985, p. 193), and several of the interviews in the Oral History Collection, Moses and Calvin (1958).

[83] See Stepka et al. (1948) for the first publication from the Berkeley Group based on paper chromatography (on amino acid separation, which was Stepka's field of expertise). Benson et al. (1950) then demonstrated how successfully paper chromatography could be applied to identify carboxylic acids and phosphate esters.

the spot of the sample dipping into a trough that had an appropriate solvent, such as water or ethanol. The solvent then moved up the paper by capillary action, met and dissolved the sample mixture and carried the compounds along, according to their solubility in the solvent (the most soluble compounds travelled fastest up the paper). The sample mixture separated out into different spots and the general structural properties of the various components could then be inferred according to their position on the paper. This could be complemented with a technique, which became known as co-chromatography: substances assumed to have been in the spot were applied to the paper at the same time as the suspension to be analysed; and if the properties of the known molecule's chromatogram coincided with those of the substance in question, one could then conclude that the two of them were identical. Having been eluted, the components of the spots could also be analysed by chemical or physical means. The Calvin–Benson group tried, for example, to apply fluorescence and ultraviolet absorption spectra to find out more about the compounds. However, most of the time there was not enough substance available for these analyses, so that the researchers had to use more traditional analytical techniques.[84]

Several improvements had to be introduced before paper chromatography revealed its full potential. *First*, the team developed a two-dimensional chromatography, in order to separate the labelled compounds more finely: after the first solvent had dried out, the paper was turned at right angles and submitted to a second chromatography run, with a different solvent. This strongly increased the chances that as many compounds as possible were separated from each other: two intermediates might have the same extent of solubility in one solvent, but it was unlikely that they would share the same extent in a completely different solvent. *Second*, paper chromatography was combined with autoradiography: since the compounds were radioactively labelled, their position on the paper could be visualised by placing an X-ray film onto the paper. This made it far easier to localise the spots and to preserve the chromatograms (since the paper was frequently destroyed at the analytical stage). For quantitative work, the amounts of radioactivity in the spots were determined with a Geiger counter. *Third*, the solvents themselves had to be adapted, since neither water nor ethanol (nor any other standard solvent) was able to separate the many phosphate compounds that by then were known to be contained in the extracts. It was chiefly Benson who developed this aspect of chromatography. He persistently experimented with one solvent after the other until he came up with a sophisticated variant that could also separate the sugar phosphates.[85]

Paper chromatography eventually became a highly powerful analytical tool. In the course of time, the Berkeley team succeeded in producing a whole series of chromatograms from which the sequence of radioactively labelled compounds could be inferred (see Fig. 6.5, for an example of chromatograms after different time periods).

[84] For a detailed description of this procedure, see also Calvin (1962, p. 881, 1989, p. 9); and Bassham and Calvin (1960, pp. 890–895).

[85] Moses and Moses (2000); Interview with Bassham, p. 7/10. See also Benson (2002a) and Bassham (2003).

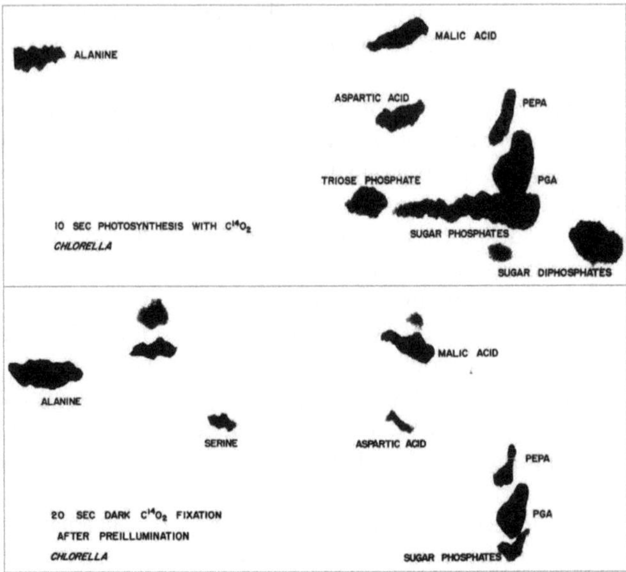

Fig. 6.5 A sample chromatogram of two-dimensional paper chromatography, after ten seconds (*top*) and twenty seconds (*bottom*) photosynthesis. (Alcoholic extract of *Chlorella pyrenoidosa*.) The rapid accumulation at the spot identified as PGA after ten seconds is evident—while ten seconds later much of the material had gone elsewhere. (Taken from Calvin and Massini (1952, pp. 15, 18).

However, it was neither a straightforward, nor a very pleasant procedure to carry out: the solvents—highly noxious organic compounds—had an extremely strong odour, so that, following increasing complaints of the group members about the constant exposure to these dreadful organic vapours, the chromatography room was eventually moved out of the laboratory into a more isolated area. This, of course, did not change the situation for those carrying out the procedure (Bassham recalled that some laboratory members who were working with the new solvents were once asked to leave a cinema because of complaints from other cinema goers seated nearby—they had shed their laboratory coats but otherwise not changed their clothes[86]). Moses, who most of the time carried out the procedure, vividly remembered how much he detested the work:

> I was always smelly-ish, of course, and when you put the solvent in [the tank], it got smellier and then you closed the lid and left the papers to let the solvent travel across the paper for however many hours it took. [. . .] Then you had to take the papers out of the tanks in order to dry them [. . .]. And you then lifted up this sodden wet piece of paper, held in place by two or three paper clips, and delicately took it to a drying rack in some sort of oven, hood really, fume cupboard. During this process, you got the full force of the vapours in your face [. . .]. And if you weren't careful, these papers would occasionally simply tear and rip off and fall on the floor and that would be the end of that one. [. . .] It was a most painful activity. [. . .]

[86] Bassham (2003, p. 41).

> Anyhow, we used to spend lots of time there counting these spots on chromatograms and it was on that, on the data from those measurements, that everything depended. And so it was very important to get it done.[87]

With the help of paper chromatography, it became apparent that even after just 30 s, the radioactively labelled carbon had become incorporated into a broad range of compounds, so that it became necessary to shorten the duration of the experiments even further—eventually down to fractions of a second.

6.5 The Second Cyclic Model (1950)

By 1950, the one thing that had been established beyond any doubt was that PGA was a central compound in the early stages of photosynthetic carbon reduction. Paper chromatography had impressively shown that in short-time photosynthesis (30 seconds of illumination) 80–90 % of the radioactively labelled carbon was present in the PGA; and this was confirmed by the use of other techniques. The other potential intermediates that were still being debated included pyruvic, malic and glycolic acids, while most of the compounds, which had been assumed by the Berkeley team to be part of the regenerative cycle in 1948, had disappeared from the array of promising candidates. Not even the existence of triose phosphates, which were highly probable intermediates on the pathway from PGA to glucose phosphate, were confirmed, at that time, to be present with sufficient reliability.[88]

In fact, the Berkeley team had to acknowledge that the labelling of the compounds of the regenerative cycle that they had originally proposed, namely succinic, fumaric, tartaric and malic acids as well as other acids, had been an artefact. With hindsight it became clear that, up to approximately 1948, most of the radioactivity measurements had been significantly influenced by the high background radiation emitted during the runs of the cyclotron in the Crocker Laboratory next door. Bassham recalled this frustrating realisation: "Further work with more careful shielding of the Geiger counter showed that radioactivity was not spreading to the central positions fast enough for these compounds to be intermediates in the primary cycle."[89] The working hypothesis—that the process of carbon dioxide fixation during photosynthesis might operate in the same way as the non-photosynthetic process—thus started to fall apart. The first wave of enthusiasm for the unlimited potentiality of tracer techniques also began to wear off. Emerson, for example, who had been delighted when he first heard about the experiments set up by Ruben and Kamen, now very much doubted the significance of the tracer studies' results. In a letter to Gaffron, written on 4 April 1950, Emerson maintained:

---

[87] Moses and Moses (2000), interview with Moses, pp. 17/12 ff.

[88] See Gaffron and Fager (1951, p. 91).

[89] Bassham (2003, p. 39).

> Provisionally, I conclude that pick-up of radioactivity, in either light or dark, is of doubtful significance as far as photosynthesis is concerned. It's going to be difficult to prove that any particular channel of pick-up is the channel of photosynthesis. I hear from California that radioactivity appears also in fats (??), 15 seconds after illumination."[90]

Fats, of course, are large molecules that could not possibly have anything to do with the early stages of photosynthetic carbon dioxide reduction; and if substances like these were identified among the early products, how could one possibly rely on anything found using this means? Yet the Berkeley Group did not lose faith in their methods, and instead tried to learn from their mistakes. In 1950, they amended their proposal of the regenerative cycle, and now based it on the first extensive findings using the paper chromatographic methods (see Fig. 6.6, for a reconstruction in graph form). In addition to the strong confirmation that PGA was one of the first products of photosynthesis, they also endorsed the assumption that the hexoses might be formed from the PGA in a reversal of the process of glycolysis, and this hypothesis henceforth was no longer seriously challenged.[91] The 1950 cycle started with a [C_2]-acceptor, the identity of which was unknown. The group suspected that acetic acid had this function but they had not yet been able to produce any evidence for this assumption. The acceptor was carboxylated (and phosphorylated) to form PGA [C_3]. Then PGA would either be supplied to form glucose-6-phosphate (in what constitutes the reversal of glycolysis), or it would be rearranged to form phosphopyruvic acid [C_3]. The latter was thought to be subject to the second carboxylation reaction, yielding oxaloacetic acid [C_4], which would regenerate the acceptor by splitting into two [C_2] fragments. Aspartic acid and alanine, two amino acids which were also among the early labelled products, could easily be formed from oxaloacetic or pyruvic acids.

Comparing this new model hypothesis with the first suggestion of 1947 and 1948, we find that the authors revised the earlier model as minimally as possible. The module "glucose formation" remained untouched although, as was mentioned earlier, the presence of triose phosphates, which were the central compounds of this path, was not undisputed. The module "carboxylation" also remained largely intact, but there was no longer any precise hypothesis on the nature of the primary acceptor. The revisions concerned only the regenerative cycle, which was punctually contracted, as, in reaction to the revised empirical evidence, malate, succinate and fumarate no longer were listed as intermediates on the path (and would not return). Yet, it was well-known that all these [C4]-compounds were easily formed starting from oxaloacetate and vice versa.

The identity of the first carbon dioxide acceptor remained a mystery. It was generally assumed to be a two-carbon compound [C_2], since the carboxylation of the latter was the easiest way to explain the formation of a molecule of PGA that had the radioactively labelled carbon incorporated into the carboxyl group. For some

[90] Emerson to Gaffron, 4 April 1950. Robert Emerson Papers, 1923-61, Record Series 15/4/28, Box 1, Folder: Gaffron, Hans, University of Illinois Archives.

[91] Benson and Calvin (1950).

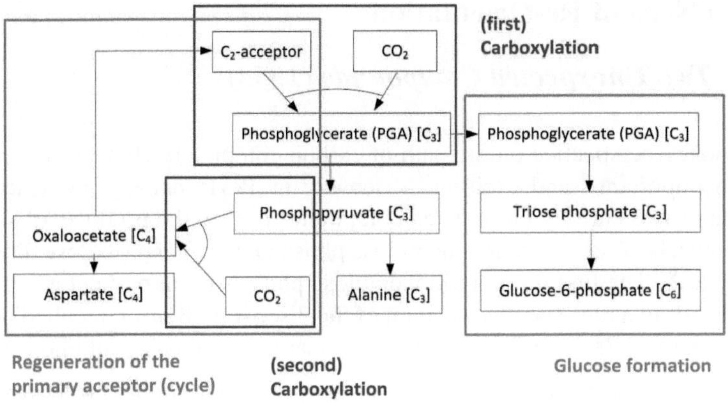

Fig. 6.6 The photosynthetic carbon cycle, as formulated by the Berkeley Group in 1950. The standard hypothesis of 1947/1948 was only minimally revised. The formation of glucose is still the reversal of glycolysis. The group still assumed two instances of carboxylation, although they left the first carbon acceptor open. The regeneration of the latter still was thought to be brought about by the split of a 4-carbon compound (oxaloacetate), although the intermediates malate, fumarate and succinate were dropped.

time, glycolic acid [C_2], which had been found to be present, was suspected to be closely related to the primary acceptor; however, the data were unclear and the issue remained unsettled. Equally unclear was the *origin* of the hypothetical two-carbon acceptor, whatever its identity might be. A priori, two possibilities prevailed: either it was the result of a [C_1] + [C_1] combination or it was the outcome of the cleavage of larger compounds, for example, [C_4]. The Berkeley team consistently voted for the second option, not the least because no one-carbon compound was ever identified in the chromatograms. This explains, at the same time, why the Berkeley team assumed that there were two carboxylation reactions, as the [C_4] compound had to be regenerated. For a long time, malic acid appeared as a promising [C_4] candidate for this role.[92] However, when the group found that inhibiting its formation did not significantly lower the amount of PGA produced, malic acid was precluded from the list of possible intermediates, without a promising candidate to replace it.[93]

---

[92] See, e.g., Badin and Calvin (1950); Calvin and Massini (1952).

[93] See Bassham et al. (1950). The idea that the "malic" enzyme should somehow also be involved in photosynthetic carbon dioxide fixation was, nevertheless, rather persistent. In Ochoa and Vishniac (1952), which was a very influential paper (see Chapter 7), the idea was not only revived but also expanded on, despite evidence to the contrary supplied by the inhibition studies.

6.6 In Need of Re-Orientation

6.6.1 Two Unexpected Compounds (1952)

The Berkeley perspective on the path of carbon completely changed when Benson made an important—and exciting—discovery in 1951: namely, evidence for the existence of two entirely unexpected compounds among the products of short-time photosynthesis. These were sedoheptulose phosphate [C_7] and ribulose diphosphate (RDP) [C_5].[94] The finding of a sedoheptulose phosphate, in particular, was hard to comprehend. In a retrospective account of the discovery, Benson recalled how these unusual compounds were first seen in a chromatogram of bacterial photosynthesis:

> Their surprising appearance was tantalizing. After preparing hundreds of radiograms from our two-dimensional paper chromatograms, the usual pattern of compounds and their relative amounts had become very familiar. But, in this case, two radioactive spots just jumped out at us, strangers among a well known group of compounds. It must have been the result of phosphatase activities liberated in preparation of the bacterial extracts.[95]

With Bassham's help the compounds were finally identified. Bassham oxidised the unknown sugar with a reagent that would yield carbon dioxide from carbonyl carbon only, that is, from carbon that was double-bonded to oxygen, such as the carbon in position 1 of most sugars. To the team's utter surprise, only 14 % of the carbon-14 in the originally eluted sugar was in this carbonyl carbon: that is, one-seventh. "That's unheard of, Al, try again!" was Benson's first reaction to Bassham's finding. Bassham tried again, and confirmed his earlier result. They had to search the literature for any precedents of such a sugar, and really found that "sedoheptulose", a seven-carbon sugar, had been described in a publication of 1917.[96] Benson was able to get hold of a sample of this compound from a Norwegian colleague, who had recently published on this topic, and on the basis of this sample the identity of the curious product was confirmed. An allegedly elusive compound, which had so far been reported to be present only in succulents (therefore the name, which refers to the genus *Sedum*), now seemed to perform a vital function in photosynthesis. It was found to be formed before the hexose phosphates, while it was too different in structure to be one of the latter's direct precursors. Thus, Benson assumed that sedoheptulose had a function somewhere in the regenerative cycle.[97]

The identification of the second mysterious compound was announced in the same issue of the *Journal of the American Chemical Society* in which the discovery of sedoheptulose was reported. Benson had managed to elucidate it through a laborious procedure. Step by step he succeeded in eliminating one improbable alternative after

[94] The first report on ribulose diphosphate (RDP) was given in Benson (1951); the first report on sedoheptulose phosphate in Benson et al. (1951).

[95] Benson (2002a, p. 39).

[96] Benson (2002a, p. 39). See also Bassham (2003, p. 42), for this episode.

[97] See Benson et al. (1951) for the pertinent publication.

the other, until he concluded that the compound had to be (the equally improbable) RDP, which today is known as "ribulose bisphosphate". Furthermore, Benson had found that, under optimum conditions, the concentration of the RDP was similar to the concentration of PGA; and that the compound's degradation yielded PGA [C_3] and phosphoglycolic acid [C_2] as the two major products. Benson concluded his discovery report by announcing that "a discussion of its [the ribulose diphosphate's] importance as a C_2 donor in the cycle for regeneration of the CO_2 acceptors will be published".[98]

A major advance had been made, which was elaborated in a publication of 1952 by Benson and a number of co-authors. Therein, they stressed the striking similarities between the five-carbon sugar and the seven-carbon sugar, and also speculated how the sugars might be related to the mysterious [C_2] primary acceptor of carbon dioxide:

> The close relationship between the structure of sedoheptulose and that of D-ribulose strongly suggests a synthetic relationship. The configuration of C-3 and C-4 of ribulose is identical with that of C-5 and C-6 of sedoheptulose. [...] None of these sugars is stereochemically related to glucose by a simple sequence of reactions. One of the functions of these compounds may be to serve as sources of 2-carbon molecules capable of accepting carbon dioxide to form phosphoglycerate during photosynthesis. [...] The fact that the two predominant carboxylations of photosynthesis result in C_3 and C_4 compounds leads one to expect a condensation of the C_3 and C_4 sugars to give sedoheptulose.[99]

Around the same time, the Berkeley Group had taken up extended kinetic studies of the radioactively labelled compounds. For this purpose, labelled bicarbonate solution was added at a certain point of time ($t = 0$) to steady-state photosynthesising algae in which the mass concentration of the compounds was assumed to be constant. Thereafter, a series of samples was taken at certain intervals, and the appearance of the radioactively labelled carbon was plotted as a function of time. In addition to the general appearance, the distribution of radioactivity in each compound was also followed. One of the main results of these kinetic studies was that no appreciable reservoir of labelled carbon between carbon dioxide and PGA [C_3] was apparently present in the cell, which again indicated that PGA was one of the first products of photosynthesis. The appearance curves of the compounds seemed to confirm that "there are two independent carboxylation reactions having different dependencies on carbon dioxide partial pressure".[100] Furthermore, it was found that fructose and sedoheptulose appeared to be labelled almost simultaneously (and rather early on), while the labelling of (newly synthesised) glucose and mannose lagged behind. In view of these findings, a new version of the photosynthetic cycle emerged, in which the seven-carbon sugar was assumed to precede the five-carbon sugar as the latter's precursor—however, this model's lifetime was extremely limited: immediately after it had been published, subsequent studies at Berkeley would produce sets of data that effectively rendered it obsolete, as the following section will show.

---

[98] Benson (1951, p. 2972).

[99] Benson and Calvin (1947, p. 713).

[100] Benson et al. (1952, p. 4481).

6.6.2 The Degradation Studies

The person who was working most intensively with the sedoheptulose was the organic chemist Lorel Kay, who was a member of the Berkeley Group from 1949 to 1954. Looking back on her work with a touch of self-irony, Kay described herself, in an interview with Moses, as the world's expert, at the time, of sedoheptulose's middle carbon atoms.[101] Kay had to decompose the sugar, step by step, carbon atom by carbon atom, in order to identify the radioactively labelled carbons and the sequence of their appearance in the chain: a laborious procedure, since it transpired that, although the principal technique was well developed, each of the carbons in the seven-carbon chain had to be separated by a different method. Although Kay did not have to develop the methods herself, she had to go and find them in the literature and then implement the procedures, which might involve, for example, preparing the necessary enzymes from rats. Another challenge was to get hold of substantial amounts of unlabelled sedoheptulose to be used as carrier molecules. As no sedoheptulose was then commercially available, Kay and other laboratory members had to go out on a *Sedum* collecting expedition—luckily, satisfactory amounts of this plant were found near Calvin's home. Back in the laboratory, the sedoheptulose had to be retrieved from the leaves: "That was kind of a fun process, to do the mashing, like making wine perhaps", Kay remembered; "Mash it down, boil it down, eventually crystallise it and then I had my carrier to put in with the sedoheptulose that had the radioactivity in it from the short-term photosynthesis."[102] This was certainly not the kind of organic chemistry that was usually carried out in university laboratories.

As prolonged as the procedure was, as uncertain were the results: "Sometimes you'd get two [carbons] together and then you'd get one of another [sugar] and your data wasn't as neat and clean as perhaps you might have wished it to be", Kay remembered.[103] This was all the more annoying as Kay knew very well how closely her work was being followed by the others in the laboratory. Kay's data, for example, were irreconcilable with the assumption that there were two different carboxylation reactions in the cycle, which was one of the principal assumptions of the group. She recalled how Calvin and others tried to talk her out of it: "So, there was almost, you know, a little pressure: 'Hey, are you sure that this data is perfectly good?' I would say, 'To the limits of how good I know it is, it says this and not that'".[104]

While Kay was the seven-carbon sugar expert, Anne Tolbert (at the time, Anne Harris) became the expert on five-carbon sugars. She joined the laboratory in 1951 as a graduate student. Originally, she had planned to become a high school teacher, but her college professor had strongly recommended that Tolbert do graduate work with Calvin. Tolbert soon became truly engrossed in her research project, which

[101] Moses and Moses (2000); interview with Kay, p. 20/5.

[102] Moses and Moses (2000); interview with Kay, p. 20/6.

[103] Moses and Moses (2000); interview with Kay, p. 20/5.

[104] Moses and Moses (2000); interview with Kay, p. 20/6.

turned out more challenging, more interesting and far more important than anybody had expected. It was assumed that, according to the model hypothesis favoured at that time, the labelling pattern of the ribulose would exactly match that of the sedoheptulose (since the latter was thought to be the precursor of the former). Tolbert said in retrospect: "They thought it was just some simple kind of mechanism where you just added two and took off two [. . .]. That was assigned to me because they knew what they wanted done."[105] Thus, to the people in the laboratory, the project given to Tolbert was a piece of standard work that was deemed to be appropriate to her skills, and of which the outcome was pretty much determined from the start. Yet, the data came out differently, and thus Tolbert also had to face the fact that her findings did not meet the group's expectations:

> It was one of the seminar sessions where I've just had to present this, you know, and everybody was really kind of uptight because these weren't the same [the data for the labelling of the five-carbon and the seven-carbon sugars]. There was not this 1:1 correlation. It was really hard for me to defend this but I said, "Well, I think I did it right". "This is the way I did it" and I told everybody. As it turned out, it was, I think, correct, and it was a more complicated cycle than they thought.[106]

Thus, Kay and Tolbert really had to fight their corner—after all, they were both only graduate students and, even worse, they were female. But when no obvious methodical mistake could be uncovered, the other members of the group sat down and thought about what this would mean for the modelling of the cycle. Tolbert stayed in the laboratory until 1953, and thus had the satisfaction of seeing how her data provided important clues to the eventual solution. However, by then she had married another graduate student of the group (Bert Tolbert) and when he got a job at Harvard, she took her Master's degree, instead of a doctorate, and left Berkeley and science (which later she regretted).[107]

The problem was not only that the labelling patterns of the five-carbon sugar and the seven-carbon sugar disagreed but they did so in such a strange way (see Fig. 6.7). In the RDP, the central carbon atom (No. 3) was labelled first, followed by carbons 1 and 2, while in the sedoheptulose monophosphate (SMP) the three central carbon atoms (3, 4 and 5) were the first to be labelled, with a slight tendency for number 4 to be labelled first. Interpreting these patterns was not straightforward. There were three possible ways of forming the seven-carbon compound: $[C_6] + [C_1]$, $[C_5] + [C_2]$ or $[C_4] + [C_3]$. The presence of PGA obviously provided a suitable $[C_3]$, so that the latter of these three options appeared the most promising at first glance. However, the labelling pattern of the three carbon atoms of PGA did not match either the top or the bottom positions of the sedoheptulose. The pattern of the RDP could not be detected in the sedoheptulose either, so that here too a direct precursor–product relationship

[105] Moses and Moses (2000); interview with Tolbert, pp. 28/2–3.

[106] Moses and Moses (2000); Interview with Tolbert, p. 28/3.

[107] See Moses and Moses (2000), interview with Tolbert, p. 28/7. In a later section of the interview (p. 28/9), Tolbert described how she took a long career break to raise a family and then eventually became an accountant.

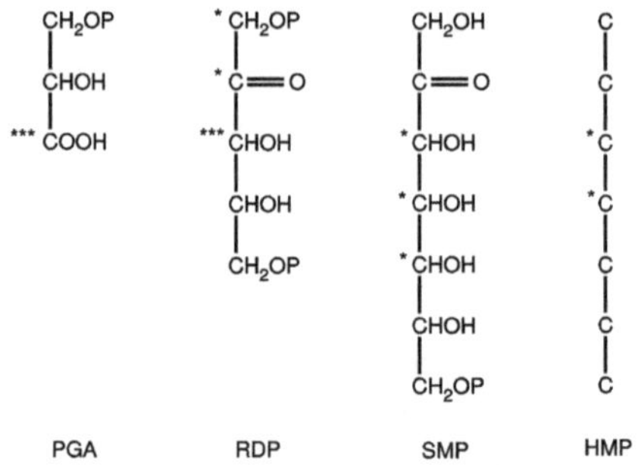

Fig. 6.7 The labelling patterns of phosphoglyceric acid (PGA), ribulose diphosphate (RDP), sedoheptulose monophosphate (SMP) and hexose monophosphate (HMP). (After Bassham et al. (1954, p. 1767)).

seemed doubtful. To find an alternative origin of the ribulose was equally puzzling. A priori several options were possible: it could be formed either from a hexose losing a carbon fragment, that is, $[C_6]-[C_1]$, or from smaller fragments combining with each other, such as $[C_1] + [C_4]$ or $[C_2] + [C_3]$. No hexose with a labelling pattern approximately equal to that of the ribulose was found, so that the first option was ruled out. Again, the PGA easily provided the $[C_3]$ fragment, and the labelling seemed sufficiently similar. But finding an appropriate $[C_2]$ remained problematic.

6.7 The Saturation Experiments

Parallel to the degradation work on the new compounds, a new type of experiment was devised in order to study the course of the pathway from a different angle. Rather than looking at the chromatograms as a static unit and trying to establish the linear sequence of events, the aim was to acquire a more dynamic understanding of the process, which included analysing the interaction of all the components. The capacity of the cycle's intermediate substances to become rapidly saturated with radioactivity was a useful starting point for determining the size of the reservoirs of these compounds and the way the system responded (by changes to the pool size of the compounds) to the variation of external variables, such as light.[108] The first of these experiments was carried out by the Swiss biochemist Peter Massini, who had come to the Berkeley laboratory in 1951 on an exchange visit for a year. This is how he remembered the experimental set-up for determining the pool sizes:

> I labelled the material for a certain time, a rather long time for what then was usual: a minute or several minutes. Then I switched off the light. Before and after switching off the light, I made [took] samples, from the lollipop, and made the chromatograms and everything. These [...] first compounds were then in a state where the labelling was all already at [a] maximum. What happened after switching off the light was an index for the concentration

[108] See Calvin and Massini (1952, p. 451).

> of these compounds, the changes of concentration of the compounds other than changes of the specific activity. That was the point.[109]

Thus, the idea was to observe the accumulation or depletion of compounds known to be involved in the pathway by cutting one of the pathway's sources. In this case, the source to be cut was the energy equivalents gathered from the photochemical reactions. Massini remembered that the whole project was of great interest to Calvin and that the two of them exchanged news on the chromatograms on a daily basis. What he found was rather exciting:

> When illumination is interrupted there appears a sudden great increase in the concentration of phosphoglyceric acid (followed by a slow decrease after 2 min), and an almost complete depletion of the diphosphate area [made up almost exclusively of ribulose diphosphate]. Analysis of the monophosphate area showed that the amount of sedoheptulose phosphate decreased also.[110]

In other words, when illumination was stopped two directly opposed responses were observed: the concentration of PGA went up, while, correspondingly, the concentration of RDP and SMP went down. This seemed to indicate that the three compounds were connected to each other in a reaction series: while the *formation* of PGA seemed to be independent of the light, the further processing of PGA "downstream" to RDP and SMP was dependent on the supply of energy provided by the photochemical reactions. Furthermore, the accumulation of one compound and the depletion of others were just the kind of effects one would expect of a cycle. And although everybody had for long suspected that there was a cycle, it was only from these experiments that the concept of a regenerative cycle was given conclusive empirical foundation.

The light experiment was complemented by a study of what happened when the supply of carbon dioxide to steady-state photosynthesising algae was reduced. This became the project of Alexander T. Wilson, a 20-year-old graduate student from New Zealand, who created a highly intricate experimental set-up: the "algal steady-state apparatus", as it became called. This apparatus allowed Wilson to keep the algae in a controlled physiological state and to collect series of samples in order to study very short exposure times (0.4–15 seconds) of the plant to $^{14}CO_2$. Wilson explained the purpose of the complicated apparatus as follows:

> Even when every attempt is made to control conditions under which algae are grown, the algae show daily variations in such properties as rates of CO_2 fixation and cell division. These considerations make it difficult to do experiments in which the results from different days must be compared on a quantitative basis. To overcome this difficulty the apparatus was designed to take small representative samples of algal suspension over short-time intervals from a system in which the external variables were under complete control. Use was made of recent advances in instrumentation to monitor continuously the variables, such as partial pressure of CO_2 and radioactivity.[111]

---

[109] Moses and Moses (2000); interview with Massini, p. 48/2. In this oral interview with Massini, his use of "switching out" the light has been corrected to "switching off" the light.

[110] Calvin and Massini (1952, p. 454).

[111] Wilson and Calvin (1955, p. 5949).

The "recent advances in instrumentation" referred to an infrared gas analyser and a corresponding recorder that continuously monitored the carbon dioxide concentration in the tanks. Before any samples could be gathered, Wilson had to cool down the whole set-up to a temperature of 6°C, which was the only way to get sufficiently fine-grained data on the reactions. Since one could not speed up the measuring process indefinitely, Wilson tried, alternatively, to slow down the reactions by having them run at a lower temperature. Samples were taken at short intervals, going from 1 to 0.03 % carbon dioxide and then in reverse. It was an ingenious apparatus, but it was impossible to run the series without the help of others. "I would set up the experiment and then everybody in the lab would help me for 30 min while I took all these measurements on the samples", Wilson recounted.[112] It was thanks to this experiment, in combination with the results of Massini and of the degradation studies, that the idea of a second carboxylation (leading to a four-carbon compound related to malic acid) was dropped. Wilson remembered that Calvin was one of the last to be convinced that the cycle really was made up of only one carboxylation reaction.[113] Yet, even more important was another finding of Wilson's:

> Perhaps the most striking result is the reciprocal relationship between PGA and RuDP [= ribulose diphosphate]. [...] As soon as the CO_2 pressure is dropped the PGA drops sharply and the RuDP rises sharply. The initial slopes of these curves, together with the fact that the other intermediates change more slowly, confirm that PGA and RuDP are related in a precursor-product relationship [...]. The results imply that RuDP is the actual CO_2 acceptor in photosynthesis, or alternatively is related to it by a vanishingly small reservoir, and that PGA is the first observable product of the carboxylation.[114]

This was the solution that finally brought to an end the relentless hunt for the two-carbon acceptor. There was no two-carbon acceptor. The substance was a five-carbon acceptor, RDP, which, immediately after carboxylation, was split into two halves, yielding two three-carbon molecules. This process had no known precedents in organic chemistry in fact, although it was easily construed on paper (carboxylation of the RDP at the second carbon atom, hydrogenation on the third carbon atom, and a subsequent splitting right between these two-carbon atoms), no direct evidence for the occurrence of these reactions was available at the time. The group simply judged that all the alternatives were even less likely to occur.[115] However, only a month after the publication of the "Path XXI" paper the group reported, in May 1954, the successful carboxylation of RDP to PGA in a cell-free system: obviously they had felt the urgent need to prove, at the very least, the actual in vitro existence of this crucial reaction step.[116]

Wilson's findings were used in the famous "Path of Carbon XXI" paper of 1954, dealt with in the next section, but a separate paper, published in 1955 in the *Journal of*

[112] Moses and Moses (2000); interview with Wilson, p. 13/7.

[113] See Moses and Moses (2000); interview with Wilson, p. 13/7.

[114] Wilson and Calvin (1955, p. 5952).

[115] Cf. Bassham et al. (1954, p. 1766).

[116] Quayle et al. (1954, p. 1766).

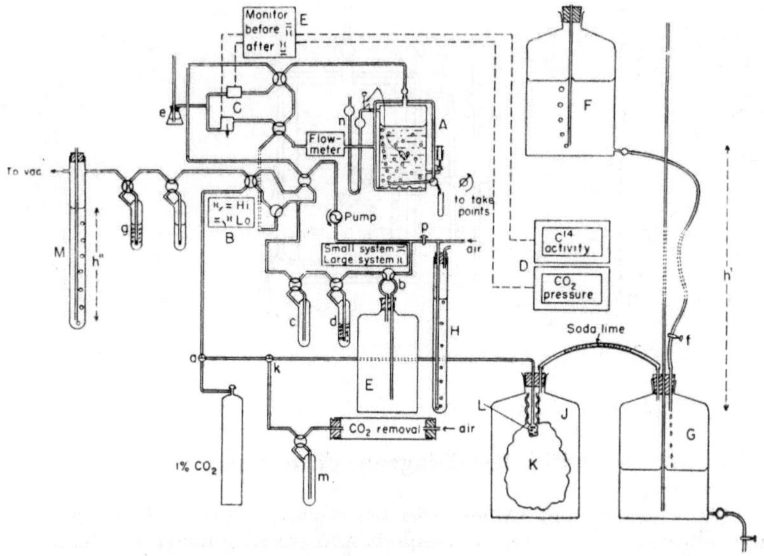

Fig. 1.—Diagram of apparatus for measuring transient phenomena.

Fig. 6.8 The drawing of Alexander T. Wilson's apparatus by Alice Holtham that was published, in 1955, in the JACS.

the American Chemical Society (JACS), gave more details on the actual technique.[117] Among other things, the paper contained a detailed illustration of the highly complex apparatus that Wilson had constructed—perhaps one of the most complicated drawings ever published in the journal (Fig. 6.8). Several members of the laboratory had unsuccessfully tried to talk Calvin out of having this "unintelligible masterpiece" published in its entirety, without being simplified.[118] The drawing was prepared by Alice Holtham (later Alice Lauber), one of the secretaries, whose office was next door to Wilson's apparatus; thus, she knew the set-up very well and had even been involved, at times, in taking measurements.[119] She remembered that completing this drawing exhausted her, and she was devastated when she learned that the journal's editors found it too big and wanted to compress it to the width of a text column. Encouraged by some members of staff, including Wilson, Holtham spontaneously drew a tiny fisherman perched with his rod on a tube leading into the algal steady-state reservoir, and converted one of the bubbles in the reservoir into a fish (Fig. 6.9). Her colleagues found this "add-on" hilarious, and urged Holtham to keep the drawing as it was when the paper was submitted to the journal, which she did. In its reduced

---

[117] See Wilson and Calvin (1955).

[118] Fuller (1999, p. 8).

[119] See Moses and Moses (2000); interview with Holtham, p. 23/8.

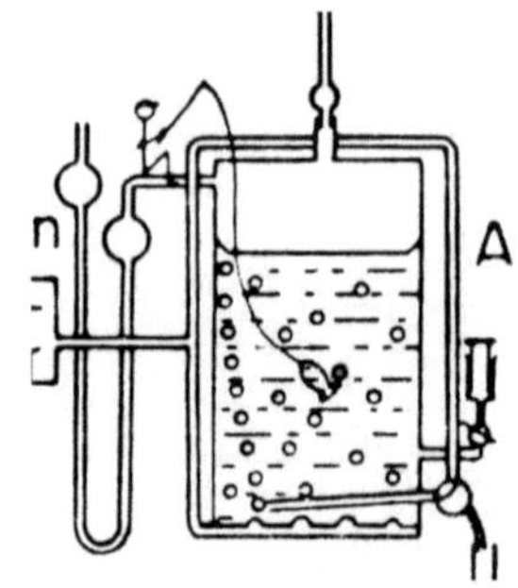

Fig. 6.9 Detail of Wilson's apparatus: the fisherman perched on a tube leading into the steady-state reservoir.

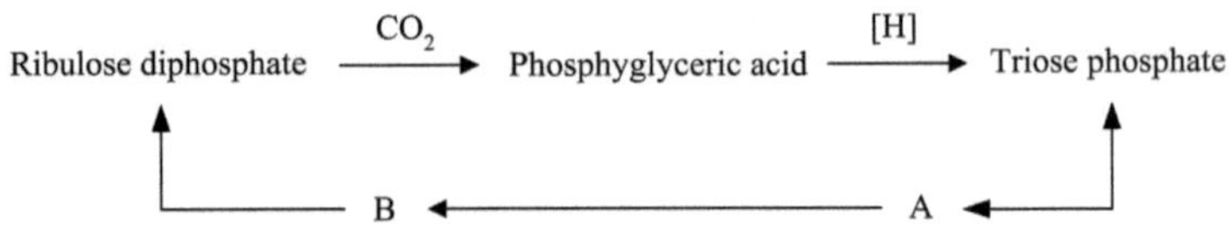

Fig. 6.10 A rough schema of the photosynthetic cycle. (After Bassham et al. 1954, p. 1766).

form the illustration came out so small that the fisherman is hardly discernible, yet it is infallibly there. As none of the reviewers complained about it, the drawing was published in its entirety by the JACS (perhaps the only humorous item ever to be published therein!), and it can still be found in many university libraries.[120]

6.8 The Path XXI Paper

The central paper in which the "definitive" model hypothesis was presented, based on the data gathered in the saturation experiments and in the degradation studies, was published in 1954. According to its number in the "path of carbon" series, it became known, within the Berkeley Group, as the "Path XXI" paper.[121] A rough scheme of the regenerative cycle, as formulated in this paper, is depicted in Fig. 6.10. RDP, the five-carbon sugar, was the acceptor of carbon dioxide, while the resulting six-carbon compound was highly unstable and immediately split into two molecules of PGA, or, alternatively, one molecule of PGA and one of phosphoglyceraldehyde. The further

[120] The story is well remembered by many members of the laboratory. See Fuller (1999, pp. 8–9), and Moses and Moses (2000), e.g., the interviews with Calvin, Holtham, Kay, Moses, Wilson, etc.

[121] See Bassham et al. (1954). Bassham thought that, although the group had had a "pretty good handle on the cycle" before 1954, he always regarded "Path XXI" as being the definitive publication. See Moses and Moses (2000), interview with Bassham, p. 7/10.

processing of the PGA required an energy input from the photochemical reactions (in the form of reducing equivalents) and resulted in the formation of triose phosphate. The latter then became the starting point both of the formation of hexose phosphates and of the cyclic regeneration of RDP.

They then needed to match this scheme with the labelling patterns that had emerged from the degradation studies. These data precluded the possibility that the RDP was entirely derived from a $[C_6] \rightarrow [C_1] + [C_5]$ split or a $[C_7] \rightarrow [C_2] + [C_5]$ split. "No five-carbon fragment of the hexose or the heptose molecules contains the same distribution of radiocarbon as ribulose" was the sober summarising statement, in accordance with the findings of Kay and Tolbert, which in effect brought to an end this part of the 1952 model.[122] The third obvious option—a three-carbon compound combined with a labelled two-carbon fragment—was discarded as well: the only way to obtain an appropriately labelled two-carbon fragment was through the breakdown of the labelled hexoses; the actual occurrence of such a breakdown, however, was considered highly unlikely in view of general biochemical knowledge. The solution that was finally arrived at was a very counterintuitive one; and it was only when all the more promising alternatives had been ruled out that the Berkeley Group came up with it. This is how it was introduced in the paper, which was authored by Bassham, Benson, Kay, Harris (later: Tolbert), Wilson and Calvin:

> Another way of accounting for the observed distribution of radioactivity, which seems quite plausible in view of the rapidly accumulating enzymatic evidence for the reverse reaction, is the formation of ribulose from sedoheptulose and triose. This reaction could result in the observed labeling. If the ribose-5-phosphate and ribulose-5-phosphate are then converted to RDP [ribulose diphosphate] the resulting distribution of labels would be that observed.[123]

Thus, the explanation for the confusing labelling patterns was that there were two different pathways that led to the formation of RDP: the first was when the two-carbon fragment from the top of the sedoheptulose was combined with triose phosphate; the second was the rearrangement of the remaining five-carbon fragment of the former sedoheptulose (see Fig. 6.11). At this point, the Berkeley team only had evidence for the occurrence of the reverse reaction of this process (see quotation above), but it was known that many enzymatic reactions are reversible, which, to some extent, justified their postulate.[124]

The next question to be solved was the point of origin of the sedoheptulose. The degradation studies had ruled out the two obvious options—that it was formed by a $[C_6] + [C_1]$ or a $[C_5] + [C_2]$ reaction. There was still the possibility that a $[C_4] + [C_3]$ pathway was constructed. However, it was clear from the labelling patterns that the reaction assumed in 1952 (the formation of a $[C_4]$ via the carboxylation of triose phosphate) was incompatible with the data, as Kay and Tolbert had eventually

[122] Bassham et al. (1954, p. 1767).

[123] Bassham et al. (1954, p. 1767).

[124] On the assumption of reverse reactions of known processes as a typical strategy within the biomedical sciences of the 1950s, see the argument in Scholl and Nickelsen (2015).

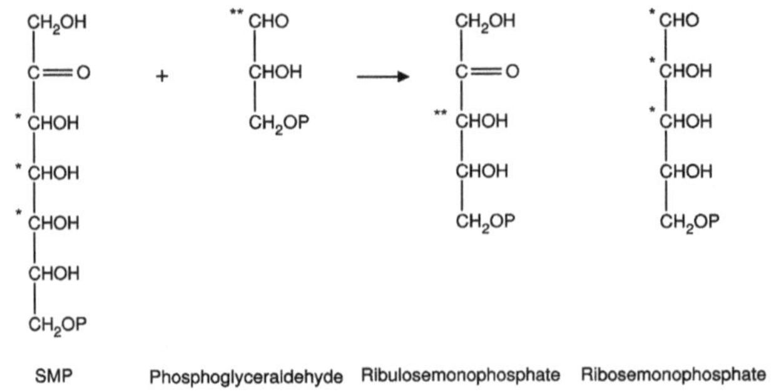

Fig. 6.11 The proposed mechanism for the formation of ribulose diphosphate (RDP): sedoheptulose monophosphate (SMP) combines with triose phosphate to form two pentoses that can easily be converted into RDP. (After Bassham et al. 1954, p. 1767).

persuaded Calvin and Benson. Therefore, by 1954 the Berkeley Group favoured a different pathway, which very nicely fitted the labelling patterns:

> The most likely source of the [C_4] fragment seems to be a [C_6] → [C_2] + [C_4] split. Trioses [C_3] could then react with [C_4] and [C_2] to give sedoheptulose and ribulose, respectively.[125]

The biochemical steps are spelled out in Fig. 6.12. The primary carbon dioxide acceptor in the cycle was a ribulose molecule [C_5], presumably RDP. Upon its reaction with carbon dioxide, two molecules of PGA [C_3] should almost immediately be formed, which were transformed into triose or triose phosphate molecules [C_3]. These were the central compounds of the cycle. This step required the input of reducing equivalents from the photochemical reactions. The triose phosphates would be the starting point for the formation of the hexose phosphates [C_6], the primary end products of photosynthesis. At the same time, they were also the starting point for the regeneration of the RDP as the acceptor of the next carbon dioxide. This was thought occur on two parallel paths. Triose phosphate, presumably phosphoglyceraldehyde, might react with glucose, upon which (through an unstable intermediate) two molecules would be released: an unknown [C_4] fragment and one molecule of ribulose monophosphate [C_5]. The [C_4] fragment would, in turn, react with another triose phosphate, presumably dihydroxyacetone phosphate, to form SMP [C_7]. The latter would also react with one molecule of triose phosphate, presumably phosphoglyceraldehyde, and (through an unstable intermediate) give rise to one molecule of ribulose monophosphate [C_5] and another of ribose monophosphate [C_5]. Both were readily converted into RDP, upon which the cycle would start again.

With this paper of 1954, the explanation of the dark reactions of photosynthesis had been reached. All the core problems had been solved: integrating the mysterious five-carbon and seven-carbon sugars; finding the carbon dioxide acceptor; and establishing a full sequence of reaction steps that was in agreement with relevant empirical

[125] Bassham et al. (1954, p. 1767).

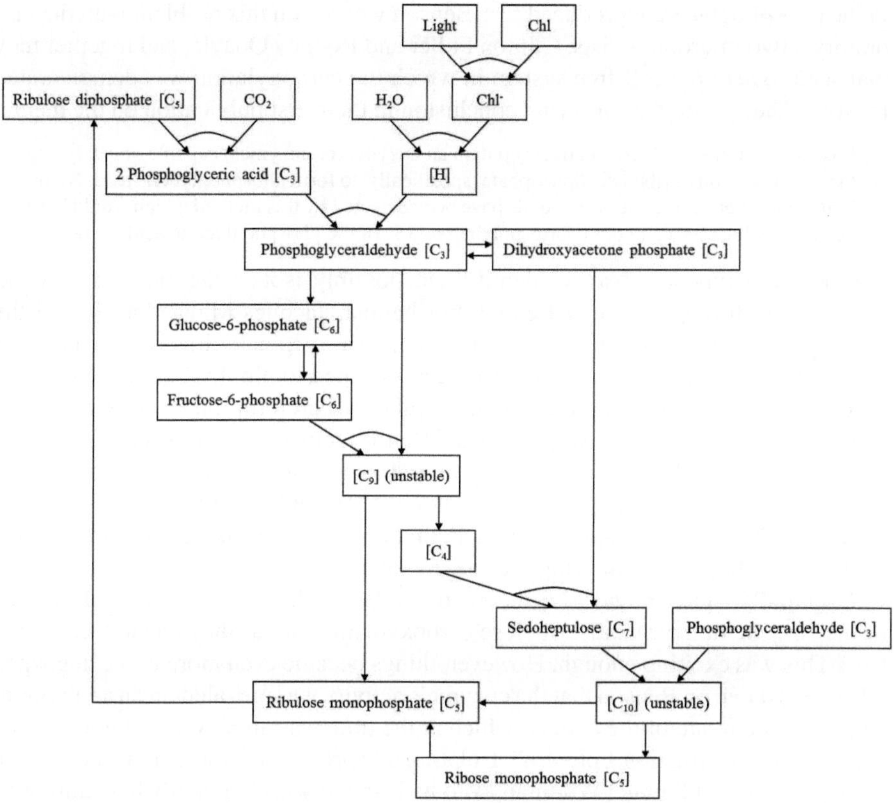

Fig. 6.12 The cyclic model of 1954. (Reconstructed from Bassham et al. 1954).

evidence. Moreover, this model, albeit with minor modifications and extensions, is still part of the common body of knowledge of the field of biochemistry today.

6.8.1 The Fraction 1 Protein

In addition to the group's effort of finding the biochemical intermediates from carbon dioxide to glucose, all major players pursued individual side projects on more specific problems. Benson's main work from 1953, for example, centred around his search for the carboxylation enzyme that catalyses the first major step of the photosynthetic carbon cycle. After a series of attempts, Benson finally succeeded in preparing sufficient amounts of the enzyme's substrate, RDP. As he later recalled, this was "the world's supply of the pure compound with which we could assay enzymatic carboxylation using $^{14}CO_2$ and measuring fixed radioactivity which would be in the phosphoglycerate produced".[126] Benson worked on this problem with the laboratory's two microbiologists, Clinton Fuller and Rodney Quayle, and together they managed to set up a cell-free system in which the carboxylation was demonstrated to work. They wrote the following conclusion in their first publication on the theme:

> It is clear that the [cell-free] extracts contain an enzyme (or enzymes) capable of catalyzing the carboxylation of ribulose diphosphate, specifically, to form phosphoglyceric acid. No intermediates between these compounds have been detected by this method which would have detected as little as an amount corresponding to 5 % of the phosphoglyceric acid formed.[127]

The next obvious step was to identify and possibly isolate the enzyme at work. With perfect timing, a young Belgian biochemist, Jacques Mayaudon, joined the Berkeley laboratory early in 1954. Although he was originally involved in a project on bacterial photosynthesis, which had been assigned to him by Calvin, Mayaudon preferred collaborating with Benson in the latter's search for the key enzyme in the cyclic path of carbon. Both Mayaudon and Benson remembered this time as a period of feverish work. Mayaudon was still required to contribute to the bacterial project, which he carried out during the day, but at night Mayaudon worked with Benson on extracts of New Zealand spinach (which, taxonomically speaking, was not spinach at all), in the hope of identifying the enzyme of enzymes.[128]

Eventually, Benson and Mayaudon succeeded in their project and had crystals of the enzyme in their hands: ribulose carboxydismutase, as they had agreed to call it.[129] This was exciting enough. However, things became even more unsettling when it began to dawn on Benson that the enzyme's activity was prevalent in an ammonium sulphate precipitate of the extract, which at the time was already well-known in the literature as the "fraction 1 protein". Isolating this protein was the achievement of the plant physiologist Samuel Wildman, who had started working on it while employed in the laboratory of James Bonner at Caltech. Benson frequently visited Caltech and was well familiar with Wildman and his work.

Wildman had been Bonner's postdoctoral student in the years 1944–1950.[130] It was known that the proteins of the leaves of green plants could be separated into two large classes: green insoluble proteins and nongreen soluble proteins. Following this division, Bonner and Wildman decided to focus on the soluble ones. The techniques for studying proteins were then rather primitive. However, by adding increasing amounts of ammonium sulphate to the leaf extracts, Wildman and Bonner found a voluminous precipitate that they called "fraction 1", in order to differentiate it

---

[126] Benson (2002a, p. 45).

[127] Quayle et al. (1954, p. 3611). On Quayle, see Kornberg (2006); see also Fuller (1999).

[128] See Moses and Moses (2000), interviews with Mayaudon and Benson. See also Benson (2002a).

[129] Ning Pon recalled that there were endless discussions about what to call the enzyme. See Moses and Moses (2000, p. 9/3). In the same vein, Benson wrote on a Christmas card to Warburg: "An enthusiastic belgian [sic], Mayaudon, and I have finally succeeded in purifying the carboxylation enzyme. Should we call it 'Photosynthase' or 'ribulose diphosphate carboxylase'? It seems to be a major leaf protein." Archive of the BBAW, NL Warburg 114. Card undated.

[130] See Wildman (1992, 1998, 2002) for autobiographical accounts.

from what remained in the supernate (which was called “fraction 2” and could be collected by making a concentrate of the extract through evaporation). Wildman was able to investigate this protein on one of the first moving-boundary electrophoresis instruments: the so-called Tiselius apparatus, which had just been constructed at Caltech in the neighbouring laboratory of Linus Pauling.[131] Wildman remembered his finding as follows:

> The result was very intriguing. Without ammonium sulfate fractionation, the cytoplasmic proteins migrated as if 70 % of their content consisted of a single protein. When the ammonium sulfate cut labeled Fraction 1 was tested, it migrated as a single, electrophoretically homogeneous component. Furthermore, the minimal spreading of the boundary during electrophoresis suggested the protein to be of high molecular weight.[132]

Four years later, the protein was studied further with the help of the ultracentrifuge, which, in the meantime, had also appeared in Pauling’s laboratory. It turned out that according to the centrifuge pattern a minimum of 50 % of the soluble spinach leaf proteins was made up of a large molecular weight component, with a weight of about 600,000 (based on a sedimentation coefficient of 18 Svedberg units). Furthermore, this very protein could be demonstrated to be present in a host of other plants. Being thus homogenous according to the most rigorous tests available at the time, the fraction 1 precipitate henceforth became known as the “fraction-1 protein”.

No wonder that Benson was thrilled when he realised that his carboxydismutase precipitated along the same lines as the mysterious protein “fraction 1”; he was to remember this moment of realisation as one of the most exciting times of his life. Wildman also clearly remembered Benson’s telephone call, in which the latter broke the good news to him. Together with Mayaudon, Benson typed out a manuscript, in the format of a “Letter to the Editor” of the JACS, in which they described their finding and mentioned that their enzyme closely resembled Wildman’s fraction 1 protein. However, this was never to be published. Benson recounted that after he had submitted the manuscript to Calvin for the usual “inhouse-review”, it disappeared. “The results of our tremendous efforts could have been published in 1954, but first appeared in print late in 1957 with no mention of the fraction 1 protein. Possibly Melvin did not recognise its importance—since he was unfamiliar with and disinterested in the work of Sam Wildman at Caltech”, Benson reminisced.[133] Very soon thereafter Calvin asked Benson to leave the laboratory by the end of 1954.[134] Benson found a position at Penn State University, and later moved on to the Scripps Institution of Oceanography in La Jolla (California). He continued to make important

---

[131] Arne Tiselius, the inventor of this apparatus, received the 1948 Nobel Prize in chemistry for analytical work done with the help of this instrument.

[132] Wildman (2002, p. 245). The results were published in Wildman and Bonner (1947).

[133] Benson (2002a, p. 46). The paper referred to is Mayaudon et al. (1957); see also Mayaudon (1957).

[134] See Benson (2010) for an account of his “Last days in the old radiation lab”. In 2012, an interview with Benson on these events was recorded and posted to the web. For a description of the interview as well as the link to the video, see Buchanan and Wong (2013).

contributions to science, while the greatest reward for having discovered the path of carbon in photosynthesis, the Nobel Prize in chemistry, was awarded to Calvin alone (in 1961).[135]

In the mean time, the research team led by the biochemist Bernard L. Horecker at the National Institutes of Health (NIH) succeeded in preparing purified RDP carboxylase, although they did not yet identify the enzyme as being the fraction 1 protein[136] (This, incidentally, demonstrates the ingenuity of Benson's conclusion, which was by no means self-evident). However, when Wildman's associates Robert Dorner and Albert Kahn learned that the enzyme in question had a sedimentation constant of 18 Svedberg units, they quickly inferred that the carboxylation enzyme and the fraction 1 protein had to be one and the same.[137] After a sequence of ever more complicated, jaw-breaking names for the enzyme had been tried out, in 1979 a name caught on, which is still used today: "RuBisCo", which is an acronym for "Ribulose-1,5-bisphosphate carboxylase".[138]

6.9 The Heuristics of the Berkeley Group

One of the reasons most frequently given for the huge success of the Berkeley Group is that the methods used by the team were first rate. The Berkeley team had an excellent infrastructure to its disposal (thanks to almost unlimited financial resources). It had the advantage of being one of the very few research groups at the time that could work with carbon-14 (provided by the cyclotron). And the group had the far-sighted intelligence to recognise the potential of paper chromatography, as well as the skills to adapt it to the compounds under study. What made this group so special at the time, however, was the highly interdisciplinary composition of the staff and the spirit of collaboration that prevailed in the laboratory. "They were tied together by a desire to solve this problem, this big, hot problem, in just one little group", the biochemist Nathan E. Tolbert recalled.[139] Tolbert and many others felt that they never again experienced such a strong sense of group identity and such collective resolve. Yet, after the decisive 21 "path of carbon" paper had been published, this unique atmosphere disappeared, although the project still continued. As Bassham pointedly

---

[135] Fuller (1999, pp. 9–10), pointedly summarised the disconcerting feeling with which one is left after having read Calvin's autobiography, Calvin (1992), in which there is no single reference to Benson. Also thereafter, Calvin consistently rejected the fact that Benson played a crucial role in the photosynthesis project; cf., e.g., Moses and Moses (2000), interview with Calvin, p. 1/30. This, of course, is a cruel distortion of history.

[136] See Horecker et al. (1954) and Weissbach et al. (1956).

[137] See Dorner et al. (1957).

[138] On the history of Rubisco, see also Portis and Parry (2007).

[139] Moses and Moses (2000), interview with N. Tolbert, p. 29/20.

1948:
PGA first product.
Ion exchange columns.
Glucose formation as the reversal of glycolysis.
Cyclic regeneration of the primary acceptor
Same process as in heterotrophic tissue
(i.e. reversal of TCA: succinic acid, fumaric acid, malic acid)
Two carboxylation events.
Acetic acid as the 2-carbon acceptor

1950:
PGA first product (confirmed).
Paper chromatography; radioautography; degradation studies.
Glucose formation as the reversal of glycolysis.
Cyclic regeneration of the primary acceptor
Unlike the process in heterotrophic tissue
(TCA intermediates precluded)
Two carboxylation events.
Unknown 2-carbon acceptor

1952:
PGA first product (confirmed).
Paper chromatography; radioautography; degradation studies.
Glucose formation as the reversal of glycolysis.
Cyclic regeneration of the primary acceptor
Process includes SMP, RuDP.
SMP assumed to split in RuDP and 2-carbon acceptor
Unknown 4-carbon compounds involved (erythronic acid?)
Two carboxylation events.
Unknown 2-carbon acceptor

1954:
PGA first product (confirmed).
Paper chromatography; radioautography; degradation studies.
Saturation experiments.
Glucose formation as the reversal of glycolysis.
Cyclic regeneration of the primary acceptor (confirmed)
Process includes SMP, RuDP.
RuDP is primary acceptor (5-carbon compound!)
One unknown 4-carbon compound involved
One carboxylation event.
Two different pathways for regenerating the acceptor
(via SMP combined with PGA)

Fig. 6.13 Synopsis of the Berkeley Group's sequence of models.

remarked: "It wasn't ever as exciting a time as [it] was during the mapping of the path of carbon in photosynthesis."[140]

According to many former members of the laboratory, the personal coherence of the group was endorsed by the highly collaborative organisation of the project. There was a clear hierarchy: a few central figures coordinated the activities and determined the general agenda—Calvin, Benson and Bassham—while a large circle of temporary or even occasional contributors was involved in the several subprojects, without any prospect to ascend within the group. This structure undeniably limited the researcher's independence, which not everybody was pleased with.[141] However, within the defined area of research, bottom-up initiatives flourished. As group members discussed the latest results and hypotheses, new subprojects emerged, and according to all reports, these initiatives were strongly encouraged, even if new equipment had to be installed in order to realise the ideas and new methods needed to be learned. This was the environment in which the path of carbon was, finally, elucidated. Figure 6.13 shows a synopsis of the four main cyclic models that were proposed in the years 1948–1954. In the following, I shall briefly recall this sequence and reflect on the group's model building heuristics.

If one reduces the models to their essential assumptions, a number of constant parameters emerge that had been assumed from the start. Among these are the assumptions that PGA is the first product of photosynthesis, that glucose formation is glycolysis run in reverse and that a cyclic regeneration of the primary carbon dioxide acceptor takes place. In terms of methods, the greatest change came around

[140] Moses and Moses (2000), interview with Bassham, p. 7/17.

[141] See Moses and Moses (2000), Introduction by Moses, p. Intro-4. This point was emphasised also by others, see, e.g., the interviews with Park, p. 25/11, and Benson, pp. 12/29–30.

1948, that is, when the group abandoned ion-exchange columns in favour of paper chromatography and radioautography.

The details of the cycle, of course, changed dramatically. To begin with, there was the standard hypothesis, which proposed that the fixation of carbon dioxide took the same path in photosynthetic and heterotrophic tissues. This was followed by a modified version of the standard path that involved oxaloacetic acid. And even the model variant of 1950 did not completely discard the close analogy between respiration and photosynthesis. Drastic changes to this first approach were required when two completely new compounds had to be incorporated: RDP and SMP, none of which were part of the respiration pathway. But the new attempt was quickly made obsolete by the crucial sets of data that became available after 1952. First, there was the fact that the straightforward hypothesis of the compounds' role in the pathway (that the sedoheptulose was the direct precursor of the ribulose) was incompatible with the curious labelling patterns that Kay and Tolbert had found. Second, there were the data from the saturation experiments of Massini and Wilson, which indicated that RDP operated as the primary acceptor of carbon dioxide. The latter resulted in the hypothesis that two carboxylation events were involved being abandoned, while it was only in 1954 that the eventual solution was developed.

Considering at these and other examples how the group generated their model hypotheses three heuristic moves stand out: (1) the transfer of knowledge from other contexts (from respiration to photosynthesis, from in vitro to in vivo); (2) the assumptions that all biochemical reactions also run in reverse (which, to some extent, is a special case of transfer); (3) the recombination of structural formula on paper.[142] Despite their simplicity, these strategies were evidently powerful.

The transfer strategy loomed large already from the very beginning. The group started with two standard assumptions: *first*, that photosynthetic carbon reduction was the reversal of respiration; and *second*, that it involved a metabolic cycle. Glucose formation was conceptualised by the group as the reversal of glycolysis; and the process of carbon dioxide fixation was conceived of as the reversal of the tricarboxylic acid cycle. This approach seemed well confirmed by the fact that non-photosynthetic carbon dioxide fixation followed exactly this path. In the end, the path turned out not to exist in algal photosynthesis. Yet, about a decade thereafter a divergent path of photosynthetic carbon reduction was found to be present in many groups of plants (predominantly grasses), which *did* involve a reversal of parts of the tricarboxylic acid cycle with malic acid as its central intermediate. The first indication of the existence of this pathway was found when investigating sugar cane plants in Honolulu, in the late 1950s, but the findings were published only years later.[143] This prompted a group in Brisbane, Australia, to pursue the question; and further research revealed the aforementioned pathway that today is called "C_4 photosynthesis", since it includes

[142] See on these heuristics Scholl and Nickelsen (2015).

[143] See Kortschak et al. (1965).

a number of [C_4] organic acids.[144] These later discoveries speak to the principal soundness of the first working hypotheses of the Berkeley Group, although for their experimental organism—unicellular freshwater algae—they proved inaccurate. Even an initially discarded hypothesis turned out to have picked out an actual pathway in nature.

The assumptions that biochemical reactions run in both directions also had a prominent place as earlier examples demonstrate. The decisive turning point for the final model suggestion in 1954 was the recognition that RDP was formed in photosynthesis on two different pathways. And although this might appear as simply piecing together a jigsaw puzzle, it is significant that the group only came up with the idea of the second pathway (combining SMP with triose phosphate) after evidence for the reverse reaction had been found by other research groups, for example, the team working around the biochemist Bernard L. Horecker at the NIH.[145] This recognition, combined with the knowledge that many enzymatic reactions were capable of running in both directions, justified, in the eyes of the Berkeley team, this unusual formation path.

As a third type of heuristics, there is the approach to recombine structural formula on paper. Consider, for example, how the group dealt with the finding of unexpected five- and seven-carbon sugars. In order to integrate these into a sensible pathway, the group turned to biochemical knowledge on structurally similar compounds and inferred possible reaction steps. This was the first time that they tentatively introduced reactions that so far had not been reported in the literature. When strong evidence suggested that the five-carbon sugar was the primary carbon dioxide acceptor, the group recombined everything they knew on other compounds of the regenerative cycle and suggested the eventual solution.

Finally, it is remarkable that even in the Berkeley Group, which was known not to be shy of bold assumptions, hypothesis generation was marked by deep conservatism, despite its highly original results. It began with attempts to transfer the causal scheme of glycolysis, a well-understood, central achievement of early twentieth century biochemistry, and other parts of the respiratory pathway. Only when this proved insufficient did it become necessary to try more complex—but qualitatively similar—transfers from other bodies of previous knowledge. In 1950, to name a second example, a contraction of the model was required in view of the continued failure to detect the expected intermediates of the tricarboxylic acid cycle in photosynthesising plants, namely malate, fumarate and succinate. However, the modifications were as minimal as possible: the compounds in doubt were simply replaced by another four-carbon compound (oxaloacetate), which in a well-known reaction series was connected to the eliminated compounds—and so the group still had the option of

---

[144] Cf. Hatch and Slack (1966); Hatch et al. (1967); retrospective accounts are given in Hatch (1992), Hatch (2002). Around the same time the occurrence of a "reductive carboxylic acid cycle" in some bacteria was reported, which in fact is a tricarboxylic acid cycle run in reverse in order to produce organic compounds from carbon dioxide in water; see Evans et al. (1966).

[145] The following references were cited in Bassham et al. (1954): Axelrod et al. (1953), Horecker and Smyrniotis (1952, 1953) and Racker et al. (1953).

returning to their original hypothesis. More radical hypotheses were only developed when empirical evidence demanded it; that is, when entirely unexpected substances turned up and needed to be integrated into the existing pathways.

But also these were eventually dealt with. It was one of the major advantages of the Berkeley Group that its composition and cooperative nature allowed Calvin and Benson to follow up almost every lead, many of which turned out to be blind alleys.[146] A great many intermediate model hypotheses were formulated, most of which never made their way onto paper. "Yeah, schemes always came up and were abandoned", Kay confirmed in her interview.[147] At any one time, a whole set of model alternatives were being circulated and heatedly discussed, and the papers that were published represented only the tip of the iceberg. The principle of plurality that was introduced in chapter 2 for a whole community was, in this case, eagerly practiced within one single research team. The intensive debate within the group was dearly needed, as the Berkeley Group very soon outmatched all other laboratories in the attempt to elucidate the photosynthetic path of carbon.[148] Interestingly, the dominance of the Calvin–Benson team contrasts greatly with the situation in elucidating the light reactions of photosynthesis, to which numerous actors from very different places contributed. This, then, forms the focus of the last chapter of this book, which deals with the discovery and elucidation of photosynthetic ATP production, the generation of reducing power and the participation of two light reactions and two pigment systems in photosynthesis.

---

[146] Perhaps the most prominent of the red herrings that were followed up in the laboratory was Calvin's "thioctic acid theory" on which he worked intensely for more than two years. See Barltrop et al. (1954) and Calvin (1954) for central papers on this theme.

[147] Moses and Moses (2000), interview with Kay, p. 20/6.

[148] For some time, the University of Chicago group, which included Hans Gaffron, Edmund Fager, Jerome Rosenberg and Allan Brown, had been competing with Berkeley; but as they made much slower progress than their highly endowed Berkeley colleagues, they eventually dropped the project.

Chapter 7
Elucidating the Light Reactions (1950–1961)

The 1950s are sometimes referred to as the Golden Age of photosynthesis research.[1] The decade saw a dramatic increase in advances, concerning all aspects of the mechanism; and by the end of the period, that is, in 1961, researchers were able to propose an outline for the "light reactions" stage of photosynthesis, which is still current today: that two photoreactions involving two different pigment systems operate in series. Melvin Calvin, meanwhile, was awarded the Nobel Prize in chemistry for having elucidated the "dark reactions" stage of photosynthesis via the Calvin–Benson cycle. This chapter looks at how the studies in the light reactions stage of photosynthesis undertaken in the 1950s finally culminated in the two photoreactions, two pigment systems model. This model was reached by way of a number of research paths, the results of which were surprisingly convergent. The 1960 paper by Robin Hill and Fay Bendall, in which they presented what became known as the "Z-scheme", is best known to the public; however, its thermodynamic argument crucially required complementary studies by others, including the two biophysicists Louis N. M. Duysens from the Netherlands and Horst T. Witt from Germany. Also noteworthy are the early contributions made by Bessel Kok; and Robert Emerson's finding of the enhancement effect in 1957 (described in chapter 5) provided not only the incentive for many of these studies but also a convincing argument for the accuracy of the model. Since photosynthesis research in this decade became highly diversified, and the entanglement of the different strands has never been reconstructed before, large parts of the chapter are devoted to describing the course of events before analysing it.

One symptom of the dramatic changes in approach and method that occurred during this period was the fact that the technique of manometry, which had been introduced to photosynthesis research by Otto Warburg in 1919, lost its dominant position in physiological and biochemical laboratories and was superseded by the technique of spectrophotometry. It did not take long before every well-equipped laboratory was able to monitor changes in the absorption spectra of biologically

[1] See, e.g., Krogmann (2000).

K. Nickelsen, *Explaining Photosynthesis,* History, Philosophy and Theory of the Life Sciences 8, DOI 10.1007/978-94-017-9582-1_7

significant molecules by means of a range of spectroscopic instruments.[2] These minute absorption changes—shifts in the wavelengths of the absorption maxima—frequently reflected electric fields arising from charge separations, that is, oxidations and reductions; and interestingly many of these changes were light induced. Thus, photochemically driven redox reactions involving chlorophyll *a*, cytochromes and other compounds could be "directly" observed. These observations became even more informative when fluorescence data were also taken into account, given the fact that fluorescence or photosynthetic utilisation are the alternative fates of excitation energy in the chlorophyll. This provided a far more detailed basis from which to draw inferences on the biochemical and biophysical foundations of the process than all other previous techniques. Eugene Rabinowitch once said that the replacement of manometry by spectroscopy "was comparable to looking under the hood of a car in order to find out about its mechanism, as compared to studying its gas exchanges".[3] (Warburg, incidentally, never acknowledged the usefulness of spectroscopic methods.)

The main challenge at the time, to which the new spectroscopic methods were employed, was to explain how the light reactions of photosynthesis provided the necessary driving force for the dark reactions of the process (if they did so at all). It transpired, for example, as described in the previous chapter, that many compounds involved in the dark reactions were phosphorylated. This indicated that energy-rich phosphate bonds, which were known to be frequently provided by small molecules of adenosine triphosphate (ATP), were involved in driving these reactions. It was known that ATP was formed in the mitochondria and was the final, energy-rich product of respiration; but it was completely unknown how this was related to the mechanism of photosynthesis, and whether or not there were alternative ways of forming ATP specific to plants. The second open question concerned the reducing agents required for carbon dioxide reduction. In view of the processes in respiration, photosynthesis researchers suspected that at least one of the two coenzymes diphospho nucleotide (DPN) and triphospho nucleotide (TPN) was involved, as they were known to play a crucial role in other hydrogen-transfer reactions of metabolism.[4] But up to 1950, no one had observed any connection between these molecules and photosynthesis. How the different parts of the jigsaw puzzle built up and were pieced together almost simultaneously in several different laboratories: these are the main threads of the narrative of this chapter.

[2] Clayton (2002); see also Reinhardt (2006) for a study of the advancement of spectroscopy in the chemical disciplines.

[3] Quoted in Duysens (1989, p. 68).

[4] The involvement of ATP and TPN or DPN was made highly probable after phosphoglyceric acid (PGA) and triose phosphate had been assumed to be the central intermediates in glucose formation: it was well-known at the time that PGA could only be reduced to triose phosphate, with the help of pyridine nucleotides, such as TPN and DPN, if it was first phosphorylated by ATP.

7.1 The Synthesis of Reducing Power

By the 1920s, it had become clear that the reduction of an organic compound could be regarded as the acceptance of either hydrogen or electrons, while a compound's oxidation was, vice versa, the loss of one or the other. This was the basis for Albert J. Kluyver's theory, which cast all metabolic reactions as hydrogen transfers (see Chapter 4). The tendency of a compound to accept electrons from other reactants became measured in terms of "oxidation reduction potentials" or "redox potentials" for short.[5] If a system A/A^+ had a more positive potential than a system B/B^+, this meant that A would have a strong affinity to electrons, and, hence, tended to be in the reduced state (A), while B would preferentially donate electrons and hence rather become oxidised (B^+). In combination of the two, A^+ would act as the oxidising agent, since it tends to withdraw electrons from B, while B would act as the reducing agent, since it tends to transfer electrons to A^+. The transfer of electrons between these two chemical systems would determine the redox potential, which was measured in "volts" (V) or, more aptly, "millivolts" (mV).

Carbon dioxide was known to be a very stable compound: the carbon is in its maximally oxidised state, so that a strong reducing agent (with a negative redox potential) was required in order to induce any changes. Obvious candidates for this role were the two "coenzymes" that were found to act in many organisms as cofactors of metabolic reactions. Originally, these compounds were called, rather unimaginatively, coenzymes I and II. In the course of time they were identified as being diphosphopyridine nucleotide (DPN) and triphosphopyridine nucleotide (TPN).[6] It was suspected, long before it could be demonstrated, that at least one of these two cofactors, DPN and TPN, also collectively termed "pyridine nucleotides", was involved in photosynthetic carbon dioxide reduction. The next obvious assumption was that these pyridine nucleotides were reduced in the photochemical reactions, utilising the incident light energy. Afterwards the resultant reducing equivalents might be used in the thermochemical part of the process, in which carbohydrates were formed. It was only in 1951, however, that evidence was presented that lent more support to this assumption than mere plausibility: no less than three American research groups independently discovered that DPN and TPN were reduced by illuminated chloroplasts.[7] In other words, the two coenzymes could be used as "Hill reagents", which are the oxidising agents in the Hill reaction (see Chapter 4). This observation was surprising, since up to then only electron acceptors with redox potentials more positive

---

[5] Whenever the term is used here, it refers to the "midpoint" or "equilibrium" redox potential.

[6] Today they are known as nicotinamide adenine dinucleotide, $NAD^+/NADH$, or nicotinamide adenine dinucleotide phosphate, $NADP^+/NADPH$. However, in order to keep the text consistent with the historical quotations, the former, outdated nomenclature has been used throughout this chapter.

[7] All three groups—headed by Wolf Vishniac and Severo Ochoa, by Leonard J. Tolmach and by Daniel I. Arnon—published their findings as so-called Notes in the journal *Nature*: Vishniac and Ochoa (1951); Tolmach (1951a) and Arnon (1951).

than about +40 mV were known to be reduced by isolated chloroplasts, while the standard redox potential of TPN/TPNH was −320 mV! Chloroplasts were obviously able to generate a much more electronegative redox potential than had previously been thought, which implied that they were able to drive many energy-requiring reactions in plants.

The first account of these findings, published on 12 May 1951, was a paper jointly authored by the microbiologist Wolf Vishniac, then still a postdoctoral student at Stanford, and the biochemist Severo Ochoa of the New York University School of Medicine, who would later win the 1959 Nobel Prize in physiology or medicine. Vishniac and Ochoa reported that they had observed the reduction of DPN and TPN by illuminated chloroplasts. At the same time, the authors wrote, molecular oxygen was released. The experimental trick that Vishniac and Ochoa applied was to "trap" the photoreduced pyridine nucleotides (which were very unstable), by coupling the photoreduction process to the formation of a stable reaction product, so that any energetically favoured back reactions were prevented from occurring. To this effect, Vishniac and Ochoa added the so-called malic enzyme to the suspension: an enzyme of the cell's cytoplasm, which catalysed the carboxylation of pyruvic acid to malic acid—a reaction that was known to be dependent on the presence of reduced DPN or TPN. The fact that this reaction ran smoothly in the illuminated chloroplasts was taken as evidence for the light-driven reduction of at least one of the two coenzymes (see Fig. 7.1).[8]

Similar results were reported four weeks later, on 9 June 1951, in a paper written by Leonard J. Tolmach of the University of Chicago. Tolmach believed that not only was TPN reduced in the chloroplasts; it also catalytically promoted oxygen production under these circumstances: "Addition of small amounts resulted in yields of oxygen 30–40 times the equivalent of the added triphosphopyridine nucleotide", he wrote. The same was observed when DPN was added, and, as Vishniac and Ochoa did, Tolmach agreed "that pyridine nucleotides might be reduced photochemically by illuminated chloroplasts and that these reduced co-enzymes could be utilized by enzyme systems for reductive fixation of carbon dioxide".[9] This was again confirmed in a third paper on this topic published on 23 June 1951 by Daniel I. Arnon of the University of California, Berkeley; in addition, Arnon had found that TPN was more easily reduced by chloroplasts than DPN.[10]

After this series of mutually reinforcing papers, there could no longer be much doubt that chloroplasts were capable of photoreducing pyridine nucleotides and, thereby, of developing a negative redox potential of more than −320 mV. Controversial opinions prevailed, however, as to the implications of this observation for the

[8] Vishniac and Ochoa (1951). Vishniac and Ochoa suspected that the prevalence of these back reactions explained the failure of earlier researchers to find compounds with a relatively negative redox potential reduced by chloroplasts.

[9] Both quotes: Tolmach (1951a, p. 947). For the idea that TPN or DPN might be used in carbon dioxide reduction, Tolmach referred to Ochoa et al. (1950) and Ochoa (1950).

[10] Arnon (1951). Jagendorf (1956) confirmed that TPN was reduced in preference to DPN.

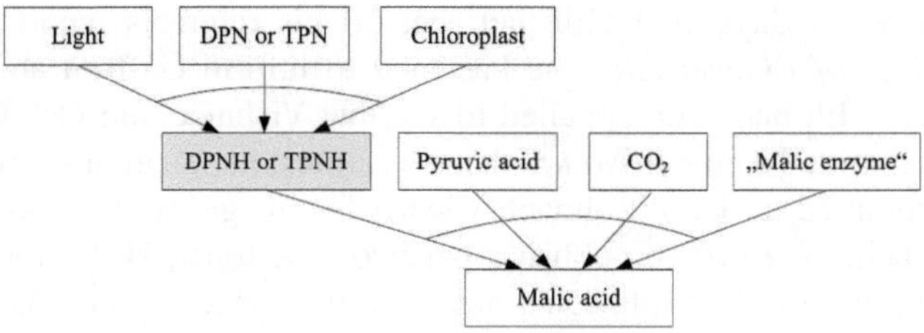

Fig. 7.1 The process underlying the experiment undertaken by Vishniac and Ochoa (1951). The production of malic acid by the "malic enzyme reaction" is dependent on the presence of reduced DPN or TPN, from which it was inferred that they were produced in illuminated chloroplasts.

mechanism of natural photosynthesis in living cells. The hope was, of course, that chloroplasts were also able to use pyridine nucleotides as oxidants for the splitting of water in plants; and that plants would thus be able, through appropriate (although still unknown) enzyme systems, to use the reduced pyridine nucleotides for the reduction of carbon dioxide to carbohydrates. This seemed all the more plausible as DPN and TPN could clearly be used to reduce carbon dioxide while malate was being formed (in the course of the "malic enzyme" reaction described above). However, one of the obvious points of criticism was the fact that, in all three laboratories, the reduction of the pyridine nucleotides was observed at rates that were about 50 times lower than would be required for the process of photosynthesis in plants to take place.

An interesting story lies behind the publication of these three papers. At the time Tolmach was working with Hans Gaffron's group at the University of Chicago (Gaffron became director of the Photosynthesis Laboratory after James Franck had retired in 1947). On 2 February 1951, Gaffron sent a draft of Tolmach's note, which had just been submitted to *Nature*, to Robin Hill, requesting the latter's opinion.[11] Hill replied 10 days later: "You can imagine how delighted I am to hear that a bridge head is now established between the chloroplasts and the coenzymes". Hill had longed to see evidence for the fact that the "Hill reaction" was, in fact, related to photosynthesis not only in terms of oxygen release but also carbon dioxide reduction. However, despite his general delight, Hill seriously advised Gaffron to withdraw the note in its current form, and write instead a more extended paper which included the precise experimental protocol and the actual data. Without these details, Hill argued, the information given in the note was almost meaningless. This blunt comment was welcomed by Gaffron, who in his next letter admitted that he had not been entirely happy with the note either.[12] Tolmach quickly wrote an extended version of the paper, submitted it to the *Archives of Biochemistry*—and Hill promptly gave it a positive review.[13]

[11] Cambridge University Library, Ms. Add. 9267/J.61. Gaffron to Hill on 2 February, 1951.

[12] Cambridge University Library, Ms. Add. 9267/J.61. Gaffron to Hill on 6 March, 1951.

[13] Hill's referee's report to the journal, dated 11 April, 1951, is preserved in the Cambridge University Library, Ms. Add. 9267/J.61. See Tolmach (1951b) for the published paper.

Yet, only a few days after Hill had sent off his referee's report to the editor of the *Archives of Biochemistry*, he hastened to inform Gaffron about the latest developments: Hill had been appalled to see that Vishniac and Ochoa had, in the meantime, submitted a note to *Nature* that was almost identical in scope and content to the one submitted earlier by Tolmach. Clearly feeling guilty about having deprived Tolmach of being the first to publish a paper on the topic, Hill informed Gaffron that he would ensure that Tolmach's note was published as soon as possible and supplemented by at least one table of data.[14] This was much appreciated by Gaffron, who had noticed how much attention Vishniac and Ochoa's results had been receiving in the USA prior to publication. Furthermore, Gaffron mentioned that "Arnon also seems to have rushed into the field".[15] The Chicago group was naturally interested in getting its share of the credit, although Gaffron still considered that Hill's original criticism was fully justified. Hill did as he had promised, and Tolmach's note appeared in *Nature* in an only slightly altered form. And despite Hills's reservations about the genre of "Notes" to *Nature*, a letter written by him at the end of April to Robert Emerson again demonstrates how enthusiastic Hill was about the general finding that DPN and TPN were reduced in chloroplast suspensions. Hill considered that this discovery finally put an end to the idea that carbon dioxide might be directly reduced through the action of light on chloroplasts:

> The new achievements of Ochoa, Tolmach & Arnon one now feels very cheerful about—the biochemical part does look encouraging. Not that it yet indicates the whole solution about CO_2 but it really is good to see this new escape from the bold "$h\nu + CO_2$" picture, even if it is to be only temporary. I think Tolmach's work gives the most interesting clues—though it needs the other two to support it, in fact each of the three has its own specific contribution.[16]

Emerson, however, was far less excited in his response. He was not at all convinced that these findings helped to explain photosynthesis in plants—after all, no one had demonstrated that reduced TPN actually was involved in the reduction of carbon dioxide. He wrote to Hill, on 8 May 1951:

> You speak optimistically about the work of Ochoa, Tolmach, etc., and I agree that it's interesting to learn more about the chemical reactions which can be brought about by illuminated chloroplasts. However, I heard yesterday that the pyruvic acid reduction previously reported by Vishniak [sic] to be carried out by reduced TPN can be just as well accomplished by illuminated chloroplasts without TPN. People here were jumping to the conclusion that the pathway of photosynthesis must be via the production of reduced TPN by illuminated chloroplasts. This may still be the case, but the experiments up to date seem to indicate only that we must add TPN (and DPN) to the growing list of substances which can be reduced by chloroplasts. I suppose it is understandable that each person finding a new reduction process

---

[14] Cambridge University Library, Ms. Add. 9267/J.61. Gaffron to Hill on 16 April, 1951.

[15] Cambridge University Library, Ms. Add. 9267/J.61. Gaffron to Hill on 21 April, 1951.

[16] Hill to Emerson, 28 April 1951, Robert Emerson Papers, 1923–1961, Record Series 15/4/28, Box 1, Folder: Hill, Robin, University of Illinois Archives. By the end of April, Hill obviously already had seen all three notes (and assumed that Emerson had, too). Thus, the notes or at least their results probably had been circulated in advance to being published.

> hopes it will prove to be the key to the mechanism of photosynthesis. But until there is real evidence of carbon dioxide reduction, and of changes in reduction level large enough to require the energy of several quanta of red light, it seems to me that only moderate claims of progress are justifiable.[17]

This rather sober evaluation of the results was in stark contrast not only to Hill's views but also to the occasionally overly optimistic tone that the publications coming from Arnon's laboratory at Berkeley were about to receive. In the next section, the work done by this Berkeley-based group will be examined more closely: namely, in the context of the discovery of the light-driven production of ATP, independent of respiration.

7.2 Photophosphorylation

7.2.1 Energy-Rich Phosphate Bonds and Photosynthesis

In 1929, the German chemist Karl Lohmann found that fermentation was linked to the formation of molecules of ATP, which could be stored in the cells for several hours.[18] Then, in the 1930s, the Soviet physiologist Vladimir A. Engelhart and his collaborators discovered that muscle contraction required ATP, the Danish biochemist Herman Kalckar established (in 1937) that the formation of ATP was linked to cell respiration, and Otto Warburg elucidated how adenosine diphosphate (ADP) was phosphorylated to ATP during glycolysis.[19] Thus, by the end of the 1930s, the central role of ATP in the organism and its intricate linkage to energy-producing processes such as fermentation and cell oxidation were beyond any doubt. In fact, more and more metabolic redox reactions were demonstrated to be linked to the cleavage and formation of ATP. It seemed that most, if not all, higher organisms were able to store the energy yielded from exergonic processes as ATP, the splitting of which could, in turn, release this energy again in order to promote the occurrence of endergonic processes.

The most influential contributions to this emerging field of phosphate metabolism studies were provided in 1941 in the form of two simultaneously and independently written reviews of the phenomenon and its bioenergetic implications. These were authored by the aforementioned Kalckar and his German colleague, friend and mentor Fritz Lipmann, both of whom would eventually find themselves working in the USA, due to political circumstances.[20] Lipmann, who left Nazi Germany in 1932

---

[17] Emerson to Hill on 8 May 1951, Robert Emerson Papers, 1923–1961, Record Series 15/4/28, Box 1, Folder: Hill, Robin, University of Illinois Archives.

[18] Lohmann (1929).

[19] See, e.g., Engelhardt (1930, 1939); Kalckar (1937) and Warburg and Christian (1939). See Maruyama (1991) for a historical account of the discovery of ATP and its structure.

[20] See Kalckar (1941) and Lipmann (1941). Information on Kalckar can be found in Kennedy (1996), and on Lipmann in Kleinkauf et al. (1988); Jencks and Wolfenden (2000) and Kennedy (2001). Autobiographical accounts are provided in Lipmann (1971); Kalckar (1974, 1991).

(and would win the 1953 Nobel Prize in physiology or medicine), had met Kalckar in 1934 in Copenhagen, in the laboratory of the Danish physiologist Ejnar Lundsgaard, where Kalckar had just started his PhD studies. This encounter was the start of a life-long friendship. When Lipmann went to Copenhagen, he was already deeply interested in the biological functions of phosphorylation reactions, and not only pursued these questions in his own research but also closely followed the work done by Kalckar on the mechanism of oxidative phosphorylation (i.e. the production of ATP in cell oxidation processes). It is one of the striking but not inexplicable coincidences in the history of science that Kalckar and Lipmann both embarked on a seminal review of the "phosphate problem" at the same time, that is, around 1940.

It was in Lipmann's 1941 paper that the notion of "energy-rich phosphate bonds" was explicitly introduced, symbolised by the famous "squiggle" that is usually attached to the phosphorus atom of a phosphate group (~Ⓟ). This symbol became the accepted notation of linkages, such as the pyrophosphate bonds in ATP, the hydrolysis of which causes a relatively large energy release. The two papers by Kalckar and Lipmann were widely received, and at least among biochemists the concept quickly met with consent, if not enthusiasm (although Lipmann recalled that organic chemists *sensu stricto* and physical chemists were outraged by the proposal).[21] It eventually transpired that ATP might be the solution to the problem of how the energy gained from the decomposition of carbohydrates could be chemically preserved and transferred in a stabilised form to other, endergonic reactions of metabolism (i.e. "unfavourable" reactions that did not run spontaneously).[22] In view of these developments, it was soon speculated that ATP was also the source of energy for carbon dioxide reduction in photosynthesis. In 1943, Samuel Ruben was the first to develop a general model of carbon dioxide fixation in this vein: he assumed that the carboxylation of an organic residue, which acted as primary acceptor, was coupled to the splitting of an energy-rich phosphate bond of the ATP type (see Chapter 6).[23]

It has already been mentioned that even his close friend and collaborator Martin Kamen was sceptical about the value of Ruben's suggestion; however, although this attitude was held by most of their colleagues, by a handful of fellow scientists, including Emerson, the idea was enthusiastically picked up. Together with the plant physiologist John F. Stauffer and the microbiologist Wayne W. Umbreit, Emerson published a paper in 1944 in which they pushed Ruben's principal suggestion a little further, although the authors admitted that it was impossible at the time to argue conclusively for the accuracy of their concept, "since undoubtedly other conceptions could equally well fit the facts observed". Emerson and his co-workers proposed the following:

[21] See Lipmann (1971, p. 37).

[22] See Fruton (1972b, p. 363), for a general account of how the importance of ATP in intracellular respiration emerged.

[23] Ruben (1943) is the paper in question.

> The function of light energy in photosynthesis is the formation of "energy-rich" phosphate bonds. According to this view, the light energy absorbed by the chlorophyll system is converted more or less immediately, into "energy-rich" phosphate bonds [...] which furnish the energy for the remainder of the photosynthetic process. [...] We therefore conceive of the entire process of photosynthesis (the fixation, reduction, and synthesis of organic molecules from carbon dioxide and, in green plants, the production of oxygen from water) as being "dark reactions" which could be accomplished without the use of light if one were able to substitute into this system the "energy-rich" phosphate compounds which actually result from the absorption of light by the chlorophyll system. [...] Therefore, the light *per se* is not essential for photosynthesis, but some result of the absorption of light is essential.[24]

Thus, while oxygen release had been dropped from the list of essential properties for the phenomenon of photosynthesis in the 1930s, as a consequence of the work done by Cornelis B. van Niel, now the action of light was also being seen as replaceable if the necessary compounds could be provided in another way. Again, the relevant clues came from comparing photosynthesis in plants with the metabolism of bacteria. The authors referred, in particular, to the recent elucidation of the mechanism of energy transport in *Thiobacillus thiooxidans*: an autotrophic, chemosynthetic sulphur bacterium, which gained the energy required for carbon dioxide reduction from oxidising sulphur to sulphuric acid. It was found that this bacterium could oxidise sulphur in the absence of carbon dioxide and store at least a portion of the energy in a form that could later be used for carbon dioxide fixation under conditions in which sulphur oxidation was impossible. Thus, the two partial processes—the oxidation of sulphur and the reduction of carbon dioxide—were coupled but could later be separated, much like the photochemical production of oxygen and carbon dioxide reduction in green plants. It was demonstrated that the formation of the unknown "energy storage material" was accompanied by a substantial uptake of inorganic phosphate, while during carbon dioxide fixation an inorganic phosphate release was observed. Hence, the authors concluded that the energy gained from sulphur oxidation was stored as ATP molecules.

In view of these findings, Emerson, Stauffer and Umbreit asked whether photosynthetic organisms might also use energy-rich phosphate bonds in the form of ATP to move the energy gained from the light-induced splitting of water to the endergonic process of carbon dioxide reduction. In order to learn more about this possibility, the authors emphasised, they would need to determine whether or not phosphorylation was an important part of metabolism of photosynthetic cells. However, they had to admit that, at the time, it was impossible to differentiate between the products of oxidative phosphorylation, which were the result of respiratory processes, and the potential products of photosynthetic, light-induced phosphorylation. Their own experimental data suggested that phosphorylation was, indeed, taking place in photosynthesising cells (which was hardly surprising, since these cells were known to respire), while there were only a few indications that the underlying mechanism of ATP formation was not entirely the same as the mechanism that had been found in animal and bacterial cells.

[24] Emerson et al. (1944, p. 107).

Emerson, Stauffer and Umbreit's proposal, just like Ruben's earlier attempt, was not favourably received. In the first volume of his seminal monograph, published in 1945, Rabinowitch rather sceptically commented on the possible role of ATP in photosynthesis. The problem of photosynthesis was a problem of energy *accumulation*, Rabinowitch underlined, since so much energy was required to reduce carbon dioxide. Bearing this in mind, it seemed like a step in the wrong direction to *dissipate* the energy of red light quanta (of 43 kcal) by storing it in high-energy phosphate bondings of only 10 kcal per mole. Rabinowitch concluded:

> To sum up: we think it unlikely that the bulk of the light energy utilized in photosynthesis (or of the oxidation energy utilized in chemosynthesis) is first converted into phosphate energy. Furthermore, if phosphorylation does play an auxiliary role in photosynthesis (e.g., in the way envisaged by Ruben)—which is by no means certain—we think it much more probable that the required high-energy phosphates are supplied by nonphotochemical oxidation processes than that light quanta are diverted for their synthesis.[25]

These objections based on energetic considerations were strengthened by the results that Samuel Aronoff and Melvin Calvin published in 1948, on their studies incorporating radioactively labelled phosphorus into ATP in isolated chloroplasts: "Using radioactive phosphorus, no direct connection between gross formation of organic phosphorus compounds and photosynthesis or photochemical reductions has been found to occur" is how they briskly summarised their finding.[26] Thus, in view of these results, it seemed (erroneously, as we know in hindsight) that searching for a unique mechanism of phosphorylation during photosynthesis would be a road to nowhere.

Far more promising appeared an alternative model that was proposed by Vishniac and Ochoa in 1952. By the early 1950s, it had transpired (due to the studies of the Calvin–Benson group at Berkeley) that phosphorylated compounds were prominently present in the photosynthetic reduction of carbon dioxide; thus, there had to be a way to account for their formation during photosynthesis. Vishniac and Ochoa suggested that, in fact, photosynthesis was accompanied by phosphorylation, while the latter took place elsewhere: namely in the mitochondria, which had recently (in 1951) been demonstrated by Albert L. Lehninger to be the site of ATP production during respiration. The ATP would be formed in these organelles, and then they would be transported to the sites of photosynthetic carbon dioxide reduction.[27]

Their evidence mostly came from an in vitro system: ATP was formed if a suspension was illuminated that contained chloroplasts, mitochondria, DPN and phosphorus-32 (which served as the tracer molecule). The assumption was that the DPN was first reduced by the illuminated chloroplasts (as earlier observations had indicated) and, subsequently, oxidised in a reaction coupled to the formation of ATP in the mitochondria. This suggestion also was able explain the limited success to

[25] Rabinowitch (1945, pp. 229–230).

[26] Aronoff and Calvin (1948, p. 357).

[27] See Vishniac and Ochoa (1952a, b) and Ochoa and Vishniac (1952). Lehninger (1951) is the paper they referred to.

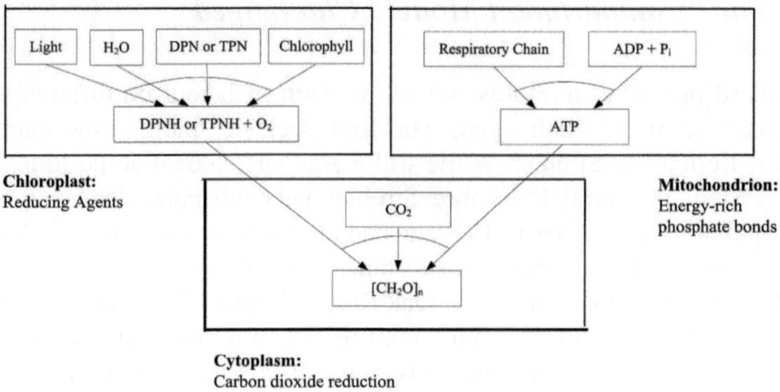

Fig. 7.2 The mechanism of photosynthesis, reconstructed from Vishniac and Ochoa (1952b). Three partial processes were identified (the production of reducing agents, the production of energy-rich phosphate bonds and the reduction of carbon dioxide), which the authors located in three different compartments of the cell (the chloroplast, the mitochondrion and the cytoplasm).

find any traces of carbon dioxide reduction in isolated chloroplasts. According to Vishniac and Ochoa's model, this was self-evident, since in Hill-type chloroplast suspensions there were no mitochondria present to provide the necessary ATP. The authors formulated their conclusion as follows:

> [T]he incorporation of P^{32} into ATP is dependent on light, oxygen, and DPN. No incorporation of P^{32} takes place in the absence of particles capable of carrying out oxidative phosphorylation. It is suggested that in photosynthesis energy-rich phosphate bonds are generated by the mechanism outlined above.[28]

For the next couple of years, this became the accepted standard hypothesis: chloroplasts were the sites of light absorption and water splitting, as well as of TPN (or DPN) reduction, while ATP was assumed to be produced in the mitochondria, and the fixation of carbon dioxide presumably occurred in the cytoplasm. The latter assumption was endorsed by the finding that the "malic" enzyme, then considered to play a central role in carbon dioxide reduction, was present in ample amounts in the cytoplasmic fluid of plant cells (but not in the chloroplasts themselves). Even when it slowly transpired, through the advances made by the Calvin–Benson group, that the "malic" enzyme itself had no bearing on photosynthetic carbon dioxide fixation, the finding of further carboxylases and other enzymes in the cytoplasm made the latter the most probable place in which other enzymatic reactions would occur. (A reconstruction of this "compartment" model is given in Fig. 7.2.)

[28] Vishniac and Ochoa (1952b, p. 502).

7.2.2 The "Compartment Model" Challenged

This standard picture of a photosynthetic division of labour on different compartments was soon to be challenged. The first decisive finding was made by the biochemist Bernard L. Strehler. While still a graduate student at the Johns Hopkins University Medical School, Baltimore, Strehler had contributed decisively to finding out what makes fireflies glow. In 1949, he had identified an enzyme, which he named "luciferin", and established that this substance gave off light when it was combined with ATP.[29] Strehler moved on to the Oak Ridge National Laboratory in Tennessee, where he met William Arnold, with whom he began a long and fruitful collaboration.[30] Starting from Strehler's previous research, the two men wanted to find out whether ATP was formed as a result of photosynthesis; the firefly enzyme seemed to provide them with a powerful and very sensitive indicator of ATP formation. Instead, they found, first of all, that, after being illuminated, the chloroplasts gave off light even without the addition of any luciferin: Strehler and Arnold had discovered the phenomenon of delayed fluorescence in photosynthesising systems.[31] However, a short while later, in 1952, Strehler established that ATP was, in fact, formed in plants immediately upon illumination, and that the site of formation was *the chloroplast*. He presented these findings at the first Gatlinburg conference on photosynthesis in 1952, where, incidentally, he was the only speaker on ATP and photosynthesis. It was clearly not yet regarded as a hot topic.[32] One year later, in 1953, Strehler presented the first detailed suggestion as to how ATP might be produced in a light-induced mechanism in plants. This was combined with the suggestion that a major portion of this ATP might immediately be used up again in order to produce reducing agents.[33] As one can take from one of the illustrations in Strehler's text (see Fig. 7.3), the concept of step-by-step oxidation and the reduction of intermediates, with ATP formation at one of the transition points, already existed in 1953, even though the details were unknown.

Around the same time, starting from 1950, evidence was found by the German botanist Otto Kandler that in Chlorella cells the inorganic phosphate went down upon illumination, while at the same time high-energy phosphor bonds were formed.

[29] Strehler and McElroy (1949).

[30] See Strehler (1996) for an autobiographical account.

[31] Strehler and Arnold (1951) and Strehler (1951); see also Strehler and Hendley (1961).

[32] According to Strehler, both Arnon and Vishniac received a copy of Strehler's papers on ATP formation at the Gatlinburg conference. However, neither of them acknowledged this source of inspiration in their later papers on the subject; cf. Strehler (1996, p. 14). See also Strehler's comment on his personal homepage, accessed in May 2014 at http://web.archive.org/web/20021207070619/fihg.org/founder1.htm.

[33] Strehler (1952, 1953). The methods were developed in Strehler and Totter (1951).

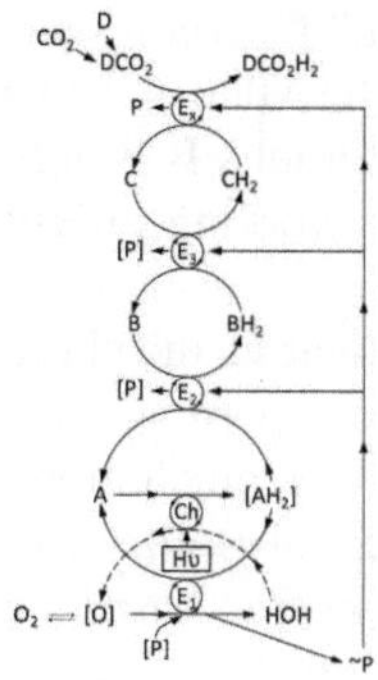

Fig. 7.3 Redone figure of Strehler (1953, p. 75). Legend of the symbols: *Ch* photochemical apparatus, *Hν* light, AH_2, BH_2, CH_2 intermediate reductants, [*O*] photochemically produced oxidant, precursor to O_2, *D* "C_2 acceptor" molecule or other acceptors, ~*P* ATP, DCO_2H_H primary fixation-reduction product, E_1 terminal oxidase, E_2, E_3, E_x intermediate transhydrogenases.

If illumination was stopped, the reverse process took place.[34] This was the first time that the process of photophosphorylation, that is, photosynthetically driven ATP formation, was clearly identified as such. These first steps were complemented in 1954 by the work carried out by Daniel I. Arnon's group at Berkeley. Born in Warsaw, Poland, Arnon had trained as a plant physiologist at the University of California, Berkeley, and received his PhD there in 1936.[35] Arnon had been a student of Dennis R. Hoagland, who was well-known for his pioneering work on plant and soil interrelations; therefore, it is not surprising that Arnon's early research focused on plant nutrition and the role of trace elements in plants. In 1941, Arnon was made an assistant professor at Berkeley, and he remained at the university for the rest of his working life. In 1961, he was appointed the first director of the Department of Cell Physiology, which developed into a centre of photosynthesis research.

In an autobiographical account, Arnon recalled how he felt increasingly uneasy with the standard hypothesis that the ATP required for photosynthesis was formed in the mitochondria. Among other things, he found this to be fundamentally at odds with well-known observations of plant cytology. Photosynthetically active tissues were so full of chloroplasts that the few mitochondria present "could not possibly supply the ATP needs of photosynthesis whose rate in saturating light can be 30 times

[34] See Kandler (1950); the theme was pursued further in Kandler (1954, 1955). On Otto Kandler, see Schleifer (2011). Strangely enough, while Arnon (see below) acknowledged this earlier paper in his first contributions, it was later dropped from the lists of references.

[35] On Arnon's life and work, see, e.g., Buchanan (1995, 2001). See Melis and Buchanan (1995) for a special issue of *Photosynthesis Research* dedicated to Arnon.

higher than the rate of respiration".[36] Arnon set out to scrutinise this problem with two close collaborators: Mary Belle Allen, who earlier in her career had worked with Charles Stacy French and with Cornelis B. van Niel; and F. Robert (Bob) Whatley, who had received his PhD while working in Hill's laboratory at the University of Cambridge (UK) in 1946.[37]

The group started off by investigating the photosynthetic capacities of the chloroplasts of green algae, the preferred experimental organism of photosynthetic studies; but very soon spinach proved to be a much more appropriate source of chloroplast suspensions. Yet, Arnon and his collaborators decided, deviating from standard procedures, not to use the membranous fraction of the chloroplasts—that is, chloroplast fragments obtained by ultracentrifugation. Instead, Arnon and his team chose to investigate *whole* chloroplasts, including their soluble content. This proved to be a fortunate decision, since it turned out later that the soluble fraction of chloroplasts contained the necessary catalysts for photophosphorylation as well as for reducing carbon dioxide.[38] Arnon's group also used isotonic salt solutions (NaCl) instead of the more usual sugar solutions, which had been developed in Hill's laboratory. Sugar was a potential confounding factor, as it was possibly used as an additional source of energy and metabolites. If one wanted to find out the *photosynthetic* capacities of chloroplasts, light ought to be the only source of energy, Arnon mused. Microscopic observations were employed to ensure that the structure of the chloroplasts thus gained was unimpaired; and they carefully double-checked that the suspension was free of mitochondria, so that any support of the process by oxidative phosphorylation was precluded.[39]

The team found, in 1954, that not only were these chloroplasts able to produce oxygen; they also accumulated ATP if supplied with the substrates of phosphorylation: adenosine monophosphate (AMP), and inorganic phosphate. When bicarbonate was then added, the system was able to reduce carbon dioxide at constant rates for about 1 h. An analysis of the products yielded an insoluble compound, which was identified as starch, and several soluble compounds, which included the major components of the photosynthetic carbon reduction cycle recently proposed by the Calvin–Benson team. "In the light of our present evidence, isolated chloroplasts emerge as remarkably complete cytoplasmic structures, equipped to carry out not only oxygen evolution but also carbon dioxide fixation and the conversion of light

[36] Arnon (1984, p. 258).

[37] The history of how the chloroplast was established as the unit of complete photosynthesis and how photosynthetic phosphorylation was discovered, has been told before; see, e.g., Arnon (1977, 1984, 1984), Frenkel (1993, 1995) and Shen and Wei (1998).

[38] See Arnon (1984, p. 259).

[39] Whatley explained, in retrospect, that, despite all the precautions they took, these "whole" chloroplasts were later found to be leaky, since they lacked the external membrane of intact chloroplasts. This alleged flaw, however, turned out to be fortuitous, as it enabled the substrate of phosphorylation, AMP, to come into contact with the thylakoids, the sites of phosphorylation, which otherwise would have been precluded. See Whatley (1995, p. 18).

into chemical energy", Arnon's group summarised their findings.[40] They consciously introduced the term "photosynthetic phosphorylation"[41] in order to distinguish this process from oxidative phosphorylation, which occurred in the course of respiration; however, it was the shorter version "photophosphorylation" that caught on.[42]

Yet, the accumulated evidence for the standard compartmental hypothesis—that all the cell's ATP was produced in the mitochondria—was too weighty to be simply swept aside. In their attempt to evaluate the new possibility, some scientists failed to reproduce Arnon's findings altogether;[43] and even those who accepted the data as such (reproduced or not) tried to provide explanations for Arnon's results within the framework of the compartment hypothesis. One line of criticism was to assume the effect of confounding factors. It was suggested, for example, that Arnon's suspensions might be contaminated with mitochondria (which was entirely plausible, since it was so difficult to get rid of these tiny structures).[44] Alternatively, it was suspected that Arnon's group had formed, as artefacts of their procedure, "a coagulation membrane of some sort which traps [cytoplasmic] enzymes in association with the chloroplast", so that the chloroplasts "may be merely floating around in a cytoplasm which itself contains all the necessary enzymes".[45] Others were suspicious of the low rates of phosphorylation and carbon dioxide reduction observed in Arnon's experiments, which made people doubt the significance of the findings.[46] In later accounts of the discovery, Arnon never failed to mention these critical reactions, not without displaying the satisfaction of having been proven right by history. He also liked to recount how the journal *Chemical and Engineering News* had invited him to give an account of his group's work, "but in the end the editor declined to publish the article because it did not pass review by three outstanding authorities in the field".[47] Even the plant physiologist André T. Jagendorf, who would later be one of the ardent supporters of the photophosphorylation hypothesis, was at first not at

---

[40] Arnon (1954a, p. 394).

[41] This lengthy term prompted Robert Emerson, who had a strong aversion to the way Arnon advertised his findings, to the following remark to his old friend James Bonner: "I'm sorry I'm not getting out to the meetings at Stanford. I would not enjoy Arnon's pompous maunderings about 'photosynthetic' phosphorylation. (As far as I understand what he has done, it is no more than 'photo'phosphorylation.)" Emerson to Bonner, 22 August 1957, Robert Emerson Papers, 1923–1961, Record Series 15/4/28, Box 1, Folder: Bonner, James, University of Illinois Archives.

[42] See Arnon (1954b) for more details. The full evidence for complete photosynthesis in isolated chloroplasts was elaborated in subsequent publications, such as Allen et al. (1955), Arnon (1955) and Whatley et al. (1956).

[43] See, e.g., Ohmura (1955).

[44] Ochoa put forward this objection; see the discussion in Arnon (1956, pp. 307–308).

[45] This objection was brought forward by Strehler at the 1955 Gatlinburg conference; see Allen et al. (1957, p. 293).

[46] With hindsight, Arnon believed that these low rates were caused by the fact that the chloroplasts were kept in salt solutions. See Arnon (1987, p. 41).

[47] Arnon (1984, p. 258).

all convinced.[48] Thus, whether *all* photosynthetic processes really did, in fact, take part in the chloroplast remained a controversial issue for some years to come.

The fact that there might be a light-induced phosphorylation mechanism was discovered independently in the same year (1954) by Albert Frenkel. According to his recollections, Frenkel was prompted to study ATP metabolism as a result of working, together with the plant physiologist Allan Brown, on an extended review of recent work in photosynthesis.[49] While Brown concentrated on those issues related to carbon metabolism, Frenkel covered phosphorus metabolism. He studied intensively, for example, the 1952 papers by Vishniac and Ochoa and decided to spend some time in a research laboratory that focused on related questions. He arranged to go and work with Fritz Lipmann at Harvard Medical School in Boston during the first half of 1954. Lipmann drew Frenkel's attention to a paper by Howard Gest and Martin Kamen, which indicated that ATP was produced in the illuminated cells of *Rhodospirillum rubrum*.[50] Building on these findings, Frenkel found that in cell-free preparations of this very organism "light induced anaerobically a pronounced disappearance of orthophosphate.[51] Frenkel was convinced that the phosphate ions disappeared as a direct result of the substantial formation of ATP by bacterial cell fragments, as he had successfully coupled this process to the ATP-dependent phosphorylation of glucose. Frenkel emphasised that the observed phosphorylation was strictly light dependent, did not require oxygen and was not inhibited by the typical inhibitors of respiration. It took another three years before further confirmation of ATP formation in chloroplasts was published by the plant scientists Mordhay Avron (who had worked on oxidative phosphorylation before) and the aforementioned Jagendorf.[52]

In the mean time, Arnon's group had continued to investigate the capacities of the chloroplast and the interrelation of the different partial processes they had identified (in a typical move of functional decomposition of the mechanism): photolysis of water, photosynthetic phosphorylation and carbon dioxide assimilation. Arnon and his collaborators argued that these were distinct phases of photosynthesis, which were linked to each other in a hierarchy of increasing complexity:

> Photolysis could be carried out by preparations incapable of photosynthetic phosphorylation and CO_2 fixation. In turn, photosynthetic phosphorylation was found to proceed unimpaired in preparations which could not fix carbon dioxide. CO_2 fixation, however, has been observed only in chloroplast preparations capable of active photolysis and phosphorylation.[53]

[48] See Jagendorf (1998, p. 219).

[49] See Frenkel (1993, 1995). The review was published as Brown and Frenkel (1953).

[50] This was Gest and Kamen (1948).

[51] Frenkel (1954, p. 5568).

[52] See Avron and Jagendorf (1957). On Avron, see Gromet-Elhanan (1992).

[53] Arnon et al. (1956, p. 458).

These results were supplemented by the surprising observation that the occurrence of complete photosynthesis outside the cell did not necessarily depend on the structural integrity of chloroplasts.[54] Chloroplast fragments, produced by osmotic shock treatment with water or dilute salt solutions (which prompted the cells to burst), were also able to carry out all three partial processes at surprisingly high rates, as long as the appropriate cofactors were added: "With proper additions, the photosynthetic activity of the reconstituted system, as measured by CO_2 fixation per unit of chlorophyll, was several times greater than that of the intact chloroplasts."[55] Restoring the carbon dioxide fixation required that the membranous fragments were supplied with pyridine nucleotides, ATP and the water-soluble portion of intact chloroplasts. This indicated that the membrane fragments lacked the necessary enzymes for carbon dioxide fixation, which instead seemed to be localised in the soluble fraction of chloroplasts. Photophosphorylation (as it came to be called), by contrast, was found to be fully restored in the fragments without the soluble chloroplast fraction, on the mere addition of either magnesium ions (Mg^{2+}), vitamin K, ascorbate or FMN (flavin mononucleotide or riboflavin-5′-phosphate). No oxygen was required for phosphorylation to take place in this system, which was in sharp contrast to the well-known oxidative phosphorylation in mitochondria. Arnon's group concluded that, since only cofactors had to be added, all the enzymes of photophosphorylation seemed to be contained in the membrane fragments.

This was the first of several papers published by Arnon's group that were concerned with the question of the appropriate "cofactors" of photophosphorylation. Not all these papers were consistent with each other, which caused confusion, occasionally consternation and even amusement among Arnon's colleagues. One researcher who was exceedingly sceptical about the importance of specific "cofactors" was Robin Hill. Hill's long-standing collaborator David Walker vividly remembered the following episode:

> At the time the word from Berkeley was of more and more co-factors. Despite our immense respect for Dan Arnon et alia [sic], we irreverently labelled a jar of marmite "Arnon's Reagent" because it worked as well as most compounds that we had tried. In the same spirit Robin suggested "spit, urine and floor sweepings". We shrank from the first two and felt that the third, given Robin's lab, would have been a bit of a forgone conclusion. Even so, there seemed to be a good reason for supposing that almost anything, with an appropriate redox potential, might suffice as a co-factor.[56]

Finally, in 1958, Arnon's group was able to demonstrate that, by making a more considered choice of experimental conditions, such as pH and chloroplast density, the rate of photophosphorylation in isolated chloroplasts, or chloroplast fragments, could

---

[54] See Whatley et al. (1956) and Arnon et al. (1956).

[55] Arnon et al. (1956, p. 462).

[56] Walker (1992b, p. 337). For a more detailed (and serious) description of the work on cofactors, going on in Cambridge at the time, see Cambridge University Library, Ms. Add. 9267/ B.275: Programme of Research c. 1959, formulated for the renewal of Fay Bendall's grant (then still referred to as "Miss Myers".

be dramatically increased (up to 170 times higher than the rates initially described).[57] Thus, the phenomenon had definitely reached significant dimensions. Before turning to Arnon's finding that there might be different types of ATP formation in the chloroplast, the contemporary developments made in a completely different field of research ought to be examined: the work on photosynthetic cytochromes, primarily undertaken by Hill in his laboratory at Cambridge (UK), and the intensive search for a photosynthetic electron transport chain.

7.3 Cytochromes and Ferredoxin

7.3.1 Background

The intricate history of respiration research, and in particular of oxidative phosphorylation, has been covered elsewhere and will not be repeated in detail.[58] The first thoughts that there might be a stepwise electron transport chain in respiration were explored in the study of specific molecules that seemed of importance for respiratory processes during the 1930s. By 1940, this had led to the idea of a chain of successively oxidised and reduced intermediates that soon were spoken of as "carriers". The details of this chain were unknown, although it was strongly suspected that it involved pyridine nucleotides, flavins and cytochromes in the order of their relative oxidation–reduction potentials. The general concept that emerged was formulated, for example, by the Harvard-based biochemist Eric G. Ball in 1942 as follows:

> The energy liberated when substrates undergo air oxidation is not liberated in one large burst, as was once thought, but is released in stepwise fashion. At least six separate steps appear to be involved. The process is not unlike that of locks in a canal. As each lock is passed in the ascent from a lower to a higher level a certain amount of energy is expended.[59]

This strong metaphor of a "downhill" cascade from high-energy to lower energy compounds has dominated the field ever since. However, decisive experimental evidence pointing to the actual sequence of electron carriers in respiration was provided only in the mid-1950s, thanks to the development of rapid and sensitive spectrophotometric techniques introduced by Britton Chance in collaboration with the biochemist G. R. Williams.[60] The underlying notion was that the absorption spectra of biologically relevant molecules, such as cytochromes, flavins and pyridine nucleotides, changed appreciably upon changes in the molecule's oxidation–reduction state. This made the recording of absorption spectra an extremely powerful way of tracing the

---

[57] See Allen et al. (1958).

[58] See Fruton (1972b) and Edsall (1974) for classic treatments of the subject; Weber (1991) deals specifically with oxidative phosphorylation.

[59] Ball (1942); quoted in Fruton (1972b, p. 387).

[60] See Chance (2004) for an autobiographical review of the development of the pertinent instruments. Yodh and Tromberg (2000) is a tribute to Chance.

oxidation and reduction states of certain compounds under certain conditions—for example, when they were supplied with molecular oxygen. It was also found that these oxidation–reduction processes were somehow linked to the formation of ATP (although the mechanism of this coupling was then still unknown).[61] Based on their spectroscopic studies, Chance and Williams finally suggested that, in the respiratory chain, electrons were transported in the sequence DPN, flavin, cytochrome *b*, cytochromes *c* and *a*, and, finally, oxygen.[62] Looking back, Chance recalled the frustrating first presentation of their results:

> The responses to the ideas were definitely "not great"; the initial presentation before the American Society of Biological Chemists at San Francisco as a ten-minute presentation was disaster-ridden. The chairman, my good friend A[lbert] Lehninger, apparently misread the clock and told me that I was to sit down after four of the ten minutes allotted for the talk.[63]

However, after the results had been confirmed by several laboratories that had used Chance's dual wavelength spectrophotometer and extended it to the study of living tissue, other scientists started to become interested in the concept. And while Chance and Williams had explicitly refrained from making any comments on photophosphorylation, and its potential coupling to analogous electron transport chains, suggestions construed along these lines soon appeared in print.[64] The principle seemed extremely suggestive, but the vexing problems had to be solved how the "uphill flow" from a redox system with a relatively negative potential (O_2/H_2O) to a redox system with a relatively positive potential (CO_2/CH_2O) was brought about. People agreed that this most probably was overcome by the absorbance of light energy and the latter's efficient use in the system; but nobody had any idea how this might work in detail. Rabinowitch pointedly formulated in 1945 (continuing the "lock" metaphor that Ball used in his formulation quoted above): "When a canal is built between two bodies of water situated at different levels, the provision of locks cannot be avoided; but whether these locks are constructed at the upper or lower end of the waterway is a purely practical problem."[65] One of the keys to solving the problem was sought in the study of components that were also involved in the respiratory chain: cytochromes. This had intensively been done, for example, in Cambridge , UK, since the 1930s; while it was successfully continued by Robin Hill and his group, as will be shown in the next section.

---

[61] This mechanism turned out to be rather sophisticated and became known as the "chemiosmotic theory" of oxidative phosphorylation; see Mitchell (1961a, b) for the seminal publications. The episode was treated, with special reference to the work of Peter Mitchell in, e.g., Prebble and Weber (2003); Prebble (2010) and Scholl and Nickelsen (2015).

[62] Chance and Williams (1956) provides the celebrated review article of these findings. A retrospective commentary on the latter is given in Chance (1983). See Chance and Williams (1955) and Chance et al. (1955) for the earlier papers.

[63] See Chance (1983).

[64] The first rudimentary ideas on the occurrence of electron transport chains in photosynthesis have been traced back to Katz (1949); Levitt (1953, 1954), although none of these were widely received or promoted any further at the time.

[65] Rabinowitch (1945, p. 151).

7.3.2 Photosynthetic Cytochromes

The 1930s were an enormously fruitful period from the viewpoint of understanding cytochromes and their role in metabolism. It was predominantly Cambridge-based biochemist David Keilin who elucidated the action of cytochromes *a*, *b* and *c* as oxidation–reduction catalysts in cellular respiration, when he found them to be reversibly reduced and oxidised by changes in the iron portion of their haeme moieties. Arguably Keilin's most important contribution was to identify the (up to then) mysterious enzyme cytochrome oxidase with a specific component of cytochrome a_3, which interacted directly with molecular oxygen.[66]

Cytochrome research was introduced to the study of photosynthesis in the 1940s and 1950s by Robin Hill at Cambridge. The latter's close association with Keilin was already mentioned as being highly significant in this context: Hill and Keilin had collaborated around 1930 on the isolation of cytochrome *c* from both yeast and muscle tissue, and even in later years Hill never lost contact with his mentor (see Chapter 4). The study of the "chloroplast reaction", as Hill insisted on calling the production of oxygen by chloroplast fragments (although everybody else called it the "Hill reaction"), developed at the end of the 1930s into a more in-depth study of the redox capacities of chloroplasts. Hill strongly suspected that cytochromes were involved in these processes as well, and started to explore a little further the presence and function of cytochromes in plants. His first success, which he made with one of his students, Kamala Bhagvat, was to demonstrate the presence and activity of cytochrome oxidase, the enzyme that Keilin had found, in plant tissues. The same tissues were also shown to contain the cytochromes *a*, *b* and *c*, which since Keilin's work were assumed to be associated with cellular respiration. Interestingly, the cytochrome system was found to be attached to small particles, which could be obtained from a variety of plant tissues—today these particles are known as "mitochondria".[67] In the course of these studies, Hill realised that leaves and the other green parts of plants had a surprisingly high concentration of haematin compounds (one species of which were cytochromes)—surprisingly high, in view of the relatively low respiratory activity of the tissue. This encouraged Hill, together with another collaborator, Richard Scarisbrick, to embark on a thorough study of the plant's haematin compounds: if respiration only accounted for a minor portion of these compounds, what explained the remainder? This project was started before 1940, but the outbreak of the Second World War forced Hill and Scarisbrick to discontinue their studies, which they were only able to resume in the late 1940s.

The first findings were thus only published in 1951: Hill and Scarisbrick reported that three different cytochromes had been obtained in a soluble form from plant

---

[66] For a more detailed account of these discoveries, see Mann (1964) and Keilin (1966).

[67] Hill and Bhagvat (1939); see also Bhagvat and Hill (1951) for an elaboration. According to Bendall and Walker (1991, p. 4), this work of Hill's, together with his student (later Professor Kamala Sohonie), "represented the first biochemical study of plant mitochondria".

tissue.[68] One of these was the well-known cytochrome *c*, while the other two had been hitherto unknown and seemed to be characteristic of plants only. The first, which Hill and Scarisbrick had extracted from the acetone powders of leaves, was tentatively called "cytochrome b_3": it was found in the green as well as in the colourless parts of plants; it was autoxidisable, did not combine with carbon monoxide and was easily denatured by heat and organic solvents. The second new cytochrome compound, however, was far more exciting: it was only found in the green parts of plants, was not autoxidisable, did not combine with carbon monoxide and was rather stable in the presence of organic solvents (despite the latter's denaturing influence on many substances). Its absorption properties in intact leaves resembled those of cytochrome *c*, although the bands were much more sharply defined, and there was a characteristic α-band of the reduced component that could be observed at 555 nm. Furthermore, the new cytochrome had definitely more oxidising redox potential than cytochrome *c*. Hill and Scarisbrick called it cytochrome *f*, after the Latin word *frons* (leaf), as it was only found in the green parts of plants.[69] In contrast to cytochrome b_3, this compound was found to be rather firmly associated with the chloroplast fraction. It proved, in fact, impossible to extract it in an unmodified form from acetone preparations, although Hill and Scarisbrick found that freshly ground leaves to which ethanol was added (which was a rather drastic treatment!) gave surprisingly high yields. This might be explained, the authors suggested, by assuming that cytochrome *f* was an integrated constituent of the insoluble parts of chloroplasts (which, incidentally, made it very unlikely that it was involved in respiration). In any event, it was the first substantial indication that there were distinct photosynthetic oxidation–reduction processes in which cytochromes played a role. Hill and Scarisbrick stated in their conclusion:

> The occurrence of this specialized cytochrome component strongly indicates the presence of oxido-reduction mechanisms in connexion with chloroplasts which differ both in nature and intensity from those characteristic of the normal respiration of green tissue.[70]

However, many questions were left open—the most pressing one of which concerned the identity of this proposed cytochrome *f*. After all, it had been isolated under unphysiological conditions to say the least, so that the whole compound might turn out to be an artefact of preparation that had arisen from the unintentional modification of other haematin compounds in the course of extraction. Therefore, together with yet another associate, Harold Davenport, Hill immediately set out to make an in-depth exploration of the new cytochrome's properties, in particular by comparing it with the better known cytochrome *c*.[71]

---

[68] Hill and Scarisbrick (1951).

[69] Hill and Scarisbrick (1951, p. 99). See also Bendall (2004) for a review of the history of the discovery and further investigation of cytochrome *f*.

[70] Hill and Scarisbrick (1951, p. 110). In view of the crude methods available at the time, isolating cytochrome *f*, which, as we know today, is an integrated membrane protein, was a remarkable achievement.

[71] The result of this investigation was Davenport and Hill (1952).

Out of a range of plant species that Davenport and Hill had tested, curled garden parsley in its first year of growth turned out to be the best suited to their purpose. In the usual down-to-earth type of analysis practised at the time in Hill's laboratory, parsley leaves were puréed in an electric meat grinder, squeezed through a cloth and centrifuged for 10 min. From the resulting suspension, a satisfactory portion of the compound in question was precipitated. Although the yield was found to depend on several details of the method, the findings were in agreement with Hill and Scarisbrick's results. Cytochrome *f* was not an artefact but in all probability a real compound. This was confirmed, only a few months later, by experiments that Davenport had conducted on his own.[72] Thus, Hill felt more than entitled to report his finding of cytochrome *f* at the 1952 Gatlinburg conference; and already then considered it a possibility that this compound might be an intermediate in a photosynthetic electron transport chain.

In addition to its absorption band at 555 nm, the most striking property of cytochrome *f* was its redox potential, determined as + 365 mV: it was more positive than the potential of any of the haematin compounds examined up to then. This implied that the reduced state of cytochrome *f* was energetically strongly favoured. In fact, Davenport and Hill were unable to find an enzyme in the leaves capable of oxidising this cytochrome, that is, an enzyme (or system of enzymes) comparable to the cytochrome *c* oxidase in the mitochondrion, the existence and function of which by then had been firmly established. In terms of electron transfer, cytochrome *f* was apparently a dead end. Thus, Hill and Davenport were unable to find any evidence of the metabolic function of this compound, which they had hoped to be analogous to the central function of cytochrome *c* in respiration. However, Davenport and Hill were still convinced that cytochrome *f* did have an important part to play in the photosynthetic electron transport process.[73] In the mean time, Hill continued to work on plant-specific cytochromes, and in 1954 he was able to report yet another discovery:

> The chloroplasts of the etiolated barley leaves were found to contain also a cytochrome *b* component present in amounts definitely larger than that of cytochrome *f*. The *b* component, which here will be designated b_6, was autoxidizable in the chloroplast suspensions and did not combine with carbon monoxide. It was not found possible to remove it from the solid material in an unmodified form. [...] An apparently identical b_6 component was found to be present, together with cytochrome *f*, in *Chlorella*.[74]

In the chloroplast suspensions made from barley leaves, cytochrome *f* was almost immediately and completely reduced on the addition of ferrous iron (Fe^{II}), while the newly identified cytochrome b_6 remained in an oxidised state for a surprisingly long period of time; the redox potential was estimated to be approximately −60 mV. "The presence of cytochrome b_6 could account for the experimentally observed reducing properties of chloroplast preparations when illuminated in the presence of certain hydrogen acceptors", Hill suggested. Thus, cytochrome b_6 was considered to be a

---

[72] Davenport (1952).

[73] Cf. also Hill and Hartree (1953), a comprehensive review of cytochromes in plants.

[74] Hill (1954, p. 502).

decisive factor in the Hill reaction. In an illuminated leaf, Hill speculated, cytochrome b_6 might be reduced, while cytochrome *f* was oxidised: "This represents obviously a definite amount of available chemical energy" (p. 503). Hill was clearly hunting for the appropriate components of a potential photosynthetic electron transport chain. The two cytochromes were promising possibilities, while their relationship to each other was far from clear.

7.3.3 Ferredoxin

In parallel to the investigation of cytochromes, another line of research that Hill and Davenport pushed ahead was the search for the natural hydrogen acceptor in chloroplasts.[75] It was already mentioned that Hill had found chloroplast suspensions to produce oxygen, if ferric oxalate or other hydrogen acceptors were added (see Chapter 4). "There must therefore be [in the chloroplast] some primary substance which is reduced [in the light], while at the same time giving oxygen", he wrote in 1939.[76] For unknown reasons, this distinctive part of the chloroplasts' oxidation–reduction system was lost or, at least, inactivated when the chloroplasts were isolated; but if one wanted to learn more about the redox system in the intact chloroplast, these inactivated elements had to be found. A first step in this direction had already been made in 1949, in studies undertaken by Davenport. He had demonstrated that, although suspensions of washed chloroplasts were unable to produce any molecular oxygen without the additional supply of hydrogen acceptors, the same chloroplasts happily released a substantial amount of oxygen, when they were suspended in an aqueous extract made from acetone-treated leaves.[77] At the time Davenport was unable to identify any specific compound in the extract that could induce this reduction.

As can be taken from Hill's correspondence, the question was of considerable concern to him and his collaborators in the following years; but it still took them a couple of years after Davenport's first attempt before they were able to obtain any substantial results. Although Hill himself was fully convinced that, in the end, photosynthesis would turn out to function in a stepwise process similar to respiration, for a long time there was no compelling evidence to support this assumption. In October 1951, Hill wrote to French:

> We are still continuing the work on the methaemoglobin factor [i.e. the natural hydrogen acceptor]; it is not very stable so that our progress is bound to be slow. As you know one of our main concerns is how the energy is to be applied to the biochemical systems and now there does seem to be a matter of two alternatives: one big hitch or a lot of little ones.

---

[75] For accounts of the discovery of what eventually became known as "ferredoxin", see Walker (2002); Besse and Buchanan (1998); Bendall (1994) and Arnon (1988).

[76] Hill (1939, p. 207).

[77] See Davenport (1949).

> Franck has always seemed to favour the one big change so I sent him our papers in the hope of finding out what he would think of this other point of view. The *total* amount of experimental evidence seems to leave the question of the two alternatives quite neutral as far as I can see.[78]

This "methaemoglobin factor" that Hill mentioned in the letter was the focus of a paper that Davenport, Hill and Whatley published in the *Proceedings of the Royal Society* in 1952 (i.e. before Whatley went to work with Arnon at Berkeley). Methaemoglobin is a derivative of haemoglobin in which the iron in the haeme group was oxidised to the ferric state (Fe^{III}). In the presence of oxygen the iron of the methaemoglobin was reduced to the ferrous state (Fe^{II}) of normal haemoglobin. This transition is measurable and, hence, a quantitative indicator of the presence of oxygen, as Hill and Scarisbrick had already established in 1940.[79] Now, Hill's group thought they had identified at least one of the natural hydrogen acceptors in the chloroplasts. They preliminarily called it "the methaemoglobin reducing factor" (MRF), after its capacity to reduce methaemoglobin (obtained from whale muscle, as can be taken from the acknowledgement) to its oxidised form.[80] According to the authors' observations, washed chloroplasts were able to reduce methaemoglobin in the light, when the soluble fraction of the chloroplast suspension was added. In the dark, by contrast, no reduction occurred. The rates observed in the light were the same as Davenport had found earlier with acetone extracts of leaves and comparable to the usual rate of the Hill reaction with artificial hydrogen acceptors. This made it highly probable that the same reaction was being observed. The factor in question seemed not to be present in the nongreen parts of the plants. Davenport, Hill and Whatley concluded from these findings "that the factor is reduced directly by the illuminated chloroplasts, that oxygen is produced in the process, and that the factor acts in the sense of a catalyst for the reduction of methaemoglobin in the illuminated system".[81]

The compound's behaviour displayed some similarities to the hydrogen donor of the cytochrome oxidase system in respiration. In particular the rapidity of the factor's reaction with illuminated chloroplasts suggested, the authors thought, that the factor was immediately concerned with the main function of the chloroplasts, even though they had no direct evidence to support this assumption.[82] But Davenport, Hill and Whatley did not hesitate to underline the general coherence of this observation with

---

[78] Cambridge University Library, Ms. Add. 9267/J.58. Hill to French on 6 October, 1951. The same concern was expressed in a letter from Hill to Whatley, on 7 October 1951: "As you may gather my concern now is to find out whether the light causes one big chemical reaction or lots of little ones. During the summer the latter seemed the most satisfying but as you can imagine when one comes to the minute physicochemical details one can get in a tangle." Cambridge University Library, Ms. Add. 9267/J. 116.

[79] See Hill and Scarisbrick (1940a, b); cf. Chapter 4.

[80] See Davenport et al. (1952). The acknowledgement reads: "We are very grateful to Dr J. G. Sharpe for his help in supplying us with whale muscle" (p. 358).

[81] Davenport et al. (1952, p. 346).

[82] Davenport et al. (1952, p. 357).

the finding of cytochrome *f*, which was reported in the same issue of the *Proceedings*. Both indicated, the authors stated, "the partial analogy [...] between the chloroplast and the respiratory system of catalysts that seem generally to be associated with mitochondria". The two analogous systems differed mainly in the direction of the electron transfer, with reference to oxygen: "The methaemoglobin reducing factor may be regarded as reacting with the illuminated chloroplast in a sense *opposite* to a cytochrome reductase reacting with the insoluble respiratory cytochrome oxidase system".[83] It is striking to see also here the fundamental assumption at work that photosynthesis was the reversal of respiration.

Yet, more precise information could not be given at the time, neither on the physiological function of the factor nor on its material identity. It was only years later, in 1960, that Davenport and Hill were able to isolate the actual compound, from leaves as well as from *Chlorella* cells, and characterise the MRF as a non-haeme protein of rather low molecular weight and of deep reddish brown colour. By then, it had been established that the MRF could also catalyse the reduction of a range of other haeme proteins, including cytochromes *c* and b_3. Strangely enough, the factor seemed not to catalyse the reduction of any artificial electron acceptor such as ferricyanide.[84] The redox potential of methaemoglobin was around +100 mV, so that the factor's position in a potential electron transport chain was suspected somewhere "below" cytochrome *f* (that is, more to the negative), while it was unclear how it was related to the redox potential of oxygen. At the time, there was no reason to assume that the MRF would be the agent of the light-induced reduction of TPN (which was later found to be its actual function). As Hill's biographer Derek Bendall pointed out: "The crucial experiment which would have shown that the 'met factor' (ferredoxin as we now know it) was, what he [Hill] was looking for was never carried out"—although Hill would have had the chance to do so: "Almost certainly he possessed a sample of NADP [i.e. TPN, in the language of the time] but there is a suspicion that he regarded it as too precious actually to use."[85]

While the MRF was being investigated at Cambridge, a few years later, in 1956, the biochemist Anthony San Pietro and his associate Helga M. Lang, at the Johns Hopkins University studied the reduction of pyridine nucleotides (DPN and TPN) in chloroplast suspensions. They found that the reduced compounds accumulated even without any "trapping" enzymes having been added, provided that the conditions were appropriately chosen.[86] San Pietro and Lang suggested that "the reduction is an enzyme-catalyzed reaction", and that "the enzyme is highly specific for the

83 Davenport et al. (1952, p. 358). Emphasis added.

84 See Davenport and Hill (1960).

85 Bendall (1994, p. 159).

86 When it was discovered, in 1951, that DPN and TPN were reduced by chloroplasts, the reduced compounds had only been demonstrated indirectly, by means of coupling the process to the enzymatic production of malate.

intact dinucleotide structure".[87] Later in 1958, San Pietro and Lang were able to isolate a soluble protein from spinach leaves that seemed to be the agent of these reductions; they consequently named it after this capacity: "*p*hotosynthetic *p*yridine *n*ucleotide *r*eductase", or PPNR for short. This, San Pietro and Lang believed, was the decisive enzyme that catalysed the light-induced reduction of pyridine nucleotides (it was later found to act specifically on TPN).[88] Still, in 1958, San Pietro and Lang were unaware of the earlier findings by Davenport, Hill and Whatley (or considered them as irrelevant to their work), so that they did not even discuss the possible relationship—which, in actual fact, was an identity relation—between PPNR and the MRF.

A third soluble factor identified in aqueous extracts of spinach leaves, which also catalysed the reduction of TPN by isolated chloroplasts, was reported in 1957 by Arnon's group at Berkeley, and named, according to its displayed capacity, the "TPN-reducing factor". It was noted that this factor was not required for the production of oxygen in those reactions in which TPN was not involved, that is, for example, when ferricyanide or other Hill reagents were added.[89] This latter feature very much resembled the behaviour of the MRF. Yet, while Hill and Davenport, based on their experiments, had no reason to suspect that the MRF had anything to do with the reduction of pyridine nucleotides, there was likewise no reason for Arnon and his colleagues to suspect that there was any connection between their newly found protein and the factor described at Cambridge.

Thus, by 1960, three different electron acceptors of a protein nature had been independently described, and they all seemed to be present in the soluble fraction of chloroplasts: the MRF, the PPNR and the TPN (NADP)-reducing factor. However, slowly evidence began to accumulate that these factors had striking resemblances and, thus, might be closely related to each other: all of them contained non-haeme iron; they had similar absorption spectra in the visible region; they had comparable redox potentials and they were devoid of flavin.[90] By 1959, Davenport had found that the MRF and PPNR were interchangeable in their effects, that is, they both catalysed the photoreduction of either methaemoglobin or TPN in the chloroplast; so that he, finally, suspected that the two compounds might eventually turn out to be one and the same. Furthermore, Davenport reported that more reduction took place when ADP, inorganic phosphate and magnesium ions were added to the reaction mixture.[91] Hill was exceedingly pleased with this discovery. The factor's capacity to form the

[87] San Pietro and Lang (1956). See also Kresge et al. (2005) on the work done by San Pietro and his group, and Pietro (2008) for an autobiographical account.

[88] San Pietro and Lang (1958).

[89] Arnon et al. (1957).

[90] See Pietro (2008, p. 191). This episode reminds one of Bruno Latour's description of how biochemical objects are "constructed" through the investigation of their properties, and how nontrivial it is to establish the identity of two objects that, in the first place, had been defined through different procedures. On this point, see Latour and Woolgar (1985, Chapter 3).

[91] Davenport (1959, 1960).

"reducing equivalents" that were required in the "dark" reactions of photosynthesis and the fact that the rate of reduction was increased upon adding the ingredients of ATP formation were most promising. Thus, Hill enthusiastically stated in his Annual Report: "[T]he original isolation of the active 'methaemoglobin reducing factor' now finally supplies an efficient link between the light driven reactions and the path of carbon in photosynthesis as elucidated by Calvin."[92]

It was only in 1962, that in Arnon's group the reducing factors were finally recognised as belonging to the same protein family: not only did the three factors resemble one another; they were also similar to an iron–sulphur protein that had recently been isolated from the prokaryote *Clostridium pasteurianum*.[93] This latter protein, named "ferredoxin", was found to mediate the transfer of hydrogen to hydrogenases. Kunio Tagawa and Arnon demonstrated that the effects of the MRF, PPNR and the TPN-reducing factor were fully comparable to the effects of ferredoxin, and suggested that the term "ferredoxin" be extended to a whole family of compounds. This was generally accepted, and ferredoxin is listed to this day as an essential element in the photosynthetic electron transport chain.

Going back to the mid-1950s, however, the latter was still a long way ahead. Having established how some of the compounds of a potential photosynthetic electron transport chain were found—cytochromes and the compound that later became known as ferredoxin—it is now time to turn to the first suggestions of how this chain worked. Few people still challenged the principal idea that, most probably, upon illumination chlorophyll became the donor of high-energy electrons, which were then transferred to acceptor molecules and utilised to form reducing equivalents or ATP. The identity of these acceptor molecules was still unclear, however, as were the other details of the mechanism.

7.4 Photosynthetic Electron Transport Chains

7.4.1 Arnon's First Model

Arnon's group in Berkeley was the first, in 1956, to present a suggestion of an explanatory scheme (which is reproduced in Fig. 7.4).[94] In this model, the photolysis of water was identified as the reaction in which light energy was first converted into chemically usable energy. This explained, the group argued, why photolysis

---

[92] See the Annual Departmental Report for the academic year 1958/1959; Cambridge University Library, Ms. Add. 9267/D. 10. In this report, Hill also mentioned that he had been working with Fay Bendall on "the action of oxidation reduction reagents required to establish the hydrogen transport necessary for photophosphorylation".

[93] Tagawa and Arnon (1962) established the convergence of the factors, while Mortenson et al. (1962) presented the *Clostridium* protein.

[94] See Arnon et al. (1956).

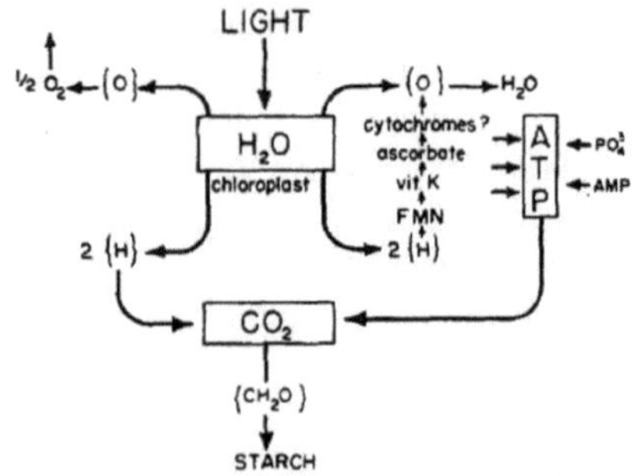

Fig. 8. Scheme for photosynthesis by isolated chloroplasts. Photolysis of water (center) leading either to ATP synthesis and the reconstitution of water (right) or to CO_2 reduction (below) linked with oxygen evolution (upper left).

Fig. 7.4 The photosynthesis model published in Arnon et al. (1956, p. 458).

was a prerequisite for both photophosphorylation and carbon dioxide assimilation. The latter two processes were thought to be in competition with each other: the photolysis of water was either linked with the formation of ATP whereupon the water molecule was immediately restored; or with carbon dioxide reduction, upon which the oxygen of water was released as a gas, while the hydrogen was used in the reduction process.[95] Each of the two processes thus required a separate expenditure of light energy. The authors then formulated the following hypothesis on the details of the phosphorylation process and the underlying electron transport chain:

> [I]t is envisaged that in photosynthetic phosphorylation the recombination of the products of photolysis of water proceeds in several successive steps, which together constitute an "electron ladder" analogous to that discussed for respiration [. . .]. Of the catalysts of photosynthetic phosphorylation Mg^{++} probably has a function in the transfer of phosphate, whereas FMN, vitamin K, and ascorbate could serve as electron carriers in the "electron ladder" shown in Fig. 8 [. . .]. The identity of the electron carriers above ascorbate is unknown, but they may very likely prove to be components of a cytochrome system.[96]

Although the individual steps differed from those envisaged for respiration, the principle was the same. Light energy was used to effect a "splitting" of a water molecule into [H] and [OH]; the "re-unification" of these molecules was an exergonic process, the free energy release of which was used to build up ATP molecules. Possible steps on the "electron ladder" included the cofactors that Arnon's group earlier had found to be necessary for photosynthetic phosphorylation to occur: FMN, vitamin

[95] This was in agreement with the suggestion made fifteen years earlier by Cornelis B. van Niel that the primary photochemical process, namely the splitting of water into [H] and [OH], was the same in all types of photosynthesis, oxygenic and anoxygenic, although in the latter case, the water molecule was immediately restituted; cf. van Niel (1941); see also Chapter 4.

[96] Arnon et al. (1956, p. 259).

K and ascorbate, while they additionally (and hypothetically) included cytochromes to cover the final step of the transport chain.

7.4.1.1 Arnon's Second Model

This scheme was not to last very long, though. In 1957, Arnon's group made several confusing observations that could not be reconciled with their hitherto concept of photophosphorylation.[97] The first surprising result was that, under certain conditions (an alkaline pH value, a rather low chlorophyll concentration level and ADP instead of AMP as substrate), TPN acted as a significant catalyst of photophosphorylation. When TPN was added to isolated chloroplasts, it was immediately reduced, while at the same time substantial amounts of oxygen developed, in agreement with the following reaction:

$$2TPN + 4H_2O \underset{\text{light, chloroplast}}{\longrightarrow} 2TPNH_2 + O_2 + 2H_2O.$$

Although this reaction had been known for several years already, it turned out that, in the presence of ADP and P_i (i.e. inorganic phosphate!), this reaction was found to be coupled to the formation of 2 moles of ATP. Arnon, Whatley and Allen were particularly struck by the stoichiometry of the participating components, which they found to be highly significant; namely that "the synthesis of 2 moles of ATP accompanies the generation of four hydrogen equivalents, which are required for the reduction of 1 mole of CO_2 to the level of carbohydrate".[98] Formulated as an equation, the overall process was the following:

$$2TPN + 4H_2O + 2ADP + 2P_iO \underset{\text{light, chloroplast}}{\longrightarrow} 2TPNH_2 + O_2 + 2H_2O + 2ATP.$$

Thus, the light quanta absorbed seemed to be used simultaneously to produce reducing equivalents *and* to form ATP, although in the model described above it had been assumed that these processes were *competing* with each other for the light quanta. Arnon's group concluded that they had hit upon a different type of photophosphorylation that had so far not been accounted for:

> Photolysis of water is now no longer regarded as resulting either in the synthesis of ATP or in the reduction of CO_2. Adenosine triphosphate [ATP] synthesis is coupled with the formation of the reductant ($TPNH_2$) required for CO_2 fixation. Thus, the same light quanta which accomplish the reduction of TPN also bring about the synthesis of ATP and generate the assimilatory power needed for the conversion of CO_2 into carbohydrates or analogous end products of photosynthesis.[99]

It was found that TPN was much more favoured than DPN; and the most interesting fact was that this kind of phosphorylation occurred without the presence of any additional cofactors, such as vitamin K or flavin mononucleotide (FMN). It had been

---

[97] The first published account of these observations was Arnon et al. (1957), while Arnon et al. (1958) provided a more comprehensive overview.

[98] Arnon et al. (1958, p. 1029).

[99] Arnon et al. (1958, p. 1030).

demonstrated, in the meantime, that only one of them had to be added, and not both, as Arnon's group had previously believed.[100] However, when either of these cofactors *was* added, the resulting picture changed dramatically: "Phosphorylation was sharply increased, whereas oxygen evolution and the accumulation of reduced TPN were abolished."[101] Thus, all the energy was again channelled into ATP synthesis, just as it had been found earlier. The most direct explanation of these results was, Arnon and his collaborators surmised, that the addition of FMN or vitamin K to the reaction mixture initialised again the already well-known path of ATP formation reported earlier, which precluded the reduction of TPN. In order to differentiate between the different types of phosphorylation, Arnon, Whatley and Allen decided to call the first type "cyclic" photophosphorylation, since all the electrons were kept within the system, while the alternative process, in which the reduction of TPN was coupled to ATP formation, was named "noncyclic" photophosphorylation.

This differentiation was taken up and elaborated in a comprehensive summary of the research done by the group during the 1950s.[102] Cyclic photophosphorylation was described as the production of ATP only, without any reducing power being accumulated, while noncyclic photophosphorylation yielded the actual "first" products of photosynthesis: reduced TPN and ATP, both of which were required to accomplish the reduction of carbon dioxide to sugar phosphates. It was believed that the purpose of cyclic photophosphorylation was to provide additional ATP—possibly the ATP produced in the noncyclic version was insufficient for the carbon reduction process.[103]

Arnon believed that, in cyclic photophosphorylation the "downhill" electron transport could proceed via two alternative pathways, one of which involved FMN as an intermediate carrier and the other required vitamin K (see Fig. 7.5; the process of cyclic photophosphorylation is marked at the top as involving only the processes on the far left of the figure). It was assumed that, along the vitamin K pathway, two electrons expelled from chlorophyll were transferred to vitamin K, thereby reducing the latter. The reduced vitamin K was, in turn, reoxidised by a cytochrome component, which then donated electrons back to the chlorophyll. Phosphorylation was thought most probably to be coupled to the oxidation of the terminal cytochrome component, while a second phosphorylation event, coupled to the reoxidation of vitamin K, was not precluded. The pathway via FMN was conceptualised as a modification of the vitamin K pathway, and in this case TPN and two cytochrome components were assumed to be involved. The actual phosphorylation was localised at the terminal cytochrome. In Fig. 7.5, this is depicted on the left-hand side (bottom): one of the two arrows starting from Cyt_I leads to the chlorophyll (Chlp) and the other to $\sim$P

---

[100] See Whatley et al. (1957).

[101] Arnon et al. (1958, p. 1030).

[102] Arnon (1959).

[103] Details of the need for the sufficient and well-balanced excitation of both cyclic and noncyclic photophosphorylation were published in Trebst et al. (1959) as well as in Losada et al. (1959); both were confirmed in Trebst et al. (1960).

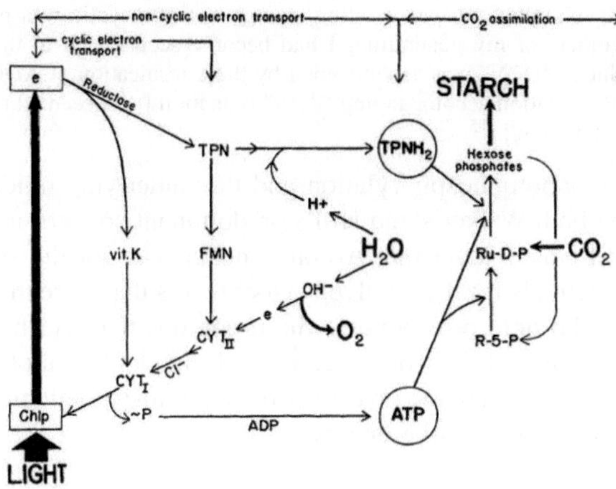

Fig. 7.5 The photosynthesis model as it was published in Arnon (1959, p. 14). The process of cyclic photophosphorylation is marked at the *top* as involving only the processes on the *left* of the figure. It proceeds either via FMN or via vitamin K. Noncyclic photophosphorylation differs from the former in that it involves the reduction of TPN.

(which indicates the formation of a high-energy phosphor bond) and eventually to ATP.

Noncyclic photophosphorylation was thought to involve the same primary photochemical reaction and the same phosphorylating site. It appeared highly unlikely that the two mechanisms should travel along completely different paths. The main difference concerned the fate of the electrons: in noncyclic photophosphorylation the electrons did not return to the chlorophyll, but were removed from the system by the reduction of TPN. The chlorophyll, in turn, was replenished again "by an interaction between hydroxyl ions (or water) and a cytochrome component peculiar to the photosynthetic apparatus of green plants but absent in photosynthetic bacteria".[104]

Explaining noncyclic photophosphorylation in this vein implied that the direction of the general process was the exact reverse of that in oxidative phosphorylation. The general idea was undisputed, as mentioned earlier: while in the latter process oxygen was consumed, in noncyclic photophosphorylation, oxygen was produced. This was well known. Yet, in respiratory, oxidative phosphorylation, reduced pyridine nucleotides (DPN) were oxidised; in non cyclic photophosphorylation now the analogous pyridine nucleotides (TPN) were reduced. This caused considerable consternation and worry among photosynthesis researchers. David A. Walker, then a postdoctoral student of Hill's at Cambridge, remembered how in the late 1950s he had battled to understand fully the curious photophosphorylation phenomena and their explanations that were coming out of Arnon's laboratory. He found the results confusing to say the least:

---

[104] Arnon (1959, p. 16). This (hypothetical) mechanism was not depicted in the figure.

> Cyclic [photophosphorylation] was exciting enough but non-cyclic was positively mind boggling. Like others of my generation, I had become accustomed to the idea that the oxidation of reduced "DPN" was accompanied by the esterification of ADP to yield ATP. Conversely, ATP formation accompanying "TPN" reduction [. . .] seemed remarkably like getting water to run up hill.[105]

The mechanism of photophosphorylation and the underlying oxidation–reduction processes became both Walker's and Hill's predominant concern in these years. To Hill and others, it was evident that Arnon's model was not the solution, since it assumed redox potentials for the involved cytochromes that were more positive than +0.81 mV, which did not correspond to the properties that had been observed.[106] However, before I turn to Hill's work once again, I shall look at another decisive area of development in these years: the investigation of photosynthetic pigments and how these pigments used incident solar energy.

7.5 Energy Migration, Chlorophyll *a* and P700

7.5.1 Early Observations

In the 1950s, it was widely assumed, as mentioned earlier, that upon illumination, chlorophyll became the donor of high-energy electrons, and it was the further fate of these electrons that Arnon, Hill and others tried to reconstruct. Yet, how the chlorophyll was able to produce these electrons in the first place—in fact, how it "absorbed" the energy of the light, was far from clear. In 1936, Hans Gaffron and Kurt Wohl had proposed that a couple of thousand of pigment molecules might "cooperate" in the absorption of light energy and in the reduction of carbon dioxide (based on Emerson and Arnold's 1932 experiments; see Chapter 4). Yet, even though this sounded like an exciting idea, by as late as the 1950s, Franck, for example, was still strongly opposed to this extravagant notion.[107] The main point of concern was exactly how this "cooperation" of light-absorbing molecules could possibly function. Given the speed of photosynthetic carbon reduction, it was impossible to imagine that the carbon dioxide moved freely around in the chlorophyll and picked up the required light quanta here and there. Carbon dioxide reduction could only occur at specific sites within the chlorophyll, which Gaffron and Wohl had tentatively called "reducing centres". The question then was how the absorbed light energy could be transported from the surrounding chlorophyll molecules to these centres.

---

[105] Walker (1995, p. 45).

[106] See, e.g., Hill and Bendall (1960b, p. 137).

[107] Franck persistently thought that energy transfers of this kind were impossible, pointing to the paper Franck and Teller (1938). The inadequacy of the latter approach was finally demonstrated in Robinson (1967). See also Chapter 4 of this book.

Around 1940, two explanatory alternatives existed: either the actual particles that carried the energy to the reducing centres might be moving (e.g. intermediate radicals), or the energy quanta themselves moved through a structurally coordinated ensemble of chlorophyll molecules in such a way that the energy of these light quanta would finally be captured by the reducing centre (which might or might not be identical with chlorophyll molecules).[108] The latter concept became known as the "optical model of the photosynthetic unit", which was strongly favoured by the physicist Wohl as the most probable explanation. "If the chlorophyll molecules are packed flatly one over the other in such a manner that all absorbing centres are in direct contact, then this crystalloid structure will be activated as a whole as soon as a light quantum is absorbed at any point", he argued. An important reference, thereby, he argued, was the analogous conception that Max Delbrück, Nikolay Timofeev-Ressovsky and Karl G. Zimmer had developed for the "gene" and its structure.[109] The close linkage between the work of Delbrück et al. in 1935 and Gaffron and Wohl's 1936 concept of the "photosynthetic unit" was already mentioned (see Chapter 4).

Closely related to how the chlorophyll molecules might be able to interact was the role of the further, so-called auxiliary or accessory pigments, which are present in all photosynthesising plants: the carotenoids, including the xanthophylls (such as fucoxanthin) or, in other organisms, the phycobiliproteins (such as phycoerythrin and phycocyanin). While it seemed beyond any doubt that these pigments also absorbed light energy, no one knew whether or not this energy was used for photosynthesis, and if so, whether it initialised a process different from the one prompted by chlorophyll. These were questions that received considerable attention towards the end of the 1930s—for example, at the University of Wisconsin in Madison, one of the first places in North America where the field of limnology had begun to thrive around 1900.[110] One of the topics of interest here was the theory of chromatic adaptation in submerged aquatic plants, which suggested that the colours of algae enabled them to use optimally the spectrum of the sun's solar radiation penetrating to different sea depths.[111]

In this local context, a study of photosynthesis in the diatom *Nitzschia closterium* (a unicellular marine alga) was undertaken in 1940 by a PhD student in plant physiology Herbert J. Dutton. In order to study the relative contribution of the accessory pigment—which in this case was mainly fucoxanthin—the photosynthetic quantum efficiency of the diatom was measured in low light intensities. We already know that this kind of experiment was extremely popular around 1940 (see Chapter 5). In particular, Dutton compared the efficiency at wavelengths where the absorption of

---

[108] A comprehensive review of this problem was given in Wohl (1940). However, at the time, Wohl still thought that: (a) photosynthesis required no more than 4 light quanta to produce one molecule of oxygen; and (b) that carbon dioxide was also firmly attached to the postulated reducing centres of unknown material identity; see Chapter 4.

[109] Wohl (1940, pp. 47–48).

[110] Cf. Beckel and Haywood (1987).

[111] Dutton (1997, p. 175).

chlorophyll was most prevalent with the efficiency at wavelengths where mainly the pigment fucoxanthin was absorbing. Efficiencies were measured in terms of oxygen release. In view of the data, Dutton and his mentor and co-author, the plant physiologist Winston Manning, maintained that "it appears necessary to conclude that light absorbed by some or all of the carotenoid pigments in *N. closterium* can be utilized in photosynthesis".[112] These results were elaborated, in order to learn whether or not the carotenoids were feeding the absorbed light energy into the chlorophyll pathway. This was determined by using chlorophyll fluorescence at different wavelengths: "If there were no transfer of energy, the yield of chlorophyll fluorescence should vary in proportion to that fraction of the absorbed light which is absorbed by chlorophyll", is how the authors explained the underlying rationale.[113] By contrast, a constant yield at various wavelengths would be strong evidence for the occurrence of the efficient transfer of absorbed energy from other pigments to the chlorophyll—and this in fact was observed: despite the low light absorption of chlorophyll in red light, the fluorescence yield of the diatom *N. closterium* was almost constant over the full range of the visible spectrum:

> [I]t may be concluded that carotenoid-sensitized photosynthesis in *N. closterium* takes place through the transfer of absorbed energy from carotenoid molecules to chlorophyll molecules with subsequent reactions the same as though chlorophyll molecules were the primary absorbers.[114]

Similar conclusions were arrived at by Emerson and Charlton M. Lewis, who, around the same time, had turned to study the role of accessory pigments in the blue-green alga *Chroococcus* and found that "light absorbed by phycocyanin is utilized in photosynthesis with an efficiency approximately equal to that of the light absorbed by chlorophyll."[115] In view of these findings, Emerson suggested to his former collaborator William Arnold, who was then working at the Hopkins Marine Station, Pacific Grove, to find out "if the energy absorbed by phycocyanin was being transferred to chlorophyll or was the phycocyanin doing photosynthesis."[116] Arnold established that the energy was being transferred to chlorophyll *a* and he went up to Berkeley to talk to the well-known physicist J. Robert Oppenheimer about the problem. The outcome was the idea that this energy transfer was analogous to the "internal conversion" of gamma rays. The suggestion was first presented by Oppenheimer at a meeting of the American Physical Society in 1941, although the concept was only developed in a paper in 1950.[117] Arnold and Oppenheimer confirmed that about 90 %

[112] Dutton and Manning (1941, p. 525). The diatom is today known as *Phaeodactylum tricornutum*. See Dutton (1997) for his memoirs of the episode.

[113] Dutton et al. (1943, pp. 308–309)

[114] Dutton et al. (1943, p. 312).

[115] Emerson and Lewis (1941b, p. 594).

[116] Arnold (1991, p. 77).

[117] See Oppenheimer (1941) and Arnold and Oppenheimer (1950). The general thought was also mentioned in a letter written by Oppenheimer to Wolfgang Pauli dated 16 April 1945, in which he offered a piece "on the analogy between the sensitization of photosynthesis on the one hand, and

of the light energy absorbed by phycocyanin was used in photosynthesis; and they additionally suggested that this energy was transferred to chlorophyll *a* by internal conversion mechanisms, that is, the nonradiative "resonance transfer of energy from one oscillator to another in resonance with it".[118] The study of the effect of auxiliary pigments was later continued by one of Emerson's students, Takuma Tanada, who in 1951 provided a fine action spectrum of the diatom *Navicula minima* (in which, incidentally, the Red Drop of photosynthetic efficiency was marvellously displayed) and confirmed the high photosynthetic efficiency of fucoxanthin.[119]

7.5.2 Fluorescence Resonance Transfer

Neither Dutton's nor Emerson and Lewis's results were received with great enthusiasm. It took another three years before the findings were confirmed,[120] while the real excitement about light energy transfer between photosynthetic pigments only got going in the 1950s. The first of these path-breaking studies was undertaken in 1952 by Charles Stacy French and his associate Violet Young, who examined the energetic side of photosynthesis in the unicellular red alga *Porphyridium* with the help of the newly developed "spectrofluorimeter". French and Young were able to demonstrate rather convincingly the transfer of energy between pigments, and suggested that phycocyanin might be an intermediate in the transfer of energy from phycoerythrin to chlorophyll. Furthermore, French and Young speculated that since "phycoerythrin and phycocyanin transfer energy to chlorophyll, it appears probable that chlorophyll plays a specific chemical role in photosynthesis in addition to acting as a light absorber".[121]

In the same year, 1952, Louis N. M. Duysens submitted his PhD thesis to the University of Utrecht in the Netherlands.[122] Later that same year, Duysens was invited to present the results of this thesis at the First Gatlinburg Conference on Photosynthesis (which was already mentioned in chapter 5): an extremely efficient way of circulating the findings. By using a newly developed sensitised fluorescence method, Duysens was able to demonstrate, in cyanobacteria and algae, the occurrence of far-reaching and highly efficient energy transfers from the auxiliary pigments, including chlorophyll *b*, to chlorophyll *a*. According to Duysens's data, all the light energy that was to be used chemically in photosynthesis had to pass through chlorophyll *a*. Whatever

the internal conversion of gamma rays on the other" as a contribution to an envisaged "Festschrift" for Niels Bohr. See Pauli (1993, p. 269) (Doc. 724).

[118] Arnold and Oppenheimer (1950, p. 424). See also Knox (1996) on this episode.

[119] See Tanada (1951).

[120] This was in Wassink and Kersten (1946).

[121] French and Young (1952, p. 889).

[122] Duysens's first results were published already one year earlier, in Duysens (1951), but see Duysens (1952) for the complete thesis.

energy was not transferred this way was lost. Analogously, Duysens found that the energy transfer in purple bacteria had to pass through a specific bacteriochlorophyll, which he named *B*890 (which was absorbing at 890 nm). Duysens also suggested a potential mechanism:

> The transfer of electronic excitation energy between pigment molecules as reported above probably takes place through inductive resonance between the excited molecules and the molecules in the ground state, a theory for which [Theodor] Förster has given a quantum-mechanical treatment, with the aid of which the probability of energy transfer can be calculated from experimental data. Estimations, based on Förster's considerations, are in accordance with, or at least do not contradict, the results recorded above.[123]

The phenomenon in question, which became known as "fluorescence resonance energy transfer", had been discovered in 1946 by the German physical chemist Theodor Förster—the same phenomenon in photosynthesis had already been discussed 1941 in the aforementioned paper by Oppenheimer, but Förster was unaware of this, and, obviously, Oppenheimer's paper had not made its way efficiently into photosynthesis circles.[124] Already in his first publications, Förster had pointed to the possible importance of these energy transfers in photosynthesis, given Gaffron and Wohl's suggestion of 1936 that a functional photosynthetic unit of chlorophyll molecules might exist.[125] Förster's theory was only widely and internationally received on the publication in 1951 of his monograph; and Duysens productively picked up the notion of induced resonance to explain his comprehensive set of data. The idea was that when a pigment molecule was excited, through the absorption of light energy, it would induce electronic vibrations in a neighbouring molecule, which at the same time would receive an electronic quantum. The latter was converted, upon transfer, from a higher into a slightly lower energy state. This process was all the more efficient, Duysens summarised in a review of 1956, "(a) the better the overlapping of the fluorescence spectrum of the transferring molecule with the absorption band of the receiving molecule, (b) the smaller the distance between the two molecules, and (c) the greater the fluorescence yield of the transferring molecule and the specific absorption of the receiving molecule".[126] Duysens compared this mechanism to a "bucket brigade", which went from those pigments that were absorbing towards the blue end of the spectrum to those absorbing at longer wavelengths, terminating at a specific type of chlorophyll *a* (or at the bacteriochlorophyll *B*890 respectively).

The speculation that there might be "specific" chlorophyll *a* molecules was another important outcome of his study (and, obviously, of high interest if one thought of Gaffron and Wohl's "reducing centres"). Duysens had obtained the confusing and

[123] Duysens (1951, p. 549).

[124] Oppenheimer (1941).

[125] See, for the original publication of Förster's results and his reflections on photosynthesis, Förster (1946, 1947, 1951). On Förster's life and work in general, see, e.g., Porter (1976). Govindjee (2004a, p. 19), provides an English translation of some central paragraphs of Förster's 1946 paper. See Clegg et al. (2010) for a historical review of the phenomenon.

[126] Duysens (1956, p. 34).

rather paradoxical observation that the light quanta absorbed by the phycobilins prompted chlorophyll *a* molecules to a stronger fluorescence intensity than the light quanta absorbed by chlorophyll *a* itself.[127] Duysens maintained:

> From these observations it may be concluded, not only that chlorophyll *a* occurs in these cells in two different modifications, differing in fluorescence yield, but also that energy is transferred from the phycobilins to the highly fluorescent part of chlorophyll *a*, probably with high efficiency, and that consequently energy is not, or only to a slight degree, transferred to the weakly fluorescent chlorophyll *a* molecules.[128]

Thus, Duysens suggested that there were two different types of chlorophyll *a*, which he identified as "active" (high fluorescence yield) or "inactive" (low fluorescence yield) in photosynthesis.[129] The light absorbed by phycoerythrin, Duysens suggested, was transferred mainly to chlorophyll *a* of the active type—he presented no convincing argument, however, as to why this should be the case. After having completed his PhD thesis, Duysens pursued these questions of reversible changes of light absorption further. Looking back at this time, Duysens wrote:

> When it had become clear that excitation energy was transferred to photosynthesis via (bacterio)chlorophyll, I began thinking about a method for studying the photochemical events following the excitation of these chlorophyllous pigments. I reasoned that if chlorophyll participated directly in the photochemical reaction, its absorption spectrum would presumably change, like that of other pigments upon oxidation or reduction. The possible absorption changes, occurring in the photosynthesizing cells upon illumination, should be small, since otherwise they would have been discovered already by the naked eye, which is rather sensitive to color changes.[130]

If the energy were transferred to a certain, presumably rather small fraction of the chlorophyll or bacteriochlorophyll, this hypothetical type of pigment P could then be assumed to be photochemically active and change its absorption spectrum upon illumination. Based on this assumption, Duysens set out to study the absorption changes of the different molecules present in the chloroplast. In a letter to the editor of *Nature*, published in April 1954, he reported, with reference to data recently obtained, that in living bacteria in the absence of oxygen and in the presence of substrate "a cytochrome pigment is oxidized by illumination and is reduced in the dark".[131] Duysens thus felt entitled to propose that according to his experiments "the photosynthetic oxidation of the substrate in *Rhodospirillum rubrum* is mediated by a cytochrome pigment". This was a stunning conclusion, given the fact that, at the

---

[127] Duysens first observed this phenomenon using the red alga *Porphyra lacineata* but it was confirmed in other organisms. His observation was in agreement with earlier findings reported in Haxo and Blinks (1950); these authors had used a polarographic method to measure oxygen, which enabled them to measure rapidly photosynthetic action spectra.

[128] Duysens (1951, p. 549).

[129] These "inactive" portions of chlorophyll *a* were later identified as part of photosystem I, in which TPN, or, more in today's nomenclature, NADP, is reduced.

[130] Duysens (1989, p. 67).

[131] Duysens (1954b, p. 692).

time, with the exception of Hill's laboratory at the University of Cambridge, very few scientists believed that cytochromes played a role in photosynthesis.

The question that naturally followed was whether the same was true of green plants: that here, too, the oxidation of the reductant—water—was mediated by a cytochrome. This would imply that in higher plants and algae light-induced changes of the cytochrome's oxidation state also occurred. At the time, Duysens had gone to spend a period of research in French's laboratory at the Carnegie Institution's Department of Plant Biology. Duysens learned that French himself, together with the biochemist-turned-biophysicist Britton Chance (who was mentioned earlier in this chapter), had been looking in vain for light-induced absorption changes in *Chlorella*. Yet, they had used Chance's apparatus, and did not then have the advantages of an absorption difference spectrophotometer. Duysens thought that he could put together a home-made instrument of that type, and French encouraged him to try.[132]

The results of this study were unfortunately less clear-cut than in the case of bacteria. In *Chlorella*, there was a peak of the difference spectrum associated with the changes of the oxidation state of a cytochrome at 420 nm; however, there was also a much higher peak at 520 nm and another at 480 nm, both of which could not possibly be associated with cytochromes.[133] The data gathered using the red alga *Porphyridium* were in better agreement with the assumption that, for example, Hill's cytochrome *f* was being reduced and oxidised; but even in this case the picture was far from conclusive.[134] In retrospect, these confusing results are no longer surprising: the action spectra of photosynthesis can only provide useful information quantitatively when no more than *one* photoreaction is involved, while in algae two photoreactions obtain (which was, of course, still unknown in 1955). In view of these difficulties with plants, Duysens went back to study the more simple systems of bacteria.

To this end, he again improved the instrumental set-up and devised an extremely sensitive differential spectrophotometer that was able to detect even minute changes of optical absorbance.[135] In 1956, Duysens suggested that his findings were compatible with the effect of oxidation–reduction processes in which bacteriochlorophyll and cytochromes were involved.[136] He formulated his proposal as a series of equations, which were (reformulated in a slightly more explicit form) the following:

$$\text{BChl} + \text{Light} \rightarrow \text{BChl}^*$$

$$\text{BChl}^* + \text{X} \rightarrow \text{BChl}^{\text{ox}} + \text{XH (photosynthetic reductant)}$$

[132] See Duysens (1989, p. 68).

[133] See Duysens (1954a).

[134] See Duysens (1955).

[135] The first findings with this instrument were reported in Duysens et al. (1956, p. 188).

[136] See Duysens et al. (1956, p. 190).

Thus, upon illumination, excited bacteriochlorophyll (BChl*) was formed. In the next step the latter was (photo)oxidised, while simultaneously an unknown substance, X, would be reduced to XH, which acted as the "photosynthetic reductant", as Duysens called it. The next steps were:

$$BChl^{ox} + cyt^{red} + proton(s) \rightarrow BChl + cyt^{ox}$$
$$cyt^{ox} + XH \rightarrow cyt^{red} + X + \text{free energy to be used in ATP formation.}$$

The oxidised bacteriochlorophyll ($BChl^{ox}$) might be replenished again by oxidising one or more reduced cytochrome compounds that were assumed to be present in the system. A fraction of the thus oxidised cytochrome, Duysens thought, might return to its reduced state by oxidising the XH, thereby releasing the energy that was required to generate ATP. Duysens believed that something similar might apply to photosynthesis in higher organisms, where the "photosynthetic reductant" would reduce TPN instead of cytochromes. The difference between the free energy of XH and TPN was sufficient, he calculated, for the phosphorylation of 0.5 ADP molecule to take place per reduction of one TPN. This hypothesis thus could actually explain Arnon's non-cyclic photophosphorylation coupled to the "mind-boggling" reduction (as Walker put it, see above) of TPN. However, even Duysens himself felt compelled to add: "It should, however, be pointed out that this is not the only possible interpretation [of the spectroscopic data]."[137]

It was in 1957 that Seymour Steven Brody and Rabinowitch, both working in the Photosynthesis Project at the University of Illinois at Urbana–Champaign (see Chapter 5), were able, for the first time, to provide evidence of the excitation energy transfer from their kinetic studies.[138] They had obtained their data by means of ultrafast fluorescence spectroscopy experiments, which were able to trace the excitation lifetime of photosynthetic pigments. These times were in the order of nanoseconds, that is, 10^{-9} seconds; and it was no trivial exercise to provide such a short light pulse in 1957. Ordinary flash lamps were far too slow, so that Brody and Rabinowitch eventually used a small hydrogen lamp for this purpose. The drawback of the latter was the relatively low intensity of this lamp, so that all the components of the experimental set-up—lamp, coloured glass filters, sample and photodetector—had to be assembled as closely together as possible. The fluorescence signal induced by the flash was detected using a photomultiplier, which was applied directly to the plates of an oscilloscope, and the display was then photographed. By this means, they eventually measured the fluorescence lifetimes of several pigments "*in vitro* [...] with a precision of $\pm$ 7 per cent, and—for the first time—also *in vivo*, with a precision of $\pm$ 20 per cent". These were excellent degrees of precision at the time. What they found was that the measured values for excited chlorophyll molecules in

137 Duysens et al. (1956, p. 190). The hypothesis was repeated and elaborated in Duysens (1957).

138 See Brody and Rabinowitch (1957) for the original publication. Brody (2002) provides a historical and partly autobiographical mini-review of fluorescence lifetime studies and energy transfer in photosynthesis. For Brody's obituary, see Hirsch et al. (2010).

vivo differed greatly from the previously calculated values (while the data in vitro were in good agreement with the calculations). The authors suggested:

> One possible interpretation of this discrepancy is to assume two forms of chlorophyll *in vivo* (a hypothesis for which some spectroscopic evidence has been obtained by other investigators); the fluorescent form must then account for about one fourth of the total, and the non-fluorescent form for about three-fourths of the total.[139]

This would have nicely confirmed Duysens's findings of two different forms in chlorophyll *a* in red algae (mentioned earlier in this chapter). Brody and Rabinowitch admitted, however, that the discrepancy could also be attributed to the special conditions of the experiment and, as always, to possible differences in the physiological state of the algae. Equally important was their finding that it proved possible to measure the times required for the transfer of excitation energy. In retrospect, Brody described the experiments and their results as follows:

> The phycoerythrin was irradiated with a nanosecond burst of green light. The excitation energy absorbed by phycoerythrin is transferred to phycocyanin and subsequently to chlorophyll. Some of the excitation energy transferred to chlorophyll is emitted as fluorescence. The time between the nanosecond burst of green light and the appearance of the red fluorescence from chlorophyll is the time required to transfer excitation energy in the red alga. [...] The measured time for energy transfer is 0.5ns.[140]

Thus, by 1958 there could be little doubt that several different pigments contributed their excitation energy to photosynthesis, while chlorophyll *a* was the only pigment directly responsible for making the excitation energy available for the further steps of photosynthesis. It not only absorbed sunlight itself; it was also the final step in a chain of highly efficient transfers of excitation energy from those pigments with absorption bands at shorter wavelengths to those with bands at longer wavelengths.[141] Whether or not there were, in fact, different types of chlorophyll *a*, only one of which was photosynthetically active, was a question that awaited further clarification.

7.5.3 Special Pigments and the Antagonistic Light Effect

While in 1956 Duysens had found certain unusual and particularly important molecules of bacteriochlorophyll, which he had dubbed *B*890, Bessel Kok, then working in Wageningen, The Netherlands, discovered that same year a comparable pigment in higher plants, algae and cyanobacteria. In studies very similar to the ones undertaken by Duysens, Kok had observed that "all plants containing chlorophyll *a* showed, after irradiation with a strong light flash, a temporary decrease in absorption

[139] Brody and Rabinowitch (1957, p. 555). The lifetime of chlorophyll *a* fluorescence in vivo was also measured, independently and using other methods, by Dmetrievsky et al. (1957).

[140] Brody (2002, p. 129). The passage refers to the results of Brody and Rabinowitch (1957).

[141] See Emerson and Chalmers (1958, p. 15), for a succinct summary of the situation up to then. See also Emerson and Rabinowitch (1960, p. 477).

(lifetime in the order of 0.01 s) in the spectral area between 620 mμ and 720 mμ".[142] Already then, Kok interpreted these changes as indicating "the photochemical transformation of a pigment that is different from chlorophyll *a* in its normal status and that occurs universally in the plant kingdom".[143] In 1957, Kok confirmed this finding of a short-lived, light-induced absorbance decrease in several photosynthetic organisms, which had its highest wavelength at 700 nm; whereupon he labelled the pigment in question "P700".[144]

In 1959, Kok presented a much improved apparatus, which was applied to the study of absorbance changes in the cyanobacterium *Anacystis nidulans*. Although the data were complex, they seemed to indicate that there was an antagonistic effect on the pigment upon illumination with different sorts of light: while far-red light yielded the decrease of background absorption that Kok had seen before, light of shorter wave-lengths (red or white) had the reverse effect.[145] Kok drew attention to the fact that an interpretation of this finding could be provided in view of the discovery of enhancement effects by Emerson and his co-workers: they had found that, at low light intensities, the severe drop in photosynthetic efficiency for wavelengths beyond 680 nm (that is, far-red light) could be compensated for by the simultaneous illumination by red light of shorter wavelengths (see Chapter 5).[146] Kok considered the following:

> The location of the absorption band of the 700 mμ pigment [...] makes it ideally suited to effectively trap all the light which is absorbed by chlorophyll *a* itself, or transferred to it by accessory pigments [...]. Suppose the 700 pigment is indeed the final light sink in photosynthesis, and its excitation by surrounding chlorophyll *a* molecules leads to a conversion, which entails disappearance of its absorption. Then, from there on, the neighboring chlorophyll *a* molecules—say those comprising a "unit"—are deprived of their outlet until the 700 pigment is restored. [...] As was described above, irradiation with red light restores the 700 mμ pigment and therefore sustains the conversion of irradiation transferred to it via chlorophyll *a*.[147]

Evidently, Kok was already then playing with the idea of two different photochemical reactions requiring light of different wavelengths, which would act in sequence upon P700. Yet, Kok hesitated to push these speculations further, and explicitly mentioned that he would not have published the results and their preliminary interpretation "if it were not to draw attention to a new road towards a better understanding of the most fundamental aspects of photosynthesis, opened up by the eminent scientist to whose memory this article is dedicated".[148] This alluded to Emerson's finding of the enhancement effect (which already then had made it very probable that two different

[142] Kok (1956, p. 399). Read "nm" for "mμ".

[143] Kok (1956, p. 401).

[144] Kok (1957).

[145] See Kok (1959, p. 190).

[146] The main papers are Emerson et al. (1957) and Emerson and Chalmers (1958).

[147] Kok (1959, pp. 191–192). Read "nm" for "mμ".

[148] Kok (1959, p. 192).

pigments with different absorption properties initiated two different photochemical reactions) and to his untimely death in an aircrash: Kok's 1959 paper was printed in a special issue of *Plant Physiology*, dedicated to the memory of Emerson. Kok continued his thinking on two different photo systems, as will be presented later in this chapter.

7.6 A Solution That Was in the Air

To sum up, by 1959 it was known that chloroplasts—at least in vitro—were able to utilise light energy to form, on the one hand, reduced TPN (which was known to act in other cases as a strong reductant), upon which a stoichiometric amount of molecular oxygen was released; and, on the other hand, ATP, provided that the chloroplasts were supplied with ADP and inorganic phosphate. The four-hydrogen equivalents required to reduce the necessary number of TPNs were, most probably, taken from water molecules. For each molecule of oxygen thus released, approximately two ATPs were found to be formed. Furthermore, it was suspected that electron transport chains with a slow and stepwise energy release were involved in the formation of ATP, and that cytochromes (and the factor that was later identified as ferredoxin) were possibly part of this chain—notably the cytochromes discovered at Cambridge to be specifically present in green plant tissues: cytochromes f and b_6. Support for the involvement of cytochrome f was provided by Duysens's finding that it was reversibly oxidised in the light. In addition, it was also suspected that there might be different types of chlorophyll *a*, which presumably were relevant to the course of the light reaction; not least, there was Kok's mysterious P700, which reacted antagonistically to light of different wavelengths. And, finally, Emerson's finding of an enhancement effect was increasingly being discussed in terms of a "short wavelength" and a "long wavelength" pigment system both of which somehow acted cooperatively to achieve photosynthesis in plant cells.

These different strands of evidence, so far presented in different sections of this chapter, were drawn together by a number of research groups, which more or less simultaneously came to similar conclusions, albeit from different starting points. Around 1960, the problem was to find a model of the photosynthetic mechanism, which was able to accommodate most of the pertinent data and, at the same time, adhere to theoretical preconditions, such as the known redox potentials of certain molecules. Hill and Fay Bendall were the first to publish, in 1960, how, from a thermodynamical point of view, the cytochromes could be arranged in a redox chain so as to provide enough energy to form ATP. At the same time, Kok elaborated his aforementioned suggestion, based on his finding of P700. Neither Hill nor Kok recognised the role of different pigments for the different light reactions, which had been discussed since the finding of the Emerson enhancement effect in 1957. The latter provided the starting point, however, for the research carried out by the groups working around Duysens and the German biophysicist Horst T. Witt; and it guided further studies of the different types of chlorophyll and the course of chlorophyll *a* fluorescence carried out at Urbana by Rabinowitch and Govindjee. All these contributions complemented each other, and all were decisive in endorsing the two photoreactions, two pigment systems model, which in outline still holds today.[149]

7.6.1 Far-Sighted Ideas at Urbana

It needs to be emphasised that what was achieved around 1960 was not the discovery of the possible *existence* of two different photochemical reactions. This idea, as a viable explanatory approach, had been around since Rabinowitch's monograph of 1945 at the latest—indeed, the notion that there might be more than one photochemical process had been a common assumption since the early 1930s. It was mentioned in chapter 3 that various actors, including Arthur Stoll, Richard Willstätter and Franck, believed that several photochemical steps were needed to achieve carbon dioxide reduction (which, at the time, was thought to be the light reaction). Franck and Herzfeld even argued, in 1941, that there were no less than eight photochemical steps (although of fundamentally the same nature), each requiring the energy of one light quantum.[150] These schemes lost their attraction when it emerged that the light reaction was concerned with the "splitting" of water and the dark reaction effected the reduction of carbon dioxide; yet, the question remained why so many light quanta seemed to be necessary to complete the full reaction series of photosynthesis.

Rabinowitch discussed the various models of the primary photochemical process in the first volume of his photosynthesis monograph in 1945. He believed that two alternatives might explain why photosynthesis seemed to require, at the very least, 8 light quanta—which he already then assumed was the accurate number (see, however, chapter 5; at this time, the real controversy on this value had not even started): either one could "activate the same four hydrogen atoms photochemically twice in succession" or one could "double the number of identical primary photochemical processes". The latter option was in line with the aforementioned suggestion brought forward by Franck and Herzfeld, in which 8 quanta were used up by eight identical photochemical processes.[151] The former option, Rabinowitch believed, had to involve photo-oxidations (in which hydrogen atoms were taken away from the water) and photo-reductions (in which the same hydrogen atoms were transferred to carbon dioxide or an intermediate acceptor). In order to clarify this option, Rabinowitch

[149] Stories of how it was established that there are two photosystems have been told repeatedly (mainly in retrospection by the actors themselves or close collaborators); see, e.g., Duysens (1989); Witt (1991) and Amesz (1998). Historical accounts were also given in Govindjee (2006); Govindjee and Björn (2012) and Govindjee et al. (2012).

[150] Stoll (1932); Willstätter (1933); Franck and Herzfeld (1941).

[151] Rabinowitch (1945, p. 161).

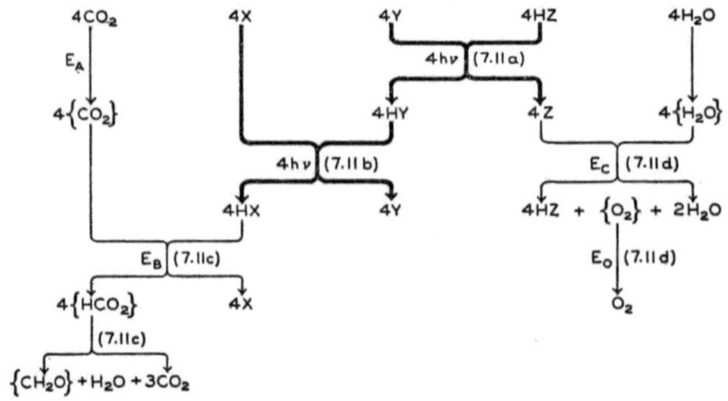

Fig. 7.6 Reproduced from Rabinowitch (1945, p. 162). Caption: "Photosynthesis with oxidation-reduction reactions between three intermediary catalysts [X, Y, Z] as the two primary photochemical processes. The central catalyst, which participates in both photochemical reactions, may be chlorophyll".

designed a scheme in which photosynthesis was framed as a series of redox reactions between three intermediary catalysts: X, Y and Z (see Fig. 7.6; the catalysts and their reduced states—HX, HY, HZ—are in the centre of the graph).[152] The first light reaction effected the transfer of hydrogen from HZ to Y, upon which Z reacted with H_2O to evolve oxygen (on the right). The second light reaction transferred hydrogen from HY to X, upon which HX reduced carbon dioxide to carbohydrates (on the left). Rabinowitch refrained from discussing the nature of the intermediates X, Y or Z, but he thought it probable that at least the central intermediate, Y, which participated in both photochemical reactions, was chlorophyll. After weighing up the arguments for each of these alternatives, Rabinowitch concluded that, at the time, the Franck–Herzfeld approach (involving eight identical primary processes) was to be preferred.[153] He emphasised, however, that the approach with two sets of different primary processes, would become first choice "if the existence of two interconvertible green modifications of chlorophyll—one a photo-oxidant and one a photo-reductant—would be definitely confirmed by experiments *in vitro*".[154]

[152] Rabinowitch (1945, p. 162), scheme 7.V.

[153] Rabinowitch also clarified the highly complicated suggestion by Franck and Herzfeld in a scheme; see Rabinowitch (1945, pp. 163–165), scheme 7.V A.

[154] Rabinowitch (1945, p. 168). Both Govindjee and Duysens have drawn attention to this remarkably far-sighted passage of Rabinowitch's book; see Duysens (1989); Govindjee (1995, p. 139), and Govindjee (2006, p. 154). Rabinowitch himself pointed to his earlier thoughts in Rabinowitch (1963, p. 113), and emphasised that by then the evidence had strongly supported the alternative of "two consecutive sets of four transfers [of hydrogen/electrons] each". See also Govindjee and Björn (2012) and Govindjee et al. (2012) on this point.

The issue had to be reconsidered in the 1950s for a number of reasons. Rabinowitch himself again brought forward the possibility of two different light reactions in the second volume (1956) of his photosynthesis monograph, when he tried to find an explanation for Duysens's experiments of 1954, that is, the observation of light-induced cytochrome *f* oxidation in photosynthesising cells. Additional confirmation was brought by similar findings in the same year of the Swedish plant physiologist Henrik Lundegårdh.[155] Rabinowitch considered two alternative interpretations: either the findings could be taken "as evidence of direct participation of these compounds in the photochemical hydrogen transfer from water to carbon dioxide (or, rather, to an organic compound into which CO_2 had been incorporated, such as PGA"; or, as suggested by Duysens himself, they could be taken "as evidence of their participation in oxidative processes (back reactions), coupled with the reduction process".[156] Rabinowitch's preference was clear:

> The first hypothesis [...] suggests photochemical transfer of electrons from reduced cytochrome to the organic acceptor (perhaps via DPN or TPN). *The transfer of hydrogen (or electrons) from H_2O to the oxidized cytochrome would then require another photochemical reaction.* To account for the observed shift, the relative probability of the two photochemical reactions would have to be such as to establish a photostationary state with most of the cytochrome in the oxidized state. The quantum requirement of the hydrogen transfer reaction as a whole would be (at least) 8, since 2 quanta will be needed to transfer each of the four required H atoms (or electrons), *first from water to the cytochrome, and then from the cytochrome to the final acceptor.*[157]

Thus, already in 1956, a fairly precise suggestion of the photosynthetic electron transport chain, with cytochrome as an intermediate, was publicly presented—if only as one of two alternative interpretations of Duysens's data (the importance of which was realised by most photosynthesis researchers only much later). All members of the Urbana laboratory were fully aware that photosynthesis involved two photochemical reactions, most probably via cytochrome *f*, so that they were astonished to observe the stir that was caused by the publication of Hill and Bendall along these lines, which will be discussed in the next section.[158]

---

[155] The relevant paper is Lundegårdh (1954). On Lundegårdh see also Larkum (2003).

[156] Rabinowitch (1956, p. 1862). See this chapter for Duysens's own interpretation.

[157] Rabinowitch (1956, p. 1862). *Govindjee's*. This passage was (re-)discovered independently by two of Rabinowitch's closest associates, Govindjee and Duysens; see Duysens (1989, p. 74), and Govindjee (2006, p. 154).

[158] Personal communication, Govindjee to the author, in September 2005. It seems that outside Urbana hardly anybody at the time had read Rabinowitch's volumes very carefully, so that most of his contemporaries were unaware of his insightful suggestions.

7.6.2 *The Thermodynamic Approach*

The main focus of Hill's work in the 1950s was solving how cytochromes *f* and b_6 functioned in a photosynthetic electron transport chain. In addition to the general suspicion that respiration and photosynthesis might operate along roughly similar lines, Hill thought that the molecular ratio of cytochromes *f* and b_6 to chlorophyll, which were found to be about 1:4000, were highly suggestive: by analogy with the relationship between cytochrome *c* concentration and respiration rate, these were just "of the right order to account for the rates of photosynthesis in terms of hydrogen transport".[159] Together with Fay Bendall, Hill eventually published a proposal on this, entitled "Function of the two cytochrome components in chloroplasts: A working hypothesis".[160] In this paper, Hill addressed the problem from the point of view of thermodynamics; and it is usually cited as the foundation of the well-known "Z-scheme" of photosynthesis.[161]

Hill and Bendall began their paper by summarising the generally accepted body of knowledge on the issues under consideration, notably the relationship between the transfer of hydrogen from water to the hydrogen acceptor and the formation of ATP from ADP and P_i in the chloroplast. The hydrogen transfer, the authors wrote, could be represented by the following set of equations:

$$H_2O + X + Y \rightarrow XH + YOH$$

$$XH \rightarrow X + (H)$$

$$YOH \rightarrow Y + \tfrac{1}{2}\, H_2O + \tfrac{1}{4}\, O_2.$$

This could be condensed to the following summary equation:

$$\tfrac{1}{2}\, H_2O \rightarrow (H) + \tfrac{1}{4}\, O_2.$$

The same process, Hill and Bendall suggested, could also be represented in the form of a diagram (see Fig. 7.7). The oxidised and reduced parts of the water would be separated from each other by the reaction with the primary acceptors, provisionally called X and Y. The (H) would then be transferred to an appropriate secondary acceptor (eventually TPN), while the (OH) moieties would react with each other and release molecular oxygen and water. Thus far, Hill and Bendall were in line with what was generally assumed to be the case.

Given the redox potentials of the hypothetical participants, notably cytochromes *f* and b_6 as intermediate reactants, Hill and Bendall then argued that, at first glance, the electron transport chain seemed to require three separate light-driven reactions on the way from YOH to XH (left-hand-side diagram of Fig. 7.8). However, as this

---

159 Hill (1965, pp. 133–134).

160 See Hill and Bendall (1960a). Very similar ideas are formulated in Hill and Bonner (1961), which was published in the proceedings of the "Light and Life" conference held in March 1960. It is unclear how much of this was already presented at the conference or elaborated for the proceedings.

161 Hill may have been influenced by the similar treatment of the subject in Rabinowitch (1945), Chapter 8, pp. 150–171. I am grateful to Govindjee who pointed me to these pages.

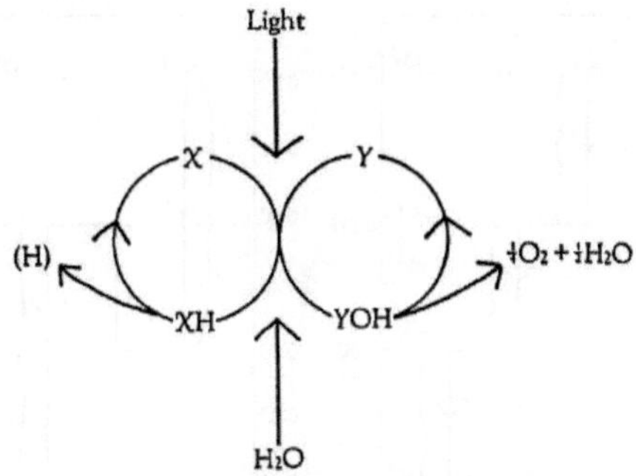

Fig. 2. *X* and *Y* are the hypothetical first acceptors of the products of the photolysis of water, forming respectively the reduced product *XH* and the oxidized product *YOH*

Fig. 7.7 The photolysis of water and the hypothetical first products. (Figure taken from Hill and Bendall (1960, p. 136).

was not in agreement with experimental evidence, Hill and Bendall suggested a system of two light-driven reactions, with a spontaneous "back reaction" in between (right-hand-side diagram of Fig. 7.8):

> The postulation of two light-driven steps, rather than three would be in better accord with present experimental results. In this case oxidized cytochrome b_6 would have to be reduced by Y to give YOH and cytochrome f would have to be oxidized by X to give XH. The reaction between cytochromes f and b_6 would then be a thermochemical process and quite analogous with a hydrogen transfer step characteristic of the mitochondrion.[162]

Under "normal" conditions in the cell, cytochrome f would tend to be in its reduced form and cytochrome b_6 in its oxidised form. When the cell was illuminated, Hill and Bendall argued, the redox states were reversed: the unknown primary acceptor, X, acted as an oxidising agent, turning cytochrome f into its oxidised state; while at the same time, the unknown reducing agent, Y, turned cytochrome b_6 into its reduced state. The two cytochromes would then spontaneously and very quickly react with each other and return to their favoured oxidation state. The energy thereby released might be used to synthesise ATP (see, for the graph representation of this model, Fig. 7.9). Hill and Bendall admitted that the nature of X and Y was still obscure, and that the whole suggestion had to be taken as a mere hypothesis.[163] However, the authors also underlined that the model had the definite advantage, in contrast to the one put forward by Arnon and his collaborators, that it was in agreement with the known redox potentials of all the components involved. Whereas Arnon's group

[162] Hill and Bendall (1960b, p. 137).

[163] This was still acknowledged in Hill and Bonner (1961): "The only evidence for the above hypothesis, in higher plants, is the observation that illumination of pale yellow-green leaves causes an oxidation of [cytochrome] f and the reduction of b_6. As yet, this sequence of reactions has not been observed, in our laboratories, with normal green leaves or chloroplasts. [...] It is our hope that future investigations will be able to elucidate more fully the role of cytochromes in the photosynthetic process of green plants." (p. 434)

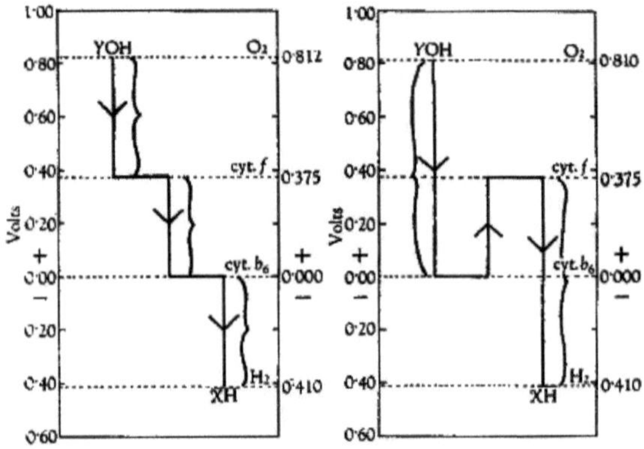

Fig. 3. Symbols as in Fig. 2 Fig. 4. Symbols as in Fig. 2

Fig. 7.8 The hypothetical electron transport chain in terms of oxidation–reduction potential. On the *left*, a potential model with three light-driven steps, which was abandoned; on the *right*, the favoured model with two light-driven steps, connected by a thermochemical reaction along the gradient between the two cytochromes. (The *downward arrows* symbolise electron transport *against* the thermochemical gradient.) (Taken from Hill and Bendall 1960, p. 137).

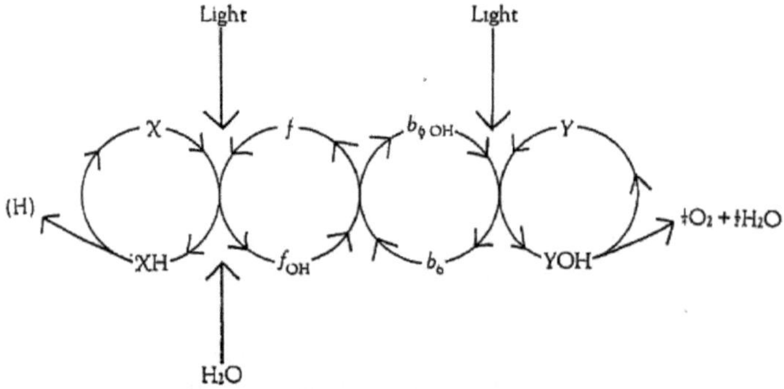

Fig. 5. X, Y, XH and YOH as in Fig. 2. f and b_6 represent ferrous and f_{OH} and b_{6OH} ferric or oxidized cytochromes

Fig. 7.9 The two photoreactions model of photosynthesis taken from Hill and Bendall (1960, p. 137).

had assumed a role for the cytochromes at one or other end of the photochemical sequence, Hill and Bendall envisaged the cytochromes in an intermediate position, between the two photoreactions, which was much more convincing.

The proposal provided a surprising solution to the problem of where the energy to produce ATP came from (namely from the energy drop between the two cytochromes).[164] It preserved the far-reaching correlation between photosynthesis and

[164] It transpired later, however, that the role of cytochrome b_6 differed from the one that Hill and Bendall had envisaged; it is involved in the process of cyclic electron transport.

respiration; and it was in agreement with the curious observation that in the light the two cytochromes tended to be in the energetically disadvantageous state: that is, when illuminated, cytochrome *f* was found in its oxidised mode, notwithstanding its highly positive redox potential; while cytochrome b_6 was found to be reduced. It is unclear, why Hill did not refer to the actual finding of this phenomenon in red algae, by Duysens in his papers of 1954, both of which Hill knew, as can be taken from his correspondence with Duysens.[165] Neither did Hill and Bendall refer to the obvious relationship of this suggestion to the Emerson enhancement effect: the conclusion that two different photochemical reactions were required in photosynthesis had also been arrived at on the basis of experiments carried out three years earlier.

7.6.3 Spectroscopy and Two Photosystems

Duysens became interested in the strange effect of different wavelengths on photosynthesis in the mid-1950s (i.e. around the same time as Emerson; see Chapter 5). Duysens later recalled that at the Second Gatlinburg Conference in 1955, Lawrence Blinks had presented data which indicated that "in red algae, illuminated alternately with red and green light for a few minutes, qualitatively different photosynthetic transients occurred, which were not caused by differences in intensity".[166] Duysens thought that these phenomena—today known as the "chromatic transients" of oxygen exchange—were most probably caused by the overlapping effects of two or more photochemical reactions with different action spectra. (Duysens did not assume, at the time, that all these reactions were actually part of photosynthesis.) He was greatly interested in the 1957 paper by Emerson and his co-workers, in which the enhancement effect was presented, although he was not very impressed by Emerson's own explanation, which assumed that accessory pigments played a specific role in photosynthesis and directly initiated a second photochemical reaction (see Chapter 5).[167] As was mentioned earlier, this approach was in conflict with Duysens's own earlier finding that all the light energy absorbed had to pass through chlorophyll *a* if it were to be used in photosynthesis.[168]

---

[165] These were Duysens (1954a, b).

[166] See Duysens (1989, p. 71). The data were published two years later in Blinks (1957); see also Blinks (1960a, b, c). Like Emerson, Blinks proposed that these effects were caused by the side reactions of respiration; this was proven to be erroneous by the findings of Govindjee et al. (1960, 1963). On Blinks's life and work, see Thorhaug and Berlyn (2009).

[167] See Emerson et al. (1957) as well as Emerson and Chalmers (1958) and several talks given at the National Academy of Sciences (documented in abstracts).

[168] Govindjee and Rabinowitch showed, in 1960, that the two pigment systems activated by the longer and the shorter wavelengths both contained species of chlorophyll *a*, so that chlorophyll *b* did not need to be involved. Cf. Govindjee and Rabinowitch (1960).

In 1957, Duysens thought that the so-called inactive (nonfluorescent) part of chlorophyll *a*, which he had identified in his fluorescence studies, might be the key to solving the puzzle: under certain conditions, this "inactive" portion of chlorophyll *a* might possibly be triggered to participate in photosynthesis, through the effect of side reactions initiated by the fluorescing, "active" portion of chlorophyll *a*. Thus, Duysens set out to investigate in more detail the functions of the two different types of chlorophyll *a*, which he had already identified in 1952. Ironically, this was exactly the same research question that Rabinowitch had originally chosen for him, when Duysens spent a year in the Urbana laboratory in 1953; at the time, however, no promising experimental approach was available, so that Duysens turned to examining the kinetics of potential electron transport components instead.[169] In 1958, the situation had changed, and Duysens decided to examine the action spectra for cytochrome oxidation and TPN reduction in the presence and absence of short and long wavelength background illumination. The first results of these studies were reported in 1960 at the Third International Congress of Photobiology held in Copenhagen, Denmark, and caused considerable excitement in the audience: contrary to expectations, at 560 nm a *low* cytochrome oxidation yield was found (although the photosynthetic yield was high), while a *high* cytochrome oxidation yield was identified in the region of 680 nm.[170] This switching of the oxidation state of a cytochrome when illuminated at different wavelengths was inexplicable, if one assumed that there was only one photoreaction in photosynthesis. Duysens recalled that he proposed the following explanation:

> I postulated the existence of two major photosystems, 1 and 2. System 1 contained the weakly fluorescent chlorophyll *a*, formerly said to be inactive, and oxidized cytochrome; system 2 contained the fluorescent chlorophyll *a*. An interaction between the two systems was shown by the different kinetics of cytochrome oxidation at different actinic wavelengths.[171]

According to Duysens's memories, at that point of time he had not yet seen the paper by Hill and Bendall; but when he was pointed to it in the discussion, he rushed back to his laboratory to finish the final experiments. His impressive findings were published in 1961, together with Jan Amesz and B. M. Kamp.[172] The authors demonstrated that Duysens's system 2 could specifically and differentially be inhibited by the herbicide DCMU. (This substance was introduced by the company Bayer in 1954 under the name "Diuron" and soon had found its way into the laboratories.[173]) By this means, the action spectrum of only system 1 could be measured, and Duysens and co-workers found it to be very similar to the spectrum of TPN reduction in cells. This strongly suggested that the latter process was driven by system 1. At the same

---

[169] See Duysens (1989, p. 69).

[170] See Duysens (1961).

[171] Duysens (1989, p. 72).

[172] Duysens et al. (1961).

[173] "DCMU" is (to this day) the abbreviated name for the herbicide 3-(3,4-di-chlorphenyl)-1,1-dimethylurea, which very effectively—and specifically—inhibits photosynthesis.

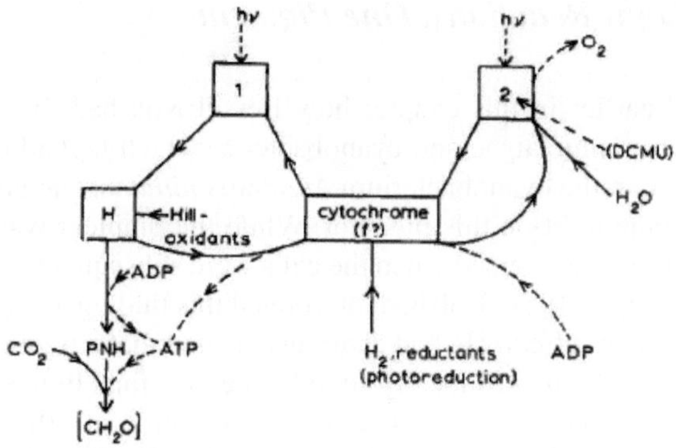

Fig. 3. Hypothetical scheme showing functions of systems 1 and 2 in photosynthesis

Fig. 7.10 Reproduced from Duysens (1961, p. 511). The two systems initiate two different light reactions, either oxidising (system 1) or reducing (system 2) the cytochrome.

time, system 1 directly or indirectly oxidised the cytochrome (with predominant activity at 680 nm), while system 2 (with predominant activity at 560 nm) reduced the cytochrome. The two functions of systems 1 and 2 were graphically summarised in a "hypothetical scheme", as the authors called it (see Fig. 7.10).

The scheme elegantly explained the enhancement effect found by Emerson: the two systems had overlapping absorption spectra, so that at simultaneous illumination with different wavelengths the excess absorptions supplemented each other. In effect, the rate of photosynthesis then would be higher than the sum of the rates at each wavelength separately. Furthermore, Duysens was able to explain that, even when photosynthesis was inhibited by DCMU (which only affected system 2), the photoreduction of either carbon dioxide or TPN (via system 1) was still possible.[174] And although the authors considered their scheme a working hypothesis, like Hill and Bendall before them, they also emphasised the explanatory strength of their proposal: "At present it explains most experiments known to us in a simple and plausible way."[175]

[174] On photoreduction, the authors cited Bishop (1958) and Vernon and Zaugg (1960).

[175] Duysens et al. (1961, p. 511). Duysens and his co-authors could not refrain from alluding to the paper by Hill and Bendall, albeit without actually citing it: "A more detailed report is in preparation, in which we will discuss partly similar but less detailed and experimentally less supported hypotheses concerning the role of the photosynthetic pigments which have been proposed by other authors." Possibly, Duysens was afraid that a paper that had failed to acknowledge earlier work of his might deprive him of his credit for prior discoveries.

7.6.4 Two Light Reactions, One Pigment

It was described earlier in this chapter how Bessel Kok had discovered a special pigment in higher plants, algae and cyanobacteria, which he had called P700, and how he had found, in the cyanobacterium *Anacystis nidulans*, an antagonistic effect of red versus orange light on this pigment. While the pigment was oxidised in red light, this oxidation was reversed when the cells were subsequently illuminated with orange light. Already in 1959, Kok had interpreted this finding as being related to the Emerson enhancement effect. He had cautiously stated that a pigment which had an absorption band at 700 nm was ideally suited to be the "final light sink in photosynthesis", and dared to speculate on a necessary alternation of different wavelengths effecting two different photochemical responses, in order to make photosynthesis work at its full efficiency.[176] Kok continued to think along these lines, and later in 1959, at the *Ninth International Botanical Congress* held in Montreal, Canada, he presented a paper entitled "Does photosynthesis require the interaction of two photochemical steps?" And his answer was decidedly positive.

In the paper's abstract Kok announced the discussion of a hypothesis which involved "a cycle of two photochemical acts in photosynthesis".[177] Kok had found that the oxidation of P700 was brought about by the light absorbed by chlorophyll *a*, whereas in the subsequent restoration and reduction processes, he (erroneously) hypothesised, it was mainly light absorbed by phycocyanin that was active.[178] Together with the biochemist George Hoch, Kok elaborated these studies and reported the outcome at the *Light and Life* symposium, held in March 1960 at the Johns Hopkins University in Baltimore.[179] The authors again asked, whether photosynthesis was actually driven by two light reactions, and their answer was, again:

> The observations discussed in the above sections strongly indicate the occurrence of two different light reactions: the first sensitized by chlorophyll *a* and a direct bleaching of "P700"; the second sensitized by accessory pigments acting indirectly via the mediation of dark steps and restoring "P700".[180]

A scheme for the cyclic process envisaged by Kok and Hoch is given in Fig. 7.11. From their data, Kok and Hoch also concluded that (in contrast to Kok's paper of 1959) phycocyanin was not essential for the process to function—chlorophyll *a* was able to sensitise photophosphorylation entirely by itself.

---

[176] See Kok (1959).

[177] See Kok (1961a) for his extended abstract. This paper had long been overlooked until attention was drawn to it in Govindjee (2006). See also Govindjee et al. (2012).

[178] Kok (1961a, p. 1072).

[179] See McElroy and Glass (1961) for the proceedings of the symposium. Therein, not only presentations but also subsequent discussions are documented. Jack Myers thought, in retrospect, that it was at this conference that the idea of two photochemical reactions being involved in photosynthesis started to dawn on photosynthesis researchers, with the contribution by Kok and Hoch being considered pivotal in this respect. See Myers (2002, p. 25).

[180] Kok and Hoch (1961, p. 407).

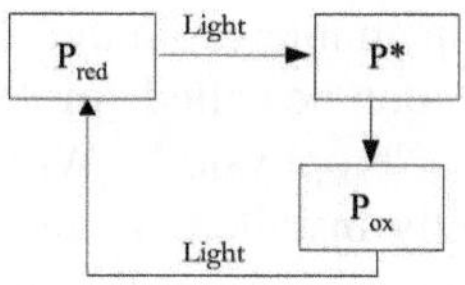

Fig. 7.11 The cyclic mechanism in which P700 is oxidised and reduced, and which involves two light reactions and one dark reaction. (After Kok and Hoch 1961, p. 408).

These findings resonated well with the conclusions arrived at by other investigators. At the aforementioned 1959 conference in Montreal, Jack Myers proposed, like Kok, the existence of two light reactions in photosynthesis;[181] while Eugene Rabinowitch went one step further and suggested that not only two (or more) different primary photochemical processes were involved, but that these were sensitised by different pigments.[182] Rabinowitch suspected the existence of two types of chlorophyll *a* in the cell that were associated with the different processes (which was soon backed up by experimental evidence).[183] These were enormous advances on the way to a better understanding of the photochemical events in photosynthesis. However, still in 1960, the reaction of the general audience at the *Light and Life* conference was rather cautious. "From the record of discussion [at the conference]", Jack Myers wrote in retrospect, "one would judge that [Kok and Hoch] had dropped an egg instead of a bomb".[184] Even though, in retrospect, it seems that by then compelling evidence for the two light reaction hypothesis had been accumulated, the hypothesis was far from being universally accepted.

7.6.5 New Flashing Light Experiments

The eventual change in attitude was not least due to the contributions to the field by the research team of the German biophysicist Horst T. Witt in Berlin. Witt had been a student of the renowned physicist Robert Pohl at the University of Göttingen in Germany, and received his PhD in solid-state physics in 1950.[185] Already at

[181] For the abstract of Myers's contribution, see Myers (1961). Related and more extended papers are Myers and French (1960a) (accepted 23 July 1959; published in March 1960) and Myers and French (1960b). See also Govindjee (2006).

[182] See Emerson and Rabinowitch (1960) for the resulting publication; here: p. 482. Govindjee (2006, p. 157), additionally cites the abstract written for the Montreal conference.

[183] The different functions of two different types of chlorophyll *a* were established in Govindjee and Rabinowitch (1960); similar findings, independently arrived at, were presented in French (1961).

[184] Myers (1987, p. 134). Myers thought that Kok and Hoch had shown too many experimental details, so that the main argument was almost lost.

[185] For biographical information on Witt, see Junge and Rutherford (2007); Jaenicke (2007) and Renger (2008). Witt (1991) is an autobiographical account of his life and work; Junge (2005)

Göttingen Witt began to develop an interest in the physical foundation of biological phenomena: a field that would soon be called "molecular biology" or, alternatively, with a slightly different focus, "biophysics". Witt recalled that in parallel to his dissertation work he had secretly installed "batteries of tubes for algae cultures", hidden behind the laboratory boards—until this undercover project was abruptly terminated: "One night the tubes burst and the algae cultures ran over the floor into Prof. Pohl's study and stained his precious carpet algae green", Witt remembered. After a rather embarrassing talk with his supervisor, Witt agreed that in future he would focus solely on his original dissertation topic. The alternative would have been to pursue instead the question of "Water Cleavage in Photosynthesis"—which, as Witt stated, was "a completely impossible topic at that time".[186] Witt started working seriously on oxygenic photosynthesis in 1952, when the German physical chemist Karl Friedrich Bonhoeffer offered him a laboratory at the Max Planck Institute of Physical Chemistry in Göttingen. In 1955, he moved to Marburg, Germany, where he completed his habilitation project, and in 1962, Witt was called on a chair at the Technische University Berlin. There he transformed the Institute for Physical Chemistry into an internationally renowned research centre for biophysical studies in photosynthesis.

Two methodical problems needed to be addressed if one wanted to find out more about the primary photochemical processes: *First*, one had to develop techniques that were able to precisely measure the absorption changes. As the largest share of the pigments were chemically inactive, the absorption changes brought about by the redox reactions were extremely small, so that the usual instruments failed to give accurate quantitative results. *Second*, these measuring techniques had to be able to deal with the extremely high speed of the reactions. Almost all the pertinent reactions were faster than 10 ms—in fact, many were suspected to be in the range of micro- and nanoseconds. Witt managed to overcome these problems with his flashing light spectroscopic methods, that greatly increased the sensitivity and the time resolution of photosynthesis studies. He presented this technique at the Second Gatlinburg Conference in 1955, and at the same time introduced himself to the most influential photosynthesis researchers of the period.[187]

With the help of this method, in 1961 Witt and his collaborators found that upon excitation with $light_1$ (710 nm) cytochrome *f* was oxidised and stayed in this state for seconds. At the same time, they established that after excitation with $light_2$ (670 nm) an unidentified component X was oxidised to XO. From these findings, the group concluded (at almost the same time as Kok and Hoch as well as Duysens, Amesz and Kamp) "that photosynthesis is triggered by two different photochemical reactions: oxidation of cytochrome by Chl*a*-680 and reduction of XO by Chl*a*-670".[188] In the

describes the later discoveries (after 1966) achieved in Witt's laboratory from the perspective of one of Witt's closest collaborators.

[186] All quotes: Witt (1991, p. 58).

[187] A survey of Witt's methods and findings up to 1959 is provided by Witt (1960).

[188] Witt et al., p. 194; see also Witt et al. Witt (1991, pp. 61–63), provides a retrospective analysis of the episode. Based on the absorption changes and other pieces of evidence, Witt et al. additionally

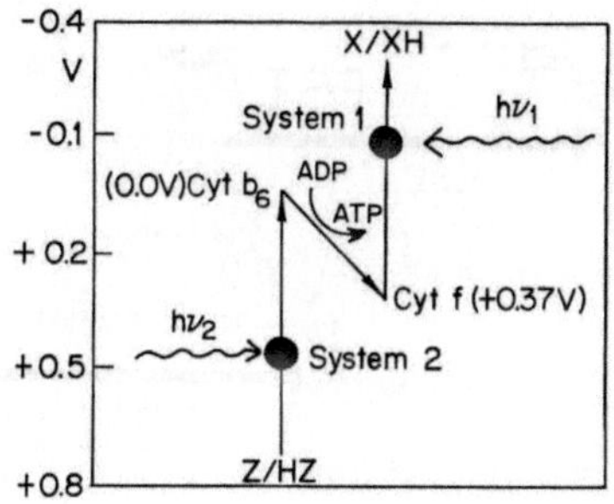

Fig. 7.12 The "Z-scheme" of photosynthesis, reproduced from Rabinowitch (1963, p. 114).

same year (1961), it was established, through independent studies undertaken by Kok on the one hand and Witt on the other, that P700 was a chlorophyll *a* molecule and the primary electron donor in the long-wavelength photosystem.[189] The different approaches finally had converged.

7.6.6 The Z-Scheme

Over the course of the years the two photochemical reactions, two photosystems model has come to be called the "Z-scheme" of photosynthesis. Today, it features prominently in most textbooks, although current standard representations often do not make it clear why it was called the "Z-scheme"—neither is this apparent from the diagrams used by the photosynthesis researchers who contributed to the development of the underlying model. In its early years of existence, rather different means were used to represent the two-photosystem model. A version that was quite close to the arrow scheme by Hill and Bendall (see Fig. 7.8), albeit with a different orientation of the vertical axis, was used by Rabinowitch in 1963, in his contribution to the conference on "Photosynthetic Mechanisms of Green Plants", held at Airlie House, Virginia (Fig. 7.12). Already then, Rabinowitch mentioned that similar schemes, "presented vertically, horizontally, in zig-zags, on circles or curlicues", had been presented by several authors.[190] Rabinowitch himself settled on a vertically oriented scheme, displaying only a minimum of information. (The vertical orientation more intuitively

---

suggested that the reaction XO → X might be the reaction of plastoquinone to hydroquinone. See Bishop (1959) for the first clear statement about plastoquinone's possible role; Klingenberg et al. (1962) was able to corroborate the theory.

[189] Kok (1961b) and Witt et al.

[190] Rabinowitch (1963, p. 115). See also David Krogmann (2000) who remembered that in the 1960s "there was heated discussion of whether the arrows depicting photoacts should point up or down and whether to rotate the Z clockwise or counterclockwise in what came to be known as the Z scheme" (p. 115).

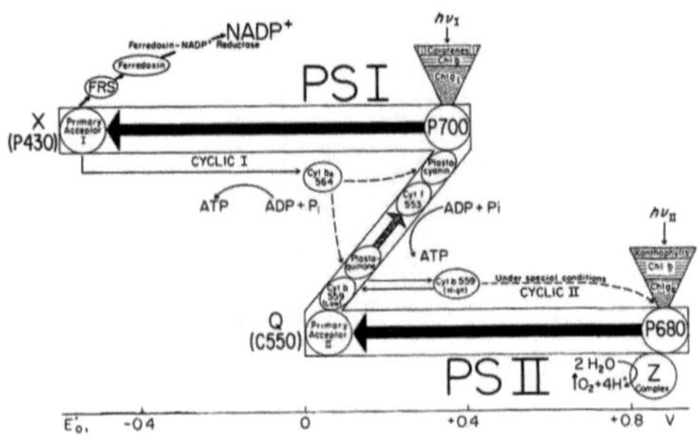

Fig. 7.13 The "Z-scheme" of photosynthesis, in a version by Govindjee and Govindjee (1975, p. 27).

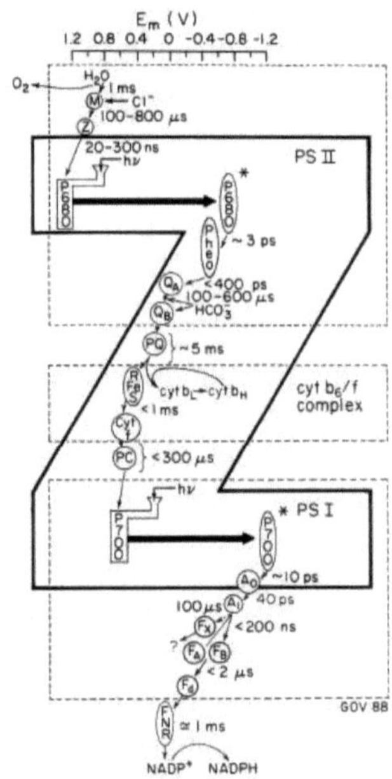

Fig. 7.14 The "Z-scheme" of photosynthesis, reproduced from Demeter and Govindjee (1989a, p. 123).

represented the spontaneous, thermodynamically "downhill" reaction between cytochromes b_6 and *f*, as the reaction also goes "downhill" in the picture.) However, later versions, which are reproduced in Figs. 7.13 and 7.14, rather chose a horizontal orientation of the axis with the redox potentials. This results in the emergence of the letter "Z" in the scheme, hence, the name.[191]

[191] The figures were taken from Govindjee and Govindjee (1975, p. 27), and Demeter and Govindjee (1989, p. 123).

This model and its current standard form of representation became a matrix that was able to accommodate most of the available evidence on how the light reactions in photosynthesis worked, and which could be expanded and completed as time went by. In this model, photosystem I (PS I) causes, on the one hand, the reduction of TPN (in the figure already named $NADP^{+}$) and eventually of carbon dioxide, as well as the oxidation of cytochrome and P700. Photosystem II (PS II) causes the reduction of the cytochrome and P700 as well as the oxidation of water, whereby oxygen is released. Both systems need to be in a sufficient and balanced excited state for photosynthesis to occur efficiently. In the years that followed, many more components of the electron transport chain were identified, and other pieces of evidence were accumulated, all of which were explained by moderately expanding the model (by inserting new intermediates, for example). This made it ever more probable that the model was on the right track and in the course of the 1960s it became almost universally accepted—although, of course, its intricate details remain controversial to this day.[192]

7.7 Convergent Research Pathways

It is clear from what has been discussed in this chapter that the establishment of the Z-scheme model of the photosynthetic light reactions, which included two sequential photochemical reactions initiated by two different pigment systems, was arrived at almost simultaneously by a number of research teams working in Europe and the USA, starting from very different angles, and using different methods and approaches.[193] This contrasts starkly with the discovery of the cyclic pathway of the dark reactions in photosynthesis, which was so strongly dominated by the Calvin–Benson team at Berkeley (see Chapter 6). No other laboratory had the manpower or infrastructure to even match the Berkeley team—rather, whoever felt that she might be able to contribute something, tried to become part of it. Thus, the divergent skills required to find the path of carbon in photosynthesis was assembled in one place. A much more decentralised approach was dominant when it came to explaining the photochemical events of the process.

Most of the work examined in this chapter was carried out in the 1950s: progress started to accelerate around 1954 and culminated in the years 1959 and 1960. Seen in retrospect, everything took off in 1951 when the Hill reaction was suddenly seen to be related to the reduction of carbon dioxide, through the finding (simultaneously in no less than three different laboratories) that chloroplasts were able to reduce TPN. Thus, they provided one of the essential components for the reduction of carbon

[192] The contributions in Tamiya (1963), which at the same time summarised the state of the art, impressively demonstrate how the field of photosynthesis continued to flourish.

[193] Perhaps this plurality of approaches, which resulted in evidence being available in many different specialties, contributed to the broad acceptance of the model; see Solomon (1994b), who demonstrated the influence of biased evaluation of data and arguments in the controversy on plate tectonics. It seems that usually new conceptual models are more easily accepted, if evidence from the own discipline is available.

dioxide in the dark cycle. Although Emerson and others were sceptical about the significance of this finding, Hill was immensely satisfied: finally, there was evidence that the Hill reaction was, in fact, not extraneous to photosynthesis but reflected an inherent part of the process. Only one year later, in 1952, the plant-specific cytochrome *f* was found in Hill's laboratory, followed by cytochrome b_6. Duysens then established, around 1956, that cytochrome *f* reversibly changed its redox state upon illumination; and Arnon construed, upon the finding that also chloroplasts produce ATP, electron transport chains that assigned a crucial role to the chloroplast's cytochromes. In 1957 Emerson found the enhancement effect of illumination with two different wavelengths. Two years later, in 1959, Kok related this effect to his finding of the antagonistic effect of red and orange light on P700; and in the same year Rabinowitch thought that the assumption of two different photoreactions, associated with two different species of chlorophyll *a* (which soon were proven to exist) was the obvious explanation of the enhancement effect. Why this was not immediately picked up by others, is not entirely clear; even in March 1960, a very similar suggestion by Kok and Hoch was still received with much reservation. Perhaps it was the more prominent place of publication (*Nature* vs. *Plant Physiology*) that granted the Hill–Bendall paper a larger audience. Perhaps it was the fact that Rabinowitch's suggestion was embedded in an extended discussion of both Emerson's findings and Franck's physical explanation—the latter, in particular, was not easy to follow. Perhaps it was the suggestive schemes that Hill and Bendall used in their paper to explain the model. In any event, by 1963, the general accuracy of a Z-scheme-like model was no longer seriously disputed. The Z-scheme was able to explain phenomena ranging from the enhancement effect to the light-induced redox changes of cytochrome *f* and the chromatic transients of oxygen exchange. This was extremely persuasive.

Duysens himself believed, in retrospect, that once the appropriate methods were known (such as absorption difference spectroscopy) and some important parts of the jigsaw puzzle pieced together (such as the discovery of light-induced redox changes in cytochromes and P700), arriving at a Z-scheme-like model was only a matter of time and could not be credited to any specific person.[194] The different approaches definitely supported each other. Hill's thermodynamics alone would have been as empty as the mere empirical data of Emerson's enhancement effect or Duysens's highly specific spectroscopy. It is significant that Kok, Hill and Duysens all believed that it was necessary to present their own suggestions as tentative "working hypotheses" that needed further clarification and modification. An enormous degree of uncertainty still prevailed around 1960. It was only when the different strands came together that the two-photosystem approach gained its persuasive power.

Instead of growing ever more self-confident in their work, photosynthesis researchers around 1960 had become cautious: time and again, in the history of the field, those solutions that at first had appeared less probable, turned out, in the end, to be accurate, whereas those alternatives that at first glance looked almost obvious

[194] See Duysens (1989, pp. 77–78).

were eventually dropped. For example, for a long time it was taken for granted that carbon dioxide was the source of photosynthetic oxygen—given the stoichiometry of the summary equation, this seemed to be evident. However, in the end, it transpired that water was the source of photosynthetic oxygen. Far into the 1930s, it was assumed that the light reaction in photosynthesis was directly related to carbon dioxide reduction; this also turned out to be erroneous. In October 1959, at a conference on bioenergetics, William Arnold was infuriated that participating physicists rejected potential mechanisms of the photochemical events in photosynthesis as they found that the mechanism under debate did not seem very *probable*. "Any statement about 'probable' might be alright in Havana, where you could make money on it; but, this is a meeting on bioenergetics, and when it comes to biology, this is simply an extraneous consideration of no importance", Arnold sharply commented.[195]

Taking stock of the entire course of events, one may want to agree. Photosynthesis is in itself a most improbable process. No chemist or physicist would ever have come up with a system that generates its energy by oxidising water and stores it by way of carbon dioxide reduction—all this at room temperature to boot. Yet, it is a process that undeniably operates extremely well in nature. In order to explain such an improbable phenomenon, scientists had to explore even the most improbable of modelling approaches; and right up until the solution was reached, it was never completely foreseeable which of the available options would, in the end, prevail.

[195] Arnold (1960, p. 324).

Epilogue

The investigation initiated by the nineteenth-century chemists to explain photosynthesis came to a preliminary conclusion around 1960, when a sophisticated, molecular-level model of the mechanism was developed. The problem—how to explain the light-driven production of carbohydrates and oxygen from carbon dioxide and water in green plants—had troubled scientists for more than a century. The solution included an intricate cyclic path, the prerequisites of which, namely, reducing equivalents and chemically usable energy in the form of adenosine triphosphate (ATP), were produced during the course of two photochemical reactions operating in series and dependent on two different pigment systems.

The story of how this solution was reached was described in this book with an emphasis placed on the internal dynamics of the modelling process; and it was suggested that several recurrent heuristic strategies can be identified that characterise the methodology of the researchers under study. It may be worthwhile to summarise some of the central features that emerged and might be of wider applicability. The modelling of a complex mechanism, such as photosynthetic carbon dioxide assimilation, was presented as being a collective enterprise. The functional decomposition of the mechanism, a widespread strategy in dealing with complexity as we know from William Bechtel and Robert Richardson, was correlated to the division of labour between several research groups, which *cooperated informally* with one another.[1] No central agency organised this process; rather, the different groups defined their own contributions to the overall project depending on their research interests and skills. Most scientists specialised in certain experimental techniques and applied them to a limited range of issues within photosynthesis research, at the same time keeping a close eye on the (complementary) work of other researchers in the field. The more complex the issue turned out to be, the more subgoals were identified (corresponding to functional subunits of the mechanism) and the more diversified the community became.

[1] Starting in the 1950s, the locations of the different processes also were considered, although it was not yet part of the primary focus of research (differing from the elucidation of other biological mechanisms, as studied by, e.g., Bechtel and Richardson).

K. Nickelsen, *Explaining Photosynthesis,* History, Philosophy and Theory of the Life Sciences 8, DOI 10.1007/978-94-017-9582-1

Decomposition of the mechanism into functional units—the *modularisation* of the process, as it is referred to in this study—helped individual researchers to focus on one specific aspect or partial process of photosynthesis, while the rest of the mechanism was left untouched. Alternatively, people imported the conception of one subprocess from the literature, while working on the rest. Early on in photosynthesis research, for example, the synthesis of carbohydrates from formaldehyde was treated as such a module: it was accepted as a self-sufficient element, which most researchers integrated into their various models of the first stages of the mechanism. The most extreme example of modularisation, which eventually resulted in the construction of two distinct partial models of photosynthesis, involved the separation of the "dark" reactions of photosynthesis from the "light" reactions. Although these two partial processes were closely interrelated, and known to depend on each other, from about 1940 they were no longer studied as a single unit. This conceptual separation coincided with a disciplinary separation: while the "dark" reactions became the domain of bio*chemistry*, the "light" reactions developed into one of the favourite research themes of bio*physics*.

The integration of experimental results into an explanatory model that adhered to accepted theory, remained a challenge. One of the most important heuristic strategies to meet this task was identified in this study as the *transfer of causal knowledge* from one field to another. This is, in effect, very similar to what Lindley Darden referred to as schema instantiation; although, it might concern only much smaller items than full schemata. The strategy is interpreted here as the tentative classification of a new situation as being similar to a type of situation that had already been elucidated. This was particularly useful in cases where it was impossible to undertake difference tests to find out more details about the involved causally relevant factors, mainly because of the lack of appropriate methods—such as in nineteenth-century chemistry, when hardly anything was known about photosynthesis except for the identity of the raw materials and the end products. The chemists, hence, started from what they knew about the reaction paths of the raw materials in vitro and tried to transfer these interaction patterns to the mechanism of photosynthesis. This strategy continued to be an extremely powerful tool, for example in elucidating the course of the dark reactions of photosynthesis around 1950: by then, intermediate compounds were identified by means of radiotracers and paper chromatography; but how these compounds interacted, had to be inferred from the body of knowledge assembled in other systems.

This type of reasoning is sometimes referred to as "reasoning by analogy" and enjoys a rather bad reputation among philosophers, due to the fact that it is obviously fallible. At the same time, it is a widespread heuristic move in scientific practice. If phenomena are observed, the underlying mechanisms of which are known in similar contexts, one might suppose that a very similar mechanism would satisfactorily explain the problem at hand. In photosynthesis research, either only a couple of factors and their interaction were transferred from one epistemic context to another or even full modules of the mechanism under study. It was frequently accompanied or followed up closely by a phase of empirical investigation in order to ascertain whether the assumed explanation did, in fact, hold. Yet, even if this empirical search did not go anywhere, scientists sometimes retained the tentatively transferred elements if their theoretical foundation and the explanatory value was persuasive enough. The long-held belief in the accuracy of the formaldehyde model of photosynthesis (which was exclusively based on the knowledge gained in artificial systems), despite the failure of scientists to demonstrate that formaldehyde was formed in plants, is a case in point.

Beyond conceptual knowledge of the mechanism, other elements of knowledge were also transferred to photosynthesis research, including new explanatory approaches, theoretical notions of a general nature or the application of new methods and instruments. Frequently, it was not the photosynthesis researchers themselves, who actively sought to import knowledge from other fields to photosynthesis studies when they found themselves in an impasse (although there were exceptions, such as the intensive search of the literature for potential chemical intermediates of photosynthesis that Martin Kamen and Sam Ruben undertook when they started their tracer studies). Rather, new pieces of knowledge were imported mostly by scientists who were experts in other areas (such as atomic physics, physiology and radiation chemistry), who had diverged from their original path of research to make a quick contribution to photosynthesis studies: *research opportunists*, as they were called in this book, in a somewhat provocative choice of term. Some of them returned to their original specialty, while others—unexpectedly for themselves—stayed in the field.

Finally, it was repeatedly emphasised that the group of photosynthesis researchers, as a whole, usually pursued a range of different modelling options concurrently. This *pluralism of alternatives* is particularly prominent in phases of high uncertainty, such as, for instance, the nineteenth-century search for a chemical pathway or the attempts to determine the maximum quantum yield of the process. Whenever important issues were at stake, while the available knowledge was meagre, a premature restriction of alternative options, in the light of what appeared promising at the time, was judged to be unwise. This strategy was applied to fundamentally diverging options (such as the chlorophyll complex model versus models including the photosynthetic unit concept) as well as to "local" alternatives (such as whether 2-phosphoglyceric acid (2-PGA) or 3-PGA was the first product of thermochemical carbon dioxide reduction). This implies that at no point there was one "winning" model, not even in 1960; it were always families of different model variants that were being discussed. The debate about the origin of photosynthetic oxygen provides a fine example of the persistent pursuit of alternatives. From the early nineteenth century onwards and well into the 1930s, the standard notion was that the oxygen originated from the carbon dioxide. However, researchers also explored the possibility that the oxygen might have, in fact, originated from the water. Not many scientists had much faith in the latter hypothesis, but it was still a viable option that they could not afford to ignore. In the end, it proved correct; but, to this day, the possibility has not been excluded that some of the photosynthetic oxygen might still result directly from carbon dioxide reduction.

How were these different options pursued? It is little surprising, although intricate in detail, that existing model suggestions were constantly modified in the light of new experimental findings or theoretical developments (although in photosynthesis research the latter initiated far less frequently a process of reconsideration than new empirical results). Researchers might add factors to the model, replace or redefine earlier ones or revise full sequences; construe alternative pathways to produce an effect or insert intermediate functional modules that had so far been neglected, and so on. Recall, for example, Robert Emerson and William Arnold's 1932 finding that in photosynthesis only one molecule of oxygen was produced per couple of thousand molecules of chlorophyll. At the time, the standard model included the assumption that oxygen was produced by the direct interaction of chlorophyll molecules with the molecules of carbon dioxide, in a one-to-one relationship, which was clearly in conflict with the new data. The community reacted in a number of ways. At first, most of the scientists seriously doubted the validity of the data. James Franck, for example, was convinced for a long time that Emerson and Arnold had not stimulated the system to operate at its maximum efficiency. Others, including Emerson himself, tried to account for the data by introducing new factors to the standard model—such as the suggestion that an enzyme which was only present in very low concentrations might be involved in oxygen evolution. And a third, more radical group, notably headed by Hans Gaffron and Kurt Wohl, postulated that fundamental changes concerning the action of chlorophyll be made to the standard model. They proposed that, instead of one chlorophyll molecule acting on one molecule of carbon dioxide, thousands of light-absorbing molecules might be "cooperating" in photosynthesis. All these alternatives were pursued, although Franck's and Emerson's more conservative approaches were strongly favoured. Gaffron and Wohl's modification implied that a previously unheard-of mechanism operated in photosynthesis; so that, although the assumption would have explained the data, most scientists considered their idea ill-founded. Nobody was able to imagine how this cooperative mechanism might function. The situation only changed when the concept of energy resonance transfer was brought up and applied to photosynthesis studies.

This reluctance on the part of researchers to revise or drop long-held model assumptions, that is, a certain *epistemic inertia*, was widespread in the episode under study. As a rule, scientists tended to respond to new data or revised theoretical knowledge by trying to modify and expand the existing models as moderately as possible. It was only when these efforts constantly failed that the group abandoned certain model families. This dropping of a model variant from the stage was, more often than not, a rather unspectacular event: nobody stopped to "falsify" the hypotheses in question, such as, for example, the existence of a complex binding of carbon dioxide to chlorophyll, which had been part of the generally accepted knowledge from 1870 until far in the 1930s. Researchers simply chose to spend their time on more productive issues, even if they were not aware of the difficulty, in fact impossibility, to prove a particular causal hypothesis implied in the mechanism definitively wrong. Take the maximum quantum yield controversy, neither Emerson nor anyone else was able to demonstrate the irrelevance of the eccentric experimental conditions that Otto Warburg and Dean Burk had advanced. Alternative pathways, incompletely understood module composition, imperfectly realised set-up and so on, could always have had an effect. All Emerson could do (and did, as far as he was able to) was to demonstrate the relevance of certain conditions for certain results and then draw inferences as to why these results of Warburg's did not reflect the maximum quantum yield of actual photosynthesis. If this proved impossible, if none of the alternatives to a contested part of the mechanism could positively be established (as was the case in the search for formaldehyde), the only option open to researchers was to wait until either the question had lost its relevance or new methods emerged. This point was reached when Kamen and Ruben used their new radioactive tracer technique to look for formaldehyde. They were as unable to find formaldehyde as their predecessors, but they found a lot of other compounds that opened up the path to more promising model alternatives.

Having so far reflected on the epistemological side of the modelling process, it needs to be underlined that the modelling of the photosynthesis mechanism was an enterprise that hinged on the mastering of experimental practice. This can most clearly be demonstrated by the fact that photosynthesis researchers spent so much time and energy on algae culturing. The more experiments they carried out using algal cells, the more intricate details they discovered about the complex metabolic reactions of these organisms. The physiological state of the algae turned out to be one of the most decisive influencing factors on the cells' photosynthetic performance. From the 1930s onwards, most researchers chose to work with a standard strain of *Chlorella* which Emerson had originally introduced to the field; and the exchange of information on experimental organisms and recipes for culturing media made up a large part of the correspondence of the actors. It is very likely that chemists such as Melvin Calvin, biophysicists such as Louis N. M. Duysens and even plant physiologists of later generations were no longer familiar with the reasons for the choice of species (*Chlorella* or *Scenedesmus*), or of specific algae cultivation techniques. They continued with established tradition not only because they trusted their predecessors' skills and decisions but also to ensure that their results could be compared with earlier findings. In this sense, experimental organisms such as *Chlorella* really did "incorporate" experimental knowledge: they were needed to satisfy the constant conditions of the experimental set-up; although, perhaps, nobody knew exactly which of their many properties influenced the mechanism in what ways. The same held true for other aspects of experimental practice. The importance of tacit knowledge is impressively demonstrated by the fact that the actors often sought to solve controversies by conducting experiments together: there might always be aspects of the know-how of an experimenter that are crucial to the outcome of the experiment; yet, they might be so self-evident to the experimenter herself that it does not even occur to her to explain these details—unless, of course, a controversy arises.

The difficulties of experimental practice and the need to master the pertinent methods explain why scientists were so conservative, not only in their support of model hypotheses but also in terms of research techniques and explanatory approaches: Warburg and many others investigated *Chlorella* cells manometrically in certain media for most of their working lives. The fact that researchers jumped from one theme or field to another, as Warburg did, was clearly prompted by the strategy to exploit the technique's possibilities in as many disciplines as possible. And Warburg, in particular, was a very successful advocate of the technique's advantages. Many of the central figures in twentieth-century photosynthesis research had learned manometry while they were fellows or members of Warburg's institute; and upon leaving they brought this technique to other places. Searching for kinetic information by manometrically measuring the process soon replaced the mainly stoichiometrically guided (and largely unsuccessful) hunt for chemical intermediates. And once the choice of technique had been made, and some standard experiments carried out with it, researchers kept to it—not the least, as to be able to compare experimental outcomes.

Yet, having emphasised the power of heuristics, methods and strategies, one must not forget that a great many of the paths thus pursued led nowhere. Warburg is a marvellous example of a protagonist whose models and hypotheses soon proved untenable, although his methods and techniques were of lasting impact. On the other hand, there were unexpected discoveries that no heuristic rule could have foreseen—for example, when Emerson set out to investigate the influence of blue light on the gas exchange of *Chlorella* cells and found instead the enhancement effect on photosynthetic efficiency. In some phases, research activity was high and methodically uncontroversial, while the actors felt they were conceptually stuck. This was the case, for example, in the 1930s, when existing model alternatives were continuously worked on, without the scientists having any idea as to where the different paths would eventually lead (if anywhere). There were experimental findings that failed to arouse the conceptional interest they would have deserved, such as the red drop of photosynthesis that Emerson and Lewis had come across in 1943; whereas there were instances of explanatory breakthroughs that were discarded as empty speculation, such as the suggestion of the photosynthetic unit by Gaffron and Wohl in 1936. The aim of this book was not to construct another grand narrative of discovery. The aim was to bring exactly those intricate details to the fore that undermine the very idea of such a narrative, that is, present the highly diverse approaches to elucidating the mechanism of photosynthesis, only some of which were to last; and to try and identify some recurrent heuristic strategies that scientists utilised when they were confronted with the inherent complexity of their subject matter: plants in light and darkness.

References

Allen, G.E.: Life Science in the Twentieth Century (Cambridge History of Science). Cambridge University Press, Cambridge (1975)

Allen, M.B., Arnon, D.I., Capindale, J.B., Whatley, F.R., Durham, L.J.: Photosynthesis by isolated chloroplasts III. Evidence for complete photosynthesis. J. Am. Chem. Soc. **77**, 4149–4155 (1955)

Allen, M.B., Whatley, F.R., Rosenberg, L.L., Capindale, J.B., Arnon, D.I.: Photosynthetic carbon dioxide fixation by broken chloroplasts. In: Gaffron, H., Brown, A.H., Stacy, F.C., Livingston, R., Rabinowitch, E.I., Strehler, B.L., Tolbert, N.E. (eds.) Research in Photosynthesis: Papers and Discussions Presented at the Gatlinburg Conference, October 25–29, 1955, Interscience Publishers, pp. 288–295 (1957)

Allen, M.B., Whatley, F.R., Arnon, D.I.: Photosynthesis by isolated chloroplasts: VI. Rates of conversion of light into chemical energy in photosynthetic phosphorylation. Biochim. Biophys. Acta **27**, 16–23 (1958)

Amesz, J.: The two photosystems of photosynthesis. In: Kung, S.-D., Yang, S.-F. (eds.) Discoveries in Plant Biology, vol. 2, pp. 1–13. World Scientific Publishing, Singapore (1998)

Andersen, H.: A field guide to mechanisms, part I & II, Philosophy Compass **9**, http://philosophy-compass.com/sections/philosophy-of-science/. Accessed 19 Feb 2015. (2014)

Angelico, F., Catalano, G.: Sulla presenza della formaldeide nei succhi delle piante verdi. Gazzetta Chimica Italiana **162**, 563–566 (1913)

Ankeny, R., Leonelli, S.: What's so special about model organisms? Stud. Hist. Philos. Biol. Biomed. Sci. **42**, 313–323 (2011)

Anonymous: (1907a) Assimilation, Meyers Grosses Konversationslexikon, vol. 1, 6th edn. Bibliographisches Institut: Leipzig., p. 890

Anonymous: (1907b) Bolometer, Meyers Grosses Konversationslexikon, vol. 3, 6th edn. Bibliographisches Institut: Leipzig., p. 184

Anonymous: Grünes Geheimnis. Der Spiegel **48**, 88–89 (1961)

Arnold, W.A.: A calorimetric determination of the quantum yield in photosynthesis. In: Franck, J., Loomis, W. E. (eds.) Photosynthesis in Plants, pp. 273–276. Iowa State College Press, Ames (1949)

Arnold, W. A.: Life: a collection of probable or improbable events. In: Augenstine, L.G. (ed.) Bioenergetics. Considerations of processes of absorption, stabilization, transfer, and utilization (Proceedings of a symposium sponsored by the U.S. Atomic Energy Commission held at the Brookhaven National Laboratory, October 12–16, 1959), Academic Press, New York/London, pp. 324–325 (1960)

Arnold, W.A.: Experiments. Photosynth. Res. **27**, 73–82 (1991)

K. Nickelsen, *Explaining Photosynthesis,* History, Philosophy and Theory of the Life Sciences 8, DOI 10.1007/978-94-017-9582-1

Arnold, W., Kohn, H.I.: The chlorophyll unit in photosynthesis. J. Gen. Physiol. **18**, 109–112 (1934)

Arnold, W., Oppenheimer, J.R.: Internal conversion in the photosynthetic mechanism of blue-green algae. J. Gen. Physiol. **133**, 423–435 (1950)

Arnon, D.I.: Extracellular photosynthetic reactions. Nature **167**, 1008–1010 (1951)

Arnon, D.I.: (1954a) Photosynthesis by isolated chloroplasts. Nature **174**, 394–401

Arnon, D.I.: (1954b) Photosynthesis by isolated chloroplasts II. Photosynthetic phosphorylation, the conversion of light into phosphate bond energy. J. Am. Chem. Soc. **74**, 6324–6329

Arnon, D.I.: The chloroplast as a complete photosynthetic unit. Science **122**, 9–16 (1955)

Arnon, D.I.: Localization of photosynthesis in chloroplasts. In: Gaebler, O.H. (ed.) Enzymes: Units of Biological Structure and Function, pp. 279–313. Academic Press, New York (1956)

Arnon, D.I.: Conversion of light into chemical energy in photosynthesis. Nature **184**, 10–21 (1959)

Arnon, D.I.: Photosynthesis 1950–1975: changing concepts and perspectives. In: Trebst, A., Avron, M. (eds.) Photosynthesis I: Photosynthetic Electron Transport and Photophosphorylation, pp. 7–56. Springer, Berlin (1977)

Arnon, D.I.: The discovery of photosynthetic phosphorylation. Trends Biochem. Sci. **9**, 258–262 (1984)

Arnon, D.I.: Photosynthetic CO_2 assimilation by chloroplasts: assertion, refutation, discovery. Trends Biochem. Sci. **12**, 39–42 (1987)

Arnon, D.I.: The discovery of ferredoxin: the photosynthetic path. Trends Biochem. Sci. **13**, 30–33 (1988)

Arnon, D.I., Allen, M.B., Whatley, F.R.: Photosynthesis by isolated chloroplasts IV. General concept and comparison of three photochemical reactions. Biochim. Biophys. Acta **20**, 449–461 (1956)

Arnon, D.I., Whatley, F.R., Allen, M.B.: Triphosphopyridine nucleotide as a catalyst of photosynthetic phosphorylation by chloroplasts. Nature **180**, 182–185 (1957)

Arnon, D.I., Whatley, F.R., Allen, M.B.: Assimilatory power in photosynthesis. Science **127**, 1026–1034 (1958)

Aronoff, S., Calvin, M.: Phosphorus turnover and photosynthesis. Plant Physiol. **23**, 351–358 (1948)

Avron, M., Jagendorf, A.T.: Some further investigations on chloroplast TPNH diaphorase. Arch. Biochem. Biophys. **72**, 17–24 (1957)

Axelrod, B., Bandurski, R.S., Greiner, C.M., Jang, R.: The metabolism of hexose and pentose phosphates in higher plants. J. Biol. Chem. **202**, 619–634 (1953)

Bach, A.: Contribution á l'étude des phénomènes chimiques de l'assimilation de l'acide carbonique par les plantes á chlorophylle. Comptes Rendus de l'Académie des Sciences **116**, 1145–1148 (1893)

Badin, E.J., Calvin, M.: The path of carbon in photosynthesis. IX. Photosynthesis, photoreduction and the hydrogen-oxygen-carbon dioxide dark reaction. J. Am. Chem. Soc. **72**, 5266–5270 (1950)

Baeyer, A.von.: Über die Wasserentziehung und ihre Bedeutung für das Pflanzenleben und die Gährung. Berichte der Deutschen Chemischen Gesellschaft zu Berlin **3**, 63–75 (1870)

Baeyer, A.von.: Erinnerungen aus meinem Leben 1835–1905, Vieweg, Braunschweig (1905)

Baeyer, A.von.: Adolf von Baeyer: Biography. Nobel Lectures, Chemistry 1901–1921. Elsevier Publishing Company, Amsterdam. http://www.nobelprize.org/nobel_prizes/chemistry/laureates/1905/baeyer-bio.html. (1966). Accessed 18 Feb 2014

Bailer-Jones, D.: Creative strategies employed in modelling. A case study. Found. Sci. **4**, 375–388 (1999)

Bailer-Jones, D.: Modelling extended extragalactic radio sources. Stud. Hist. Philos. Mod. Phys. **31**, 49–74 (2000)

Bailer-Jones, D.: Scientific Models in Philosophy of Science. University of Pittsburgh Press, Pittsburgh (2009)

Baker, S.M.: Quantitative experiments on the effects of formaldehyde on living plants. Ann. Bot. **27**, 411–442 (1913)

Ball, E.G.: Oxidative Mechanisms in Animal Tissues, A Symposium on Respiratory Enzymes, pp. 16–32. University of Wisconsin Press, Madison (1942)

Ballou, C., Barker, H.A.: William Zev Hassid (1899–1974). Biogr. Mem. Natl. Acad. Sci. **50**, 196–231 (1979)

Bannister, T.T.: The careers and contributions of Eugene Rabinowitch. Biophys. J. **12**, 707–718 (1972)

Barker, H.A.: Radioactive carbon in bacterial metabolism: something old and something new. In: Kaplan, N.O., Robinson A. (eds.) From Cyclotrons to Cytochromes: Essays in Molecular Biology and Chemistry, Academic Press, New York, pp. 67–75 (1982)

Barker, H.A., Hungate, R.E.: Cornelis Bernardus Van Niel (1897–1985). Biogr. Mem. Natl. Acad. Sci. **59**, 389–423 (1990)

Barltrop, J.A., Hayes, P.M., Calvin, M.: The chemistry of 1,2-dithiolane (trimethylene disulfide) as a model for the primary quantum conversion act in photosynthesis. J. Am. Chem. Soc. **76**, 4348–4367 (1954)

Bartha, P.: By Parallel Reasoning. The Construction and Evaluation of Analogical Arguments, Oxford University Press, Oxford (2010)

Bassham, J.A.: Melvin Calvin (1911–1997). Photochem. Photobiol. **65**, 605–606 (1997)

Bassham, J.A.: Mapping the carbon reduction cycle: a personal retrospective. Photosynth. Res. **76**, 35–52 (2003)

Bassham, J.A., Calvin, M.: The path of carbon in photosynthesis. In: Pirson, A. (ed.) Handbuch der Pflanzenphysiologie, pp. 884–922. Springer, Berlin (1960)

Bassham, J.A., Benson, A.A., Calvin, M.: The path of carbon in photosynthesis VIII. The rôle of malic acid. J. Biol. Chem. **185**, 781–787 (1950)

Bassham, J.A., Benson, A.A., Kay, L.D., Harris, A.Z., Wilson, A.T., Calvin, M.: The path of carbon in photosynthesis. XXI. The cyclic regeneration of carbon dioxide acceptor. J. Am. Chem. Soc. **76**, 1760–1770 (1954)

Baumgartner, M., Graßhoff, G.: Kausalität und kausales Schliessen. Eine Einführung mit interaktiven Übungen. Bern Studies in the History and Philosophy of Science, Bern (2004)

Baur, E.: Über die Genesis der Kohlenhydrate. Naturwissenschaften **1**(20), 474–477 (1913)

Bay, Z., Pearlstein, R.M.: A theory of energy transfer in the photosynthetic unit. Proc. Natl. Acad. Sci. U S A **50**, 1071–1078 (1963)

Bechtel, W.: Biochemistry: a cross-disciplinary endeavour that discovered a distinctive domain. In: Bechtel, W. (ed.) Integrating Scientific Disciplines, pp. 77–100. Martinus Nijhoff, Dordrecht (1986a)

Bechtel, W.: The nature of scientific integration. In: Bechtel, W. (ed.) Integrating Scientific Disciplines, pp. 3–52. Martinus Nijhoff, Dordrecht (1986b)

Bechtel, W.: Discovering Cell Mechanisms: The Creation of Modern Cell Biology. Cambridge University Press, Cambridge (2006)

Bechtel, W., Abrahamsen, A.: Explanation: a mechanist alternative. Stud. Hist. Philos. Biol. Biomed. Sci. **36**, 421–441 (2005)

Bechtel, W., Richardson, R.C.: Discovering Complexity. Decomposition and Localization as Strategies in Scientific Research. Princeton University Press, Princeton (1993)

Beckel, A.L., Haywood, C.N.: Breaking new waters. A century of limnology at the University of Wisconsin, vol. (Special Issue), Transactions of the Wisconsin Academy of Sciences, Arts and Letters (1987)

Bendall, D.S.: Robert Hill. Biogr. Mem. Fellows R. Soc. **40**, 141–171 (1994)

Bendall, D.S.: The unfinished story of cytochrome *f*. Photosynth. Res. **80**, 265–276 (2004)

Bendall, D.S., Walker, D.A.: Obituary: Robert (Robin) Hill (1899–1991). Photosynth. Res. **30**, 1–5 (1991)

Benson, A.A.: Identification of ribulose in $^{14}CO_2$ photosynthesis products. J. Am. Chem. Soc. **73**, 2971–2972 (1951)

Benson, A.A.: Tracers in biology. In: Kaplan, N.O., Robinson, A. (eds.) From Cyclotrons to Cytochromes: Essays in Molecular Biology and Chemistry, pp. 59–65. Academic Press, New York (1982)

Benson, A.A.: Following the path of carbon in photosynthesis: a personal story. Photosynth. Res. **73**, 29–49 (2002a)

Benson, A.A.: Paving the path. Annu. Rev. Plant Biol. **53**, 1–25 (2002b)

Benson, A.A.: Last days in the Old Radiation Laboratory (ORL), Berkeley, California, 1954. Photosynth. Res. **105**, 209–212 (2010)

Benson, A.A., Calvin, M.: The dark reductions of photosynthesis. Science **105**, 648–649 (1947)

Benson, A.A., Calvin, M.: Carbon dioxide fixation by green plants. Annu. Rev. Plant Physiol. **1**, 25–42 (1950)

Benson, A.A., Calvin, M., Haas, V.A., Aronoff, S., Hall, A.G., Bassham, J.A., Weigl, J.W.: C^{14} in photosynthesis. In: Franck, J., Loomis, W.E. (eds.) Photosynthesis in Plants, pp. 381–401. State College Press, Ames (1949)

Benson, A.A., Bassham, J.A., Calvin, M., Goodale, T.C., Haas, V.A., Stepka, W.: The path of carbon in photosynthesis. V. Paper chromatography and radioautography of the products. J. Am. Chem. Soc. **72**, 1710–1718 (1950)

Benson, A.A., Bassham, J.A., Calvin, M.: Sedoheptulose in photosynthesis by plants. J. Am. Chem. Soc. **73**, 2970 (1951)

Benson, A.A., Bassham, J.A., Calvin, M., Hall, A.G., Hirsch, H.E., Kawaguchi, S., Lynch, V., Tolbert, N.E.: The path of carbon in photosynthesis. XV. Ribulose and sedoheptulose. J. Biol. Chem. **196**, 703–716 (1952a)

Benson, A.A., Kawaguchi, S., Hayes, P., Calvin, M.: The path of carbon in photosynthesis. XVI. Kinetic relationships of the intermediates in steady state photosynthesis. J. Am. Chem. Soc. **74**, 4477–4482 (1952b)

Bergia, S., Navarro, L.: Recurrences and continuity in Einstein's research on radiation between 1905 and 1916. Arch. Hist. Exact Sci. **38**, 79–99 (1988)

Besse, I., Buchanan, R.B.: Plant ferredoxin: a multifunctional electron carrier protein with an unusual genealogy. In: Kung S.-D., Yang S.-F. (eds.) Discoveries in Plant Biology, vol. 2, pp. 137–156. World Scientific Publishing, Singapore (1998)

Beyerchen, A.D.: Emigration from country and discipline: the journey of a german physicist into American photosynthesis research. In: Ash, M.G., Söllner, A. (eds.) Forced Migration and Scientific Change, pp. 71–85. Cambridge University Press, Cambridge (1996)

Bhagvat, K., Hill, R.: Cytochrome oxidase in higher plants. New Phytol. **50**, 112–120 (1951)

Bishop, N.I.: The influence of the herbicide, DCMU, on the oxygen-evolving system of photosynthesis. Biochim. Biophys. Acta **27**, 205–206 (1958)

Bishop, N.I.: The reactivity of a naturally occurring quinone (Q-255) in photochemical reactions of isolated chloroplasts, Proc. Natl. Acad. Sci. U S A **45**, 1696–1702 (1959)

Blackman, F.F.: Optima and limiting factors. Ann. Bot. **19**, 281–295 (1905)

Blinks, L.R.: Chromatic transients in photosynthesis of red algae. In: Gaffron, H., Brown A.H., French C.S., Livingston R., Rabinowitch E.I., Strehler B.L., Tolbert N.E. (eds.) Research in Photosynthesis, pp. 444–449. Interscience Publishers, New York (1957)

Blinks, L.R.: Action spectra of chromatic transients and the Emerson effect in marine algae. Proc. Natl. Acad. Sci. U S A **46**, 327–333 (1960a)

Blinks, L.R.: Chromatic transients in the photosynthesis of green, brown, and red algae. In: Allen, M.B. (ed.) Comparative Biochemistry of Photoreactive Systems, pp. 329–341. Academic Press, New York (1960b)

Blinks, L.R.: Relation of photosynthetic transients to respiration. Science **131**, 1316 (1960c)

Blinks, L.R.: Winthrop John Vanleuven Osterhout (1871–1964). Biogr. Mem. Natl. Acad. Sci. **44**, 213–249 (1974)

Boberlin, U.: Photochemische Untersuchungen von R. Bunsen und H. Roscoe im Vergleich mit den Arbeiten J. W. Drapers und W. C. Wittwers: Die Anfänge der quantitativen Photochemie im 19. Jahrhundert, Köster. Berlin (1993)

Bogen, J.: Causally productive activities. Stud. Hist. Philos. Sci. Part A **39**, 112–123 (2008)

Boitreux, R.: Sur la nutrition du Trichoderma viride (Pers.) á partir du formol libre. Comptes Rendus des Séances de la Société de Biologie **83**, 737–738 (1920)

Bonner, J.: James Bonner (1910–1996). Interviewed by Graham Berry, March 13–14, 1980 (1980)

Boorse, H.A., Motz, L., Weaver, J.H.: The Atomic Scientists: A Biographical History. Wiley (1989)

Born, M., Heisenberg, W., Jordan, P.: Zur Quantenmechanik. II., Zeitschrift für Physik **35**, 557–615. (1926) (English translation in: van der Waerden, B.L. (ed.) Sources of Quantum Mechanics. Dover Publications, 1968))

Borsook, H.: Interview by Mary Terrall. Pasadena, California. (Oral History Project, California Institute of Technology Archives; permanent URL: http://resolver.caltech.edu/CaltechOH:OH_Borsook_H). Accessed May 2014 (1978)

Bourdieu, P.: The peculiar history of scientific reason. Sociol. Forum **6**, 3–26 (1991)

Bredig, G.: Katalytische Ameisensäuresynthese unter Druck und ein photochemisch-katalytisches Assimilationsmodell. Vortrag, Chemiker-Zeitung **38**(57), 186–187 (1914a)

Bredig, G.: Welche Zwischenprodukte entstehen bei der Assimilation der Kohlensäure durch die Pflanze? Die Umschau **18**, 362–364 (1914b)

Bredig, G.: Gleichgewicht der Kohlensäurereduktion. Vortrag, Chemiker-Zeitung **12**(72), 72–73 (1915)

Brock, W.H.: Justus von Liebig: The Chemical Gatekeeper. Cambridge University Press, Cambridge (1997)

Brodhun, E.: Die Physikalisch-Technische Reichsanstalt. Fünfundzwanzig Jahre ihrer Tätigkeit. Naturwissenschaften **1**(14), 321–325 (1913)

Brody, S.S.: We remember Eugene. Photosynth. Res. **43**, 67–74 (1995)

Brody, S.S.: Fluorescence lifetime, yield, energy transfer and spectrum in photosynthesis, 1950–1960. Photosynth. Res. **73**, 127–132 (2002)

Brody, S.S., Rabinowitch, E.I.: Excitation lifetime of photosynthetic pigments in vitro and in vivo. Science **125**, 555 (1957)

Brown, H.T.: The reception and utilisation of energy by a green leaf. Nature **71**, 522–526 (1905)

Brown, H.T., Escombe, F.: Static diffusion of gases and liquids in relation to the assimilation of carbon and translocation in plants. Philos. Trans. R. Soc. Lond., Series B **193**, 223–291 (1900)

Brown, H.T., Escombe, F.: Researches on some of the physiological processes of green leaves, with special reference to the interchange of energy between the leaf and its surroundings. Proc. R. Soc. Lond., Series B **76**(507), 29–111 (1905)

Brown, A.H., Frenkel, A.W.: Photosynthesis. Ann. Rev. Biochem. **22**, 423–458 (1953)

Brown, A.H., Fager, E.W., Gaffron, H.: Kinetics of a photochemical intermediate in photosynthesis. In: Franck J., Loomis W.E. (eds.) Photosynthesis in Plants, pp. 403–422. Iowa, State College Press (1949)

Buchanan, R.B.: The life of Daniel I. Arnon. Photosynth. Res. **46**, 3–6 (1995)

Buchanan, R.B.: Daniel I. Arnon (1910–1994). Biogr. Mem. Fellows R. Soc. **80**, 2–3 (2001)

Buchanan, B.B., Wong, J.H.: A conversation with Andrew Benson: reflections on the discovery of the Calvin-Benson cycle. Photosynth. Res. **114**, 207–214 (2013)

Buchanan, R.B., Douce, R., Lichtenthaler, H.K.: Andrew A. Benson. Photosynth. Res. **92**, 143–144 (2007)

Burian, R.M.: Comments on Rheinberger's "From experimental systems to cultures of experimentation". In: Wolters G., Lennox, J.G., McLaughlin, P. (eds.) Concepts, Theories and Rationality in the Biological Sciences, pp. 123–136. University of Pittsburgh Press (1995)

Burk, D., Hearon, J.Z., Caroline, L., Schade, A.L.: Reversible complexes of cobalt, histidine, and oxygen gas. J. Biol. Chem. **165**, 723–724 (1946)

Burk, D., Hearon, J.Z., Levy, H., Schade, A.L.: Reversible oxygenation of cobaltodihistidine to oxy-bis(cobaltodihistidine) and comparisons with other metal-amino acids, oxyhemoglobin, and oxyhemocyanin. Fed. Proc. **6**, 242–242 (1947)

Burk, D., Hendricks, S., Korzenovsky, M., Schocken, V., Warburg, O.: The maximum efficiency of photosynthesis: a rediscovery. Science **110**, 225–229 (1949)

Burk, D., Cornfield, J., Schwartz, M.: The efficient transformation of light into chemical energy in photosynthesis. Sci. Mon. **73**, 213–223 (1951)

Butlerov, A.M.: Bildung einer zuckerartigen Substanz durch Synthese. Annalen der Chemie und Pharmazie **120**, 295–298 (1861)

Cahan, D.: An Institute for an Empire: The Physikalisch-Technische Reichsanstalt, 1871–1918. Cambridge University Press, New York (1989)

Callon, M.: Four models for the dynamics of science. In: Jasanoff, S., Markle, G.E., Petersen, J.C., Pinch, T. (eds.) Handbook of Science and Technology Studies. Sage, London, pp. 29–63 (1994)

Calvin, M.: Chemical and photochemical reactions of thioctic acid and related disulfides. Fed. Proc. pp. 697–711 (1954)

Calvin, M.: The path of carbon in photosynthesis. Science **135**, 879–889 (1962)

Calvin, M.: The Path of Carbon in Photosynthesis, Nobel Lectures, Chemistry 1942–1962, pp. 618–644. Elsevier Publishing Company, Amsterdam (1964)

Calvin, M.: Forty years of photosynthesis and related activities. Photosynth. Res. **21**, 3–16 (1989)

Calvin, M.: Following the Trail of Light. A Scientific Odyssey. American Chemical Society, Washington, D.C (1992)

Calvin, M., Bassham, J.A.: The photosynthesis of carbon compounds. W. A. Benjamin Inc., New York (1962)

Calvin, M., Benson, A.A.: The path of carbon in photosynthesis. Science **107**, 476–480 (1948)

Calvin, M., Massini, P.: The path of carbon in photosynthesis. XX: the steady state. Experientia **8**, 445–457 (1952)

Caneva, K.L.: Robert Mayer and the Conservation of Energy. Princeton University Press, Princeton (1993)

Carson, S.F., Foster, J.W., Ruben, S., Barker, H.A.: Radioactive carbon as an indicator of carbon dioxide utilization. V. Studies on the propionic acid bacteria. Proc. Natl. Acad. Sci. U S A **27**, 229–235 (1941)

Chance, B.: Commentary on the paper Chance & Wlliams (1956). Cit. Class. Comment. **49**, 232 (1983)

Chance, B.: The stopped-flow method and chemical intermediates in enzyme reactions—a personal essay. Photosynth. Res. **80**, 387–400 (2004)

Chance, B., Williams, G.R.: Respiratory enzymes in oxidative phosphorylation. I-IV., J. Biol. Chem. **217**, 383–393, 395–407, 409–427, 429–438 (1955)

Chance, B., Williams, G.R.: The respiratory chain and oxidative phosphorylation. Adv. Enzymol. **17**, 65–134 (1956)

Chance, B., Williams, G.R., Holmes, W.F., Higgins, J.: Respiratory enzymes in oxidative phosphorylation. V. J. Biol. Chem. **217**, 439–451 (1955)

Chodat, R., Schweizer, K.: Nouvelles recherches sure les ferments oxydants. IX. De l'emploi de la peroxydase comme réactif de la photolyse par la chlorophylle. Archives des Sciences Physiques et Naturelles **39**, 334–338 (1915)

Clayton, R.K.: Molecular Physics in Photosynthesis. Blaisdell, New York (1965)

Clayton, R.K.: Research on photosynthetic reaction centers from 1932 to 1987. Photosynth. Res. **73**, 63–71 (2002)

Clegg, R.M., Sener, M., Govindjee: From Foerster resonance energy transfer to coherent resonance energy transfer and back. In: Alfano, R.R. (ed.) Optical Biopsy VII. Proceedings of SPIE, SPIE Press, pp. 7561–82 (2010)

Consden, R., Gordon, A.H., Martin, A.J.P.: Qualitative analysis of proteins: a partition chromatographic method using paper. Biochem. J. **38**, 224–232 (1944)

Craig, P.: Centennial History of the Carnegie Institution of Washington. Vol. IV: The Department of Plant Biology, Cambridge University Press, Cambridge (2005)

Craver, C.F.: When mechanistic models explain. Synthese **153**, 355–376 (2006)

Craver, C., Darden, L.: In Search of Mechanisms. Discoveries Across the Life Sciences. University of Chicago Press, Chicago/London (2013)

Creager, A.N.: Nuclear energy in the service of biomedicine: The Atomic, U.S. Energy Commission's Radioisotope Program, 1946–1950. J. Hist. Biol. **39**, 649–684 (2006)

Creager, A.N.: Life Atomic. A History of Radioisotopes in Science and Medicine. University of Chicago Press, Chicago/London (2013)

Creager, A.N.H., Lunbeck, E., Wise, M.N. (eds.): Science Without Laws: Model Systems, Cases, Exemplary Narratives. Duke University Press (2007)

Czapek, F.: Biochemie der Pflanzen, 2., revised edn. Fischer, Jena (1913)

Darden, L.: Theory Change in Science. Strategies from Mendelian Genetics, Oxford University Press, New York/Oxford (1991)

Darden, L.: Strategies for discovering mechanisms: schema instantiation, modular subassembly, forward/backward chaining. Philos. Sci. **69**, S354–S365 (2002)

Darden, L.: Thinking again about biological mechanisms. Philos. Sci. **75**, 958–969 (2008)

Darden, L., Craver, C.F.: Strategies in the interfield discovery of the mechanism of protein synthesis. Stud. Hist. Philos. Biol. Biomed. Sci. **33**, 1–28 (2002)

Dascal, M.: The study of controversies and the theory and history of science. Sci. Context **11**, 147–154 (1998)

Davenport, H.E.: Ascaris Haemoglobin as an indicator of the oxygen produced by isolated chloroplasts. Proc. R. Soc. Lond., Series B **136**(511), 281–290 (1949)

Davenport, H.E.: Cytochrome components in chloroplasts. Nature **170**, 1112–1114 (1952)

Davenport, H.E.: Coenzyme reduction by illuminated chloroplasts. Biochem. J. **73**, 45–46. (1959) (Proceedings of the Biochemical Society)

Davenport, H.E.: A protein from leaves catalysing the reduction of metmyoglobin and triphosphopyridine nucleotide by illuminated chloroplasts. Biochem. J. **77**, 471–477 (1960)

Davenport, H.E., Hill, R.: The preparation and some properties of cytochrome *f*. Proc. R. Soc. Lond., Series B **139**, 327–345 (1952)

Davenport, H.E., Hill, R.: A protein from leaves catalysing the reduction of haem-protein compounds by illuminated chloroplasts. Biochem. J. **74**, 493–501 (1960)

Davenport, H.E., Hill, R., Whatley, F.R.: A natural factor catalysing reduction of methaemoglobin by isolated chloroplasts. Proc. R. Soc. Lond., Series B **139**, 346–358 (1952)

Deichmann, U.: The expulsion of German-Jewish chemists and biochemists and their correspondence with colleagues in Germany after 1945: the impossibility of normalization? In: Szöllösi-Janze M. (ed.) Science in the Third Reich, pp. 243–280. Berg, Oxford/New York (2001a)

Deichmann, U.: Flüchten, Mitmachen, Vergessen. Chemiker und Biochemiker in der NS Zeit. Wiley VCH, Weinheim (2001b)

Delbrück, M.: Oral History Interview with Carolyn Harding, CIT, p. 55 (Permanent URL: http://resolver.caltech.edu/CaltechOH:OH_Delbruck_M). Accessed 19 Feb 2015 (1978)

Demeter, S., Govindjee: Thermoluminescence in plants. Physiologia Plantarum **75**, 121–130 (1989)

Dent, C.E., Fink, K., Fink, R.M.: Application of filter paper partition chromatography to radioactive tracer studies. Nature **160**, 801–803 (1947a)

Dent, C.E., Stepka, W., Steward, F.C.: Detection of free amino acids of plant cells by partition chromatography. Nature **160**, 682–683 (1947b)

Dmetrievsky, O.D., Ermolaev, V.L., Terenin, A.N.: Direct measurement of the lifetimes of chlorophyll and analogous pigments in various media. Doklady Akademii Nauk **114**, 468–470 (1957)

Dohrn, A.: Aus Vergangenheit und Gegenwart der Zoologischen Station zu Neapel. Deutsche Rundschau **72**, 275–298 (1892)

Dole, M., Jenks, G.: Isotopic composition of photosynthetic oxygen. Science **100**, 409 (1944)

Dorner, R.W., Kahn, A., Wildman, S.G.: The proteins of green leaves. VII. Synthesis and decay of the cytoplasmic proteins during the life of a tobacco leaf. J. Biol. Chem. **229**, 945–952 (1957)

Dutrochet, H.R.J.: Mémoires pour servir á l'Histoire anatomique et physiologique des Végétaux et des Animaux, J.-B. Baillière, Paris (1837)

Dutton, H.J.: Carotenoid-sensitized photosynthesis: quantum efficiency, fluorescence and energy transfer. Photosynth. Res. **52**, 175–185 (1997)

Dutton, H.J., Manning, W.M.: Evidence for carotenoid-sensitized photosynthesis in the diatom *Nitzschia closterium*. Am. J. Bot. **28**, 516–526 (1941)

Dutton, H.J., Manning, W.M., Duggar, B.M.: Chlorophyll fluorescence and energy transfer in diatom *Nitzschia closterium*. J. Phys. Chem. **47**, 308–313 (1943)

Duysens, L.N.M.: Transfer of light energy within the pigment systems present in photosynthesizing cells. Nature **168**, 548–550 (1951)

Duysens, L.N.M.: Transfer of excitation energy in photosynthesis, Ph.d. University of Utrecht, Netherlands (1952)

Duysens, L.N.M.: Reversible changes in the absorption spectrum of Chlorella upon irradiation. Science **120**, 353–354 (1954a)

Duysens, L.N.M.: Reversible photo-oxidation of a cytochrome pigment in photosynthesizing *Rhodospirillum rubrum*. Nature **173**, 692–693 (1954b)

Duysens, L.N.M.: Role of cytochrome and pyridine nucleotide in algal photosynthesis. Science **121**, 210–211 (1955)

Duysens, L.N.M.: Energy transformations in photosynthesis. Annu. Rev. Plant Physiol. **7**, 25–50 (1956)

Duysens, L.N.M.: Investigations in the photosynthetic mechanism of purple bacteria by means of sensitive absorption spectrophotometry. In: Gaffron, H., Brown, A.H., Stacy, F.C., Livingston, R., Rabinowitch, E.I., Strehler, B.L., Tolbert, N.E. (eds.), Research in photosynthesis: Papers and discussions presented at the Gatlinburg Conference, October 25–29, 1955, Interscience Publishers, pp. 164–173 (1957)

Duysens, L.N.M.: Cytochrome oxidation by a second photochemical system in the red alga *Porphyridium cruentum*. In: Christensen, B.C., Buchmann, B. (eds.) Progress in photobiology. Proceedings of the 3rd international Congress in photobiology, Copenhagen, 1960 Elsevier, Amsterdam, pp. 135–142 (1961)

Duysens, L.N.M.: The discovery of the two photosynthetic systems: a personal account. Photosynth. Res. **21**, 61–79 (1989)

Duysens, L.N.M., Huiskamp, W.J., Vos, J.J., van der Hart, J.M.: Reversible changes in bacteriochlorophyll in purple bacteria upon illumination. Biochim. Biophys. Acta **19**, 188–190 (1956)

Duysens, L.N.M., Amesz, J., Kamp, B.M.: Two photochemical systems in photosynthesis. Nature **190**, 510–511 (1961)

Eaton-Rye, J.: Govindjee at 80: more than 50 years of free energy for photosynthesis. Photosynth. Res. **116**, 111–144 (2013)

Edge, D.: Competition in modern science. In: Frängsmyr T. (ed.) Solomon's House Revisited. The Organization and Institutionalization of Science, pp. 208–232. Science History Publications, Canton (1990)

Edsall, J.T. (ed.): History of Bioenergetics. Special issue of the journal *Molecular and Cellular Biochemistry*. Papers from a Conference held at the American Academy of Arts and Sciences, Oct. 11–13, 1973 (1974)

Einstein, A.: Über einen die Erzeugung und Verwandlung des Lichtes betreffenden heuristischen Gesichtspunkt. Annalen der Physik **17**, 132–148 (1905)

Einstein, A.: Über die Entwicklung unserer Anschauungen über das Wesen und die Konstitution der Strahlung. Verhandlungen der Deutschen Physikalischen Gesellschaft **11**, 482–500 (1909)

Einstein, A.: Nachtrag zu meiner Arbeit: "Thermodynamische Begründung des photochemischen Aequivalentgesetzes". Annalen der Physik **38**, 881–884 (1912a)

Einstein, A.: Thermodynamische Begründung des photochemischen Aequivalentgesetzes. Annalen der Physik **37**, 832–838 (1912b)

Emerson, R.: A review of recent investigations in the field of chlorophyll photosynthesis. Ergebnisse der Enzymforschung **5**, 305–347 (1936)

Emerson, R., Arnold, W.: The photochemical reaction in photosynthesis. J. Gen. Physiol. **16**, 191–205 (1932a)

Emerson, R., Arnold, W.: A separation of the reactions in photosynthesis by means of intermittent light. J. Gen. Physiol. **15**, 391–420 (1932b)

Emerson, R., Chalmers, R.: Transient changes in cellular gas exchange and the problem of maximum efficiency of photosynthesis. Plant Physiol. **30**, 504–529 (1955)

Emerson, R., Chalmers, R.: Speculations concerning the function and phylogenetic significance of the accessory pigments of algae. Phycol. Soc. Am. News Bull. **11**, 51–56 (1958)

Emerson, R., Green, L.: Nature of the Blackman reaction in photosynthesis. Plant Physiol. **12**, 537–545 (1937)

Emerson, R., Lewis, C.M.: Factors influencing the efficiency of photosynthesis. Am. J. Bot. **26**(10), 808–822 (1939)

Emerson, R., Lewis, C.M.: Carbon dioxide exchange and the measurement of the quantum yield of photosynthesis. Am. J. Bot. **28**, 789–804 (1941a)

Emerson, R., Lewis, C.M.: The photosynthetic efficiency of phycocyanin in Chroococcus and the problem of carotenoid participation in photosynthesis. J. Gen. Physiol. **25**, 579–595 (1941b)

Emerson, R., Lewis, C.M.: The dependence of the quantum yield of Chlorella photosynthesis on wave length of light. Am. J. Bot. **30**, 165–178 (1943)

Emerson, R., Rabinowitch, E.I.: Red drop and role of auxiliary pigments in photosynthesis. Plant Physiol. **35**, 477–485 (1960)

Emerson, R., Stauffer, J.F., Umbreit, W.W.: Relationships between phosphorylation and photosynthesis in Chlorella. Am. J. Bot. **31**, 107–120 (1944)

Emerson, R., Chalmers, R.V., Cederstrand, C.N.: Some factors influencing the long-wave limit of photosynthesis. Proc. Natl. Acad. Sci. U S A **43**, 133–143 (1957)

Engelhardt, V.A.: Ortho- und Pyrophosphat im aeroben und anaeroben Stoffwechsel der Blutzellen. Biochemische Zeitung **227**, 16–38 (1930)

Engelhardt, V.A.: Myosine and adenosinetriphosphatase. Nature **144**, 668–669 (1939)

Erlenmeyer, E.: Das Wasser als Oxydations- und Reductionsmittel. Berichte der Deutschen Chemischen Gesellschaft **10**, 634–635 (1877)

Euler, H., Euler, A.: (1906a) Über die Bildung von i-Arabinoketose aus Formaldehyd. Berichte der deutschen chemischen Gesellschaft **39**, 45–51

Euler, H., Euler, A.: (1906b) Zur Kenntniss der Zuckerbildung aus Formaldehyd. Berichte der deutschen chemischen Gesellschaft **39**, 39–45

Evans, M.C., Buchanan, R.B., Arnon, D.I.: A new ferredoxin-dependent carbon reduction cycle in a photosynthetic bacterium. Proc. Natl. Acad. Sci. U S A **55**, 928–934 (1966)

Ewart, A.J.: On the function of chlorophyll. Proc. R. Soc. Lond., Series B. **89**, 1–17 (1915)

Fager, E.W.: Investigation of the chemical properties of intermediates in photosynthesis. In: Franck, J., Loomis, W.E. (eds.) Photosynthesis in Plants, pp. 423–436. State College Press, Ames (1949)

Fangerau, H.: Spinning the Scientific Web. Jacques Loeb (1859–1924) und sein Programm einer internationalen biomedizinischen Grundlagenforschung. Akademie Verlag, Berlin (2010)

Fantini, B.: The history of the stazione zoologica Anton Dohrn. An outline, S.Z.A.D. Napoli: Activity report 1998/99, Zoological Station Anton Dohrn, Naples, pp. 71–105 (2002)

Fenton, H.J.H.: The reduction of carbon dioxide to formaldehyde in aequous solution. J. Chem. Soc., Trans. **91**, 687–694 (1907)

Fink, R.M., Fink, K.: Radiocarbon and filter paper chromatography. Science **107**, 253 (1948)

Finlay, M.R.: Growing American Rubber: Strategic Plants and the Policies of National Security. Rutgers University Press, New Brunswick (2009)

Fischer, E.: Über die Verbindungen des Phenylhydrazins mit den Zuckerarten. III. Berichte der deutschen chemischen Gesellschaft **21**, 988–991 (1888)

Fischer, E.: Synthese der Mannose und Lävulose. Berichte der deutschen chemischen Gesellschaft **23**, 370–394 (1890a)

Fischer, E.: Synthese des Traubenzuckers. Berichte der deutschen chemischen Gesellschaft **23**, 799–805 (1890b)

Fischer, E.: Synthesen in der Zuckergruppe. Berichte der deutschen chemischen Gesellschaft **23**, 2114–2141 (1890c)

Fischer, E.: Synthesen in der Zuckergruppe I (Vortrag, gehalten in der Sitzung der Deutschen chemischen Gesellschaft am 23. Juni 1890), Untersuchungen über Kohlenhydrate und Fermente, Springer, Berlin, pp. 1–29. [Originally published in *Berichte der deutschen chemischen Gesellschaft* 23, 2114 (1890)] (1909)

Fischer, E., Passmore, F.: Bildung von Acrose aus Formaldehyd. Berichte der deutschen chemischen Gesellschaft **22**, 359–361 (1889)

Florkin, M.: A History of Biochemistry. Part IV. Early Studies on Biosynthesis, Vol. 32 der Comprehensive Biochemistry. Elsevier, Amsterdam (1977)

Florkin, M.: A History of Biochemistry. Part V: The Unravelling of Biosynthetic Pathways, Vol. 33A der Comprehensive Biochemistry. Elsevier, Amsterdam (1979)

Franck, J.: Elementary processes of photochemical reactions. Trans. Faraday Soc. **21**, 536–542 (1925)

Franck, J.: Emil Warburg zum achzigsten Geburtstage. Naturwissenschaften **14**(11), 203–207 (1926)

Franck, J.: Emil Warburg zum Gedächtnis. Naturwissenschaften **19**(50), 993–997 (1931)

Franck, J.: Beitrag zum Problem der Kohlensäure-Assimilation. Naturwissenschaften **23**, 226–229 (1935a)

Franck, J.: Remarks on photosynthesis. Chem. Rev. **17**, 433–438 (1935b)

Franck, J.: Photosynthetic activity of isolated chloroplasts. Rev. Mod. Phys. **17**, 112–119 (1945)

Franck, J.: An interpretation of the contradictory results in measurements of the photosynthetic quantum yield and related phenomena. Arch. Biochem. **23**, 297–314 (1949)

Franck, J., Gaffron, H.: Photosynthesis, facts and interpretation. Adv. Enzymol. **1**, 199–262 (1941)

Franck, J., Hertz, G.: Über ZusammenstöÃŸe zwischen Elektronen und den Molekülen des Quecksilberdampfes und die Ionisierungsspannung desselben. Verhandlungen der Deutschen Physikalischen Gesellschaft **16**, 457–467 (1914)

Franck, J., Herzfeld, K.F.: An attempted theory of photosynthesis. J. Chem. Phys. **5**, 237–251 (1937)

Franck, J., Herzfeld, K.F.: Contributions to a theory of photosynthesis. J. Phys. Chem. **45**, 978–1025 (1941)

Franck, J., Levi, H.: Beitrag zur Untersuchung der Fluoreszenz in Flüssigkeiten. Zeitschrift für Physikalische Chemie **B27**, 409–420 (1935a)

Franck, J., Levi, H.: Zum Mechanismus der Sauerstoff-Aktivierung durch fluoreszenzfähige Farbstoffe. Naturwissenschaften **23**, 229–230 (1935b)

Franck, J., Rabinowitch, E.I.: Some remarks about free radicals and the photochemistry of solutions. Trans. Faraday Soc. **30**, 120–130 (1934)

Franck, J., Rosenberg, J.L.: A theory of light utilization in plant photosynthesis. J. Theor. Biol. **7**, 276–301 (1964)

Franck, J., Teller, E.: Migration and photochemical action of excitation energy in crystals. J. Chem. Phys. **6**, 861–872 (1938)

Franck, J., French, C.S., Puck, T.T.: The fluorescence of chlorophyll and photosynthesis. J. Phys. Chem. **45**, 1268–1300 (1941)

French, C.S.: The rate of CO_2 assimilation by purple bacteria at various wave lengths of light. J. Gen. Physiol. **21**, 71–87 (1937)

French, C.S.: Robert Emerson, investigator of photosynthesis. Science **130**(3373), 437–438 (1959)

French, C.S.: Light, pigments and photosynthesis. In: McElroy, W.D., Glass, B. (eds.) Light and Life, pp. 447–474 . The Johns Hopkins Press, Baltimore (1961)

French, C.S.: Fifty years of photosynthesis. Annu. Rev. Plant Physiol. **30**, 1–26 (1979)

French, C.S., Anson, M.L.: Oxygen production by isolated chloroplasts. Am. J. Bot. **28**, 12s (abstract) (1941)

French, C.S., Young, V.K.: The fluorescence spectra of red algae and the transfer of energy from phycoerytherin to phycocyanin and chlorophyll. J. Gen. Physiol. **35**, 873–890 (1952)

Frenkel, A.W.: Light-induced phosphorylation by cell-free preparations of photosynthetic bacteria. J. Am. Chem. Soc. **76**, 5568–5569 (1954)

Frenkel, A.W.: Reflections. Photosynth. Res. **35**, 103–116 (1993)
Frenkel, A.W.: Photosynthetic phosphorylation. Photosynth. Res. **46**, 73–77 (1995)
Freudenthal, G.: A rational controversy over compounding forces. In: Machamer, P., Pera, M., Baltas, A. (eds.) Scientific Controversies. Philosophical and Historical Perspectives, pp. 125–142. Oxford University Press (2000)
Friedmann, H.C.: From "Butyribacterium" to "E. coli": an essay on unity in biochemistry. Perspect. Biol. Med. **47**, 47–66 (2004)
Frigg, R., Hartmann, S.: Models in science. In: Zalta, E.N. (ed.) The Stanford Encyclopedia of Philosophy. http://plato.stanford.edu/archives/spr2006/entries/models-science/. (2012)
Förster, T.: Energiewanderung und Fluoreszenz. Naturwissenschaften **33**, 166–175 (1946)
Förster, T.: Ein Beitrag zur Theorie der Photosynthese. Zeitschrift für Naturforschung **2b**, 174 (1947)
Förster, T.: Fluoreszenz organischer Verbindungen. Vandenhoeck & Ruprecht, Göttingen (1951)
Fruton, J.S.: Molecules and Life. Historical Essays on the Interplay of Chemistry and Biology. Wiley Interscience, New York (1972a)
Fruton, J.S.: Molecules and Life. Historical Essays on the Interplay of Chemistry and Biology. Wiley-Interscience, New York (1972b)
Fruton, J.S.: Proteins, Enzymes, Genes: The Interplay of Chemistry and Biology. Yale University Press, New Haven (1999)
Fuller, R.C.: Forty years of microbial photosynthesis research: where it came from and what it led to. Photosynth. Res. **62**, 1–29 (1999)
Gaffron, H.: Biochemische Zeitschrift **264**, 251–271 (1933a)
Gaffron, H.: Über den Stoffwechsel der schwefelfreien Purpurbakterien. Biochemische Zeitschrift **260**, 1–17 (1933b)
Gaffron, H.: Über die Kohlensäure-Assimilation der roten Schwefelbakterien. Biochemische Zeitschrift **269**, 447–453 (1934)
Gaffron, H.: Inwiefern ist Sauerstoff für die Kohlensäureassimilation unentbehrlich? Naturwissenschaften **23**, 528–528 (1935)
Gaffron, H.: Studies on the induction period of photosynthesis and light respiration in green algae. Am. J. Bot. **27**, 204–216 (1940)
Gaffron, H.: Van Niel's theory: thirty years Aater. In: Gest, H., San Pietro A., Vernon, L.P. (eds.) Bacterial Photosynthesis, pp. 3–14. Antioch Press, Yellow Springs (1963)
Gaffron, H.: Resistance to knowledge. Plant Physiol. **20**, 1–40 (1969)
Gaffron, H., Fager, E.W.: The kinetics and chemistry of photosynthesis. Annu. Rev. Plant Physiol. **2**, 87–114 (1951)
Gaffron, H., Wohl, K.: Zur Theorie der Assimilation. Naturwissenschaften **24**, 81–90, 103–107 (1936)
Gaffron, H., Brown, A.H., Stacy, F.C., Livingston, R., Rabinowitch, E.I., Strehler, B.L., Tolbert, N.E. (eds.): Research in photosynthesis: papers and discussions presented at the Gatlinburg conference, October 25–29, 1955, Interscience Publishers, Baltimore (1957)
Gamboa, S.: In defense of analogical reasoning. Inf. Logic **28**, 229–241 (2008)
Gest, H.: History of the word 'photosynthesis' and evolution of its definition. Photosynth. Res. **73**, 7–10 (2002)
Gest, H.: Samuel Ruben's contributions to research on photosynthesis and bacterial metabolism with radioactive carbon. Photosynth. Res. **80**, 77–83 (2004)
Gest, H.: A personal tribute to an eminent photosynthesis researcher Martin D. Kamen (1913–2002). In: Govindjee, Beatty, J.T., Gest, H., Allen, J.F. (eds.) Discoveries in Photosynthesis, pp. xxvii–xxviii. Springer, Dordrecht (2005)
Gest, H., Blankenship, R.E.: Time line of discoveries: anoxygenic bacterial photosynthesis. Photosynth. Res. **80**, 59–70 (2004)
Gest, H., Kamen, M.D.: Studies on the phosphorus metabolism of green algae and purple bacteria in relation to photosynthesis. J. Biol. Chem. **176**, 299–318 (1948)
Gibson, R.J.H.: A photoelectric theory of photosynthesis. Ann. Bot. **22**, 117–120 (1908)

Giere, R.N.: Controversies involving science and technology: a theoretical perspective. In: Machamer, P., Pera, M., Baltas, A. (eds.) Scientific Controversies. Philosophical and Historical Perspectives, pp. 125–150. Oxford University Press, Oxford (2000)

Glennan, S.: Mechanisms and the nature of causation. Erkenntnis **44**, 49–77 (1996)

Glennan, S.: Rethinking mechanistic explanation. Philos. Sci. **69**, S342–S353 (2002)

Glennan, S.: Modeling mechanisms. Stud. Hist. Philos. Biol. Biomed. Sci. **36**, 443–464 (2005)

Goenner, H.: Einstein in Berlin 1914–1933. Beck, Munich (2005)

Goenner, H., Castagnetti, G.: Einstein and the Kaiser Wilhelm Institute for physics: Institutional aims and scientific results, Preprint No. 261, MPI for the History of Science, Berlin (2004)

Goldman, A.: Social epistemology. In: Zalta, E.N. (ed.) The Stanford Encyclopedia of Philosophy. http://plato.stanford.edu/archives/spr2006/entries/models-science/. Accessed 16 March 2014 (2010)

Goldman, A.: A Guide to Social Epistemology, Social Epistemology: Essential Readings, pp. 11–37. Oxford University Press, Oxford/New York (2011)

Govindjee: Sixty-three years since Kautsky chlorophyll *a* fluorescence. Aust. J. Plant Physiol. **22**, 131–160 (1995)

Govindjee: On the requirement of minimum number of four versus eight quanta of light for the evolution of one molecule of oxygen in photosynthesis: a historical note. Photosynth. Res. **59**, 249–254 (1999)

Govindjee: Lighting the path: a tribute to Robert Emerson (1903–1959), PS2001 Proceedings: 12th International Congress on Photosynthesis, Brisbane, CSIRO Publishing, Melbourne (2001)

Govindjee: Chlorophyll *a* fluorescence: a bit of basics and history. In: Papageorgiou, G.C., Govindjee (eds.) Chlorophyll *a* Fluorescence: A Signature of Photosynthesis, Vol. 19 der Advances in Photosynthesis and Respiration, pp. 1–42. Springer (2004a)

Govindjee: Robert Emerson and Eugene Rabinowitch: understanding photosynthesis. In: Hoddeson, L. (ed.) No Boundaries: University of Illinois Vignettes, pp. 181–194. University of Illinois Press, Urbana (2004b)

Govindjee: Celebrating 20 years of historical papers in photosynthesis research. Photosynth. Res. **87**, 151–158 (2006)

Govindjee: Sam Aronoff. Am. Soc. Plant Biol. Newsl. **37**, 21–22 (2010)

Govindjee, Björn, L.O.: Dissecting oxygenic photosynthesis: the evolution of the Z-scheme for thylakoid membranes. In: Itoh, S., Mohanty, P., Guruprasad, K.N. (eds.) Photosynthesis: Overviews on Recent Progress and Future Perspective, pp. 1–27. I.K. International Publishers, New Delhi (2012)

Govindjee, Fork, D.: Charles Stacy French (1907–1995). Biogr. Mem. Natl. Acad. Sci. **88**, 1–29 (2006)

Govindjee, Govindjee, R.: Introduction to photosynthesis. In: Govindjee (ed.) Bioenergetics of Photosynthesis, pp. 1–50. Academic Press, New York (1975)

Govindjee, Krogmann, D.: Discoveries in oxygenic photosynthesis (1727–2003): a perspective. Photosynth. Res. **80**, 15–27 (2004)

Govindjee, Rabinowitch, E.I.: Two forms of chlorophyll *a* in vivo with distinct photochemical functions. Science **132**, 355–356 (1960)

Govindjee, R., Thomas, J.B., Rabinowitch, E.I.: "Second Emerson effect" in the Hill reaction of Chlorella cells with quinone as oxidant. Science **132**, 421–421 (1960)

Govindjee, Owens, O., Hoch, G.: A mass spectroscopic study of the Emerson enhancement effect. Biochimica et Biophysica Acta **75**, 281–284 (1963)

Govindjee, R., Rabinowitch, E.I., Govindjee: Maximum quantum yield and action spectra of photosynthesis and fluorescence in Chlorella. Biochim. Biophys. Acta **162**, 530–544 (1968)

Govindjee, Amesz, J., Knox, R.S.: Photosynthetic unit: Antenna and Reaction Centers (Special issue dedicated to William Arnold). Photosynth. Res. (1996)

Govindjee, Beatty, J.T., Gest, H., Allen, J.F. (eds.): Discoveries in Photosynthesis. Springer, Dordrecht (2005)

Govindjee, Björn, L.O., Nickelsen, K.: Evolution of the Z-Scheme of electron transport in oxygenic photosynthesis. In: Lu, C. (ed.) Research for Food, Fuel and Future: 15th International Conference on Photosynthesis, pp. 835–841. Zhejiang University Press (2012)

Grafe, V.: Über ein neues specifisches Formaldehydreagenz. Oesterreichische botanische Zeitschrift **56**, 289–291 (1906)

Grafe, V., Vieser, E.: Untersuchungen über das Verhalten grüner Pflanzen zu gasförmigem Formaldehyd. Berichte der deutschen chemischen Gesellschaft **27**, 431–446 (1909)

Grafe, V., Vieser, E.: Untersuchungen über das Verhalten grüner Pflanzen zu gasförmigem Formaldehyd II. Berichte der deutschen chemischen Gesellschaft **29**, 19–26 (1911)

Graßhoff, G.: Die Kunst wissenschaftlichen Entdeckens. Grundzüge einer Theorie epistemischer Systeme, habilitation treatise, University of Hamburg (1994)

Graßhoff, G.: Modelling the astrophysical object SS433: methodology of model construction by a research collective. Philosophia Naturalis **35**, 161–199 (1998)

Graßhoff, G.: Normal star observation in late astronomical babylonian diaries. In: Swerdlow, N.M. (ed.) Ancient Astronomy and Celestial Divination (Dibner Institute Studies in the History of Science and Technology), pp. 97–147. MIT Press, Cambridge (1999)

Graßhoff, G.: Inferences to causal relevance from experiments. In: Dieks, D., González, W.J., Hartmann, S., Uebel, T., Weber, M. (eds.) Explanation, Prediction and Confirmation, Vol. 2 der The Philosophy of Science in a European Perspective, pp. 167–182. Springer, Berlin (2011)

Graßhoff, G., May, M.: Causal regularities. In: Spohn, W., Ledwig, M., Esfeld, M. (eds.) Current Issues in Causation, pp. 85–114. Mentis, Paderborn (2001)

Graßhoff, G., May, M.: Hans Krebs and Kurt Henseleit's laboratory notebooks and their discovery of the urea cycle: reconstructed with computer models. In: Holmes, F.L., Renn, J., Rheinberger, H.-J. (eds.) Reworking the Bench: Research Notebooks in the History of Science, pp. 269–294. Kluwer Academic Publishers, Dordrecht (2003)

Graßhoff, G., Casties, R., Nickelsen, K.: Zur Theorie des Experiments. Untersuchungen am Beispiel der Entdeckung des Harnstoffzyklus, Educational Materials, Bern Studies for the History and Philosophy of Science, Bern (2000)

Groeben, C.: The Naples Zoological Station at the Time of Anton Dohrn: Exhibition and Catalogue. In Collaboration with Irmgard Müller. Stazione Zoologica, Naples (1975)

Groeben, C.: Catalysing science: The *Stazione Zoologica di Napoli* as a place for the circulation of scientific ideas. In: Groeben, C., Kaasch, J., Kaasch, M. (eds.) Places of Biological Research, pp. 53–64. Verlag für Wissenschaft und Bildung, Berlin (2005)

Gromet-Elhanan, Z.: Mordhay Avron, 1931–1991. J. Bioenerg. Biomembr. **24**, 520–521 (1992)

Guttman, L., Hess, D.C., Myers, F.E., Wolfe, H.C.: Foster Frederick Rieke 1905–1970. J. Appl. Phys. **41**, 3911 (1970)

Haber, F., Willstätter, R.: Unpaarigkeit und Radikalketten im Reaktionsmechanismus organischer und enzymatischer Vorgänge. Berichte der Deutschen Chemischen Gesellschaft **64**, 2844–2856 (1931)

Haberlandt, G.: Vergleichende Anatomie des assimilatorischen Gewebesystems der Pflanzen. Jahrbuch für wissenschaftliche Botanik **13**, 74–188 (1881)

Haberlandt, G.: Physiologische Pflanzenanatomie. Engelmann, Leipzig (1884)

Hall, T.S.: History of General Physiology, 2 vols. University of Chicago Press, Chicago. Originally published as: Ideas of Life and Matter, 2 vols. (University of Chicago Press, 1969) (1975)

Hall, V.M.D.: The role of force or power in Liebig's physiological thought. Med. Hist. **24**, 20–59 (1980)

Hatch, M.D.: I can't believe my luck. Photosynth. Res. **33**, 1–14 (1992)

Hatch, M.D.: C_4 photosynthesis: discovery and resolution. Photosynth. Res. **73**, 251–256 (2002)

Hatch, M.D., Slack, C.R.: Photosynthesis in sugar cane leaves: a new carboxylation reaction and the pathway of sugar formation. Biochem. J. **101**, 103–111 (1966)

Hatch, M.D., Slack, C.R., Johnson, H.S.: Further studies on a new pathway of photosynthetic CO_2 fixation in sugarcane and its occurrence in other plant species. Biochem. J. **102**, 417–422 (1967)

Haxo, F.T., Blinks, L.R.: Photosynthetic action spectra of marine algae. J. Gen. Physiol. **33**, 389–421 (1950)

Hearon, J.Z.: The configuration of cobaltodihistidine and oxy-bis(cobaltodihistidine). J. Natl. Cancer Inst. **9**, 1–11 (1948)

Hearon, J.Z., Burk, D., Schade, A.: Physicochemical studies of reversible and irreversible complexes of cobalt, histidine, and molecular oxygen. J. Natl. Inst. **9**, 337–377 (1949)

Heilbron, J.L., Seidel, R.W.: Lawrence and his Laboratory: A History of the Lawrence Berkeley Laboratory. University of California Press, Berkeley (1989)

Heilbron, J.L., Seidel, R.W., Wheaton, B.R.: Lawrence and his Laboratory: Nuclear Science at Berkeley. University of California Press, Berkeley (1981)

Hendricks, S.B.: A discussion of photosynthesis. Science **117**, 370–373 (1953)

Henning, E.: Otto Heinrich Warburg—Der "Kaiser von Dahlem". In: Treue, W., Hildebrandt, G. (eds.) Berlinische Lebensbilder. Bd. 1: Naturwissenschaftler, pp. 299–316. Colloquium Verlag, Berlin (1987)

Hentschel, K. (ed.): Physics and National Socialism. An Anthology of Primary Sources. Birkhäuser, Basel (1996)

Hentschel, K.: Die Mentalität deutscher Physiker in der frühen Nachkriegszeit (1945–1949). Synchron, Heidelberg (2005)

Herken, G.: Brotherhood of the Bomb. The Tangled Lives and Loyalties of Robert Oppenheimer, Ernest Lawrence, and Edward Teller. Henry Holt, New York (2002)

Hesse, M.B.: Models and Analogies in Science. Sheed and Ward, London (1963)

Hill, R.: Oxygen evolved by isolated chloroplasts. Nature **139**, 881–882 (1937)

Hill, R.: Oxygen produced by isolated chloroplasts. Proc. R. Soc. Lond., Series B **127**(847), 192–210 (1939)

Hill, R.: The cytochrome *b* component of chloroplasts. Nature **174**, 501 (1954)

Hill, R.: The biochemists' green mansions: the photosynthetic electron-transport chain in plants. In: Campbell, P.N., Greville, G.D. (eds.) Essays in Biochemistry, vol. 1, pp. 121–151. Academic Press, London (1965)

Hill, R., Bendall, F.: Crystallisation of a photosynthetic reductase from a green plant. Nature **187**, 417 (1960a)

Hill, R., Bendall, F.: Function of the two cytochrome components in chloroplasts: a working hypothesis. Nature **186**, 136–137 (1960b)

Hill, R., Bhagvat, K.: Cytochrome oxidase in flowering plants. Nature **143**, 726 (1939)

Hill, R., Bonner, W.D.: The nature and possible function of chloroplast cytochromes. In: McElroy, W.D., Glass, B. (eds.) Light and Life, pp. 424–435. Johns Hopkins Press, Baltimore (1961)

Hill, R., Hartree, E.F.: Hematin compounds in plants. Annu. Rev. Plant Physiol. **4**, 115–150 (1953)

Hill, R., Scarisbrick, R.: (1940a) Production of oxygen by iulluminated chloroplasts. Nature **146**, 61–62

Hill, R., Scarisbrick, R.: (1940b) The reduction of ferric oxalate by isolated chloroplasts. Proc. R. Soc. Lond., Series B **129**(855), 233–255

Hill, R., Scarisbrick, R.: The haematin compounds of leaves. New Phytol. **50**, 98–111 (1951)

Hirsch, R.E., Rich, M., Govindjee: A tribute to Seymour Steven Brody: in memoriam (November 29, 1927 to May 25, 2010). Photosynth. Res. **106**, 191–199 (2010)

Hofman, U., Baumgartner, M.: Determinism and the method of difference. Theoria **71**, 155–176 (2011)

Hofmann, K.A., Schumpelt, K.: Bildung von Formaldehyd und anderen organischen Stoffen aus Ameisensäure. I. Mitteilung: Selbstreduktion der Formiate. Berichte der Deutschen Chemischen Gesellschaft zu Berlin **49**, 303–317 (1916)

Holmes, F.L.: Lavoisier and the Chemistry of Life: An Exploration of Scientific Creativity. University of Wisconsin Press, Madison (1985)

Holmes, F.L.: Intermediary Metabolism in the Early Twentieth Century. In: Bechtel, W. (ed.) Integrating Scientific Disciplines, pp. 59–76. Martinus Nijhoff (1986)

Holmes, F.L.: Hans Krebs: The Formation of a Scientific Life, 1900–1933. Oxford University Press, New York (1991)

Holmes, F.L.: Investigative Pathways: Patterns and Stages in the Careers of Experimental Scientists. Yale University Press, New Haven (2004)

Holt, A.S., French, C.S.: Isotopic analysis of the oxygen evolved by illuminated chloroplasts in normal water and in water enriched with ^{18}O. Arch. Biochem. **19**, 429–435 (1948)

Homann, P.H.: Hydrogen metabolism of green algae: discovery and early research – a tribute to Hans Gaffron and his coworkers. Photosynth. Res. **76**, 93–103 (2002)

Hon, G., Goldstein, B.R.: Centenary of the Franck-Hertz experiments. Annalen der Physik **12**, A179–A183 (2013)

Hopkins, F.G.: On current views concerning the mechanisms of biological oxidation (with a foreword on the institutional needs of biochemistry). Skandinavisches Archiv für Physiologie **49**, 33–59 (1926)

Hopkins, F.G.: The dynamic side of biochemistry. In: Needham, J., Dunn, W., Baldwin, E. (eds.) Hopkins and Biochemistry: 1861–1947, pp. 136–159. Hefner and Sons, Cambridge (1949)

Hoppe, B.: Structural development of experimental methods in the researches on gas-exchange and photosynthesis. In: Hoppe, B. (ed.) Biology Integrating Scientific Fundamentals: Contributions to the History of Interrelations Between Biology, Chemistry, and Physics from the 18th to the 20th Centuries, Vol. 21 der Algorismus: Studien zur Geschichte der Mathematik und der Naturwissenschaften, pp. 10–30 . Institut für Geschichte der Naturwissenschaften, Munich (1997a)

Hoppe, B. (ed.): Biology Integrating Scientific Fundamentals: Contributions to the History of Interrelations between Biology, Chemistry, and Physics from the 18th to the 20th Centuries. Institut für Geschichte der Naturwissenschaften, München (1997b)

Horecker, B.L., Smyrniotis, P.Z.: The enzymatic formation of sedoheptulose phosphate from pentose phosophate. J. Am. Chem. Soc. **74**, 2123 (1952)

Horecker, B.L., Smyrniotis, P.Z.: The coenzyme function of thiamine pyrophosphate in pentose phosophate metabolism. J. Am. Chem. Soc. **75**, 1009–1010 (1953)

Horecker, B.L., Hurwitz, J., Weissbach, A.: Pentose phosphate and carbon dioxide fixation with spinach extracts. J. Am. Chem. Soc. **76**, 3611–3612 (1954)

Hungate, R.E.: Obituary: Cornelis B. van Niel (1897–1985). Photosynth. Res. **10**, 139–142 (1986)

Huzisige, H., Ke, B.: Dynamics of the history of photosynthesis research. Photosynth. Res. **38**, 185–209 (1993)

Höxtermann, E.: Photosynthese- und Stoffwechselforschung in der Geschichte der Botanik an der Berliner Universität (1810 bis 1945), number 27 in Beiträge zur Geschichte der Humboldt-Universität zu Berlin, Humboldt University, Berlin (1991)

Höxtermann, E.: Fundamental discoveries in the history of photosynthesis research. Photosynthetica **26**, 485–502 (1992)

Höxtermann, E.: Otto Heinrich Warburg (1883–1970). In: Jahn, I., Schmitt, M. (eds.) Darwin & Co. Eine Geschichte der Biologie in Portraits, vol. 2, pp. 251–274. Beck, Munich (2001)

Höxtermann, E.: (2007a) A comment on Warburg's early understanding of biocatalysis. Photosynth. Res. **92**, 121–127

Höxtermann, E.: (2007b) Der Stoffwechsel: die Chemie des Lebens. Zur Geschichte der Biochemie. In: Höxtermann E., Hilger, H.H. (eds.) Lebenswissen. Eine Einführung in die Geschichte der Biologie, pp. 143–177. Natur & Text, Rangsdorf

Höxtermann, E.: unpublished, Berliner Botaniker in der Geschichte der Biochemie, Submitted as habilitation thesis, 1993; Biology Department, Friedrich Schiller University Jena (1993)

Höxtermann, E., Sucker, U.: Otto Warburg, number 91 in Biographien hervorragender Naturwissenschaftler, Techniker und Mediziner. Teubner, Leipzig (1989)

Illari, P.M., Williamson, J.: Mechanisms are real and local. In: Illari, P.M., Russo, F., Williamson, J. (eds.) Causality in the Sciences, pp. 818–844. Oxford University Press, Oxford/New York (2011)

Illari, P., Williamson, J.: What is a mechanism? Thinking about mechanisms across the sciences. Eur. J. Philos. Sci. **2**, 119–135 (2012)

Illari, P.M., Russo, F., Williamson, J.: Why look at causality in the sciences? A manifesto. In: Illari, P.M., Russo, F., Williamson, J. (eds.) Causality in the Sciences, pp. 3–22. Oxford University Press, Oxford/New York (2011)

Jacob, C.: Philosophy and biochemistry: research at the interface between chemistry and biology. Found. Chem. **4**, 97–125 (2002)

Jacoby, M.: Über den Formaldehyd als Übergangsstufe zwischen der eigentlichen Assimilation und der Kohlenhydratbildung in der Pflanze. Biochemische Zeitschrift **101**, 1–6 (1919)

Jacoby, M.: Über den Formaldehyd als Übergangsstufe zwischen der eigentlichen Assimilation und der Kohlenhydratbildung in der Pflanze II.. Biochemische Zeitschrift **128**, 119–121 (1922)

Jaenicke, L.: Horst Tobias Witt (1922–2007). Ahnherr der biophysikalischen Photosyntheseforschung in Deutschland. BIOspektrum **7**, 815–816 (2007)

Jagendorf, A.T.: Oxidation and reduction of pyridine nucleotides by purified chloroplasts. Arch. Biochem. Biophys. **62**, 141–150 (1956)

Jagendorf, A.T.: Chance, luck and photosynthesis research: an inside story. Photosynth. Res. **57**, 215–229 (1998)

Janko, J., Strbánová, S.: Interrelations of biology, physics, and chemistry in historical perspective, number 23 in Acta historiae rerum naturalium necnon technicarum, Prague (1991)

Jencks, W.P., Wolfenden, R.F.: Fritz Albert Lipmann (1899–1986). Biogr. Mem. Fellows R. Soc. **46**, 335–344 (2000)

Johnson, J.A.: The Kaiser's Chemists. Science and Modernization in Imperial Germany. University of North Carolina Press, Chapel Hill (1990)

Johnston, H.: A Bridge not Attacked. Chemical Warfare Civilian Research During World War II. World Scientific Publishing, River Edge (2003)

Jost, W.: Kurt Wohl. Berichte der Bunsengesellschaft **67**, 347–348 (1963)

Jörgensen, I., Stiles, W.: A review of recent work on the pigments of the green leaf and the processes connected with them. New Phytol. **15**, 176–193, 205–232 (1916a)

Jörgensen, I., Stiles, W.: A review of recent work on the pigments of the green leaf and the processes connected with them. Part II. New Phytol. **16**, 24–45, 77–104 (1916b)

Junge, W.: Protons, proteins and ATP. In: Govindjee, Beatty, J.T., Gest, H., Allen, J.F. (eds.) Discoveries in Photosynthesis, pp. 571–595. Springer (2005)

Junge, W., Rutherford, A.W.: Obituary: Horst Tobias Witt (1922–2007). Nature **448**, 425 (2007)

Kalckar, H.: Phosphorylation in kidney tissues. Enzymologia **2**, 47–52 (1937)

Kalckar, H.: Origins of the concept oxidative phosphorylation. Mol. Cell. Biochem. **5**, 55–62 (1974)

Kalckar, H.: 50 years of biological research: from oxidative phosphorylation to energy requiring transport regulation. Annu. Rev. Biochem. **60**, 1–37 (1991)

Kalckar, H.M.: The nature of energetic coupling in biological systems. Chem. Rev. **28**, 71–178 (1941)

Kamen, M.D.: Some remarks on tracer researches in photosynthesis. In: Franck, J., Loomis, W.E. (eds.) Photosynthesis in Plants, pp. 365–380. State College Press, Ames (1949)

Kamen, M.D.: Early days in CO_2 fixation: some brief comments on the Berkeley experience. Mol. Cell. Biochem. **5**, 99–101 (1974)

Kamen, M.D.: Radiant Science, Dark Politics. A Memoir of the Nuclear Age. University of California Press, Berkeley (1985)

Kamen, M.D.: Onward in a fabulous half century. Photosynth. Res. **21**, 139–144 (1989)

Kamp, A.F., Riviére, J.W. M.L., Verhoeven, W. (eds.): Albert Jan Kluyver: His Life and Work. Biographical Memoranda, Selected Papers, Bibliography and Addenda. North-Holland Publishing Company, Amsterdam (1959)

Kandler, O.: Über die Beziehung zwischen Phosphathaushalt und Photosynthese: I. Phosphatspiegelschwankungen bei *Chlorella pyrenoidosa* als Folge des Licht-Dunkel-Wechsels. Zeitschrift für Naturforschung **5b**, 423–437 (1950)

Kandler, O.: Über die Beziehung zwischen Phosphathaushalt und Photosynthese: II. Gesteigerter Glucose-Einbau im Licht als Indikator einer lichtabhängigen Phosphorylierung. Zeitschrift für Naturforschung **9b**,625–644 (1954)

Kandler, O.: Über die Beziehung zwischen Phosphathaushalt und Photosynthese: III. Hemmungs-analyse der lichtabhängigen Phosphorylierung. Zeitschrift für Naturforschung **10b**, 38–46 (1955)

Katz, E.: Chlorophyll fluorescence as an energy flow meter for photosynthesis. In: Franck, J., Loomis, W.E. (eds.) Photosynthesis in Plants, pp. 287–292. State College Press, Ames (1949)

Kautsky, H., Hirsch, A.: Neue Versuche zur Kohlensäureassimilation. Naturwissenschaften **19**, 964–964 (1931)

Kautsky, H., Hirsch, A., Davidshöfer, F.: Energie-Umwandlungen an Grenzflächen. VI. Mitteilung: Kohlensäure-Assimilation (1.). Berichte der Deutschen Chemischen Gesellschaft **65**, 1762–1770 (1932)

Kautsky, H., de Bruijn, H., Neuwirth, R., Baumeister, W.: Energie-Umwandlungen an Grenzflächen. VII. Mitteilung: photo-sensibilisierte Oxydation als Wirkung eines aktiven, metastabilen Zustandes des Sauerstoff-Moleküls. Berichte der Deutschen Chemischen Gesellschaft **66**, 1588–1600 (1933)

Kautsky, H., Hirsch, A., Davidshöfer, F.: Der Einfluß der Kohlensäure auf das Fluoreszenz-Verhalten lebender Blätter. Biochemische Zeitschrift **277**, 250–260 (1935)

Kay, L.E.: The Molecular Vision of Life: Caltech, The Rockefeller Foundation and The Rise of the New Biology. Oxford University Press, New York (1993)

Keilin, D.: A comparative study of turacin and haematin and its bearing on cytochrome. Proc. R. Soc. Lond., Series B **100**, 129–151 (1925a)

Keilin, D.: On cytochrome, a respiratory pigment, common to animals, yeast, and higher plants. Proc. R. Soc. Lond., Series B **98**, 312–339 (1925b)

Keilin, D.: The History of Cell Respiration and Cytochrome. Cambridge University Press, Cambridge (1966)

Keilin, D., Dixon, M., Hill, R.: The absorption spectrum of the component *c* of cytochrome. Proc. R. Soc. Lond., Series B **109**, 29–34 (1931)

Keilin, D., Hill, R.: Estimation of haematin iron and the oxidation-reduction equivalent of cytochrome *c*. Proc. R. Soc. Lond., Series B **114**, 104–109 (1933)

Kennedy, E.P.: Herman Moritz Kalckar (1908–1991). Biogr. Mem. Natl. Acad. Sci. **69**, 148–165 (1996)

Kennedy, E.P.: Hitler's gift and the era of biosynthesis. J. Biol. Chem. **276**, 42619–42631 (2001)

Kimpflin, G.: Présence du menthal dans les végétaux vertes. Comptes Rendus de l'Académie des Sciences **144**, 148–150 (1907)

Kimpflin, G.: Essai sur l'assimilation photochlorophyllienne du carbone. A. Rey Imprimeur-éditeur de l'Université, Lyon (1908)

Kitcher, P.: The division of cognitive labour. J. Philos. **87**, 5–22 (1990)

Kitcher, P.: Patterns of scientific controversy. In: Machamer, P., Pera, M., Baltas, A. (eds.) Scientific Controversies. Philosophical and Historical Perspectives, pp. 21–39. Oxford University Press, Oxford/New York (2000)

Klein, M.J., Kox, A.J., Schulmann, R.E.: The Swiss Years: Correspondence, 1902–1914, Vol. 5 der The Collected Papers of Albert Einstein. Princeton University Press, Princeton (1993)

Klein, M.J., Kox, A.J., Renn, J., Schulmann, R.: The Swiss Years: Writings, 1912–1914, Vol. 4 der The Collected Papers of Albert Einstein. Princeton University Press, Princeton (1995)

Kleinkauf, H., von Döhren, H., Jaenicke, L.: The Roots of Modern Biochemistry: Fritz Lipmann's Squiggle and its Consequences. Walter de Gruyter, Berlin (1988)

Klemm, F.: Baeyer, Adolf Johann Friedrich Wilhelm Ritter von, Neue Deutsche Biographie (NDB), vol. 1. Duncker & Humblot, Berlin (1953)

Klingenberg, M., Müller, A., Schmidt-Mende, P., Witt, H.T.: Changes in absorption during photosynthesis in the ultra-violet spectrum. Nature **194**, 379–380 (1962)

Kluyver, A.J., Donker, H.J.L.: Die Einheit in der Biochemie. Chemie der Zelle und Gewebe: Zeitschrift für die Probleme der Gärung, Atmung und Vitaminforschung **13**, 134–139 (1926)

Knox, R.S.: Electronic excitation transfer in the photosynthetic unit: reflections on work of William Arnold. Photosynth. Res. **48**, 35–39 (1996)

Kohler, R.E.: The background to Otto Warburg's conception of the Atmungsferment. J. Hist. Biol. **6**(2), 171–192 (1973a)

Kohler, R.E.: The enzyme theory of life and the origins of biochemistry. Isis **64**, 171–192 (1973b)

Kohler, R.E.: Rudolf Schoenheimer, isotopic tracers, and the dynamic state. Hist. Stud. Phys. Sci. **8**, 257–298 (1977)

Kohler, R.E.: From Medical Chemistry to Biochemistry: The Making of a Biomedical Discipline. Cambridge University Press, Cambridge (1982)

Kohler, R.E.: Partners in Science. Foundations and Natural Scientists 1900–1945. University of Chicago Press, Chicago (1991)

Kok, B.: On the reversible absorption change at 705 nm in photosynthetic organisms. Biochimica et Biophysica Acta **22**, 399–401 (1956)

Kok, B.: Light induced absorption changes in photosynthetic organisms. Acta Botanica Neerlandica **6**, 316–336 (1957)

Kok, B.: Light induced absorption changes in photosynthetic organisms II. A split beam difference spectrophotometer. Plant Physiol. **34**, 184–192 (1959)

Kok, B.: Efficiency of photosynthesis. In: Pirson, A. (ed.) Handbuch der Pflanzenphysiologie, pp. 566–633. Springer, Berlin (1960)

Kok, B.: Does photosynthesis require the interaction of two photochemical steps?, Recent advances in botany, Volume II: From lectures and symposia presented to the IXth International Botanical Congress, Montreal, August, 1959, University of Toronto Press, Toronto, Canada, pp. 1071–1072 (1961a)

Kok, B.: Partial purification and determination of oxidation reduction potential of the photosynthetic chlorophyll complex absorbing at 700 mμ. Biochimica et Biophysica Acta **48**, 527–532 (1961b)

Kok, B., Hoch, G.: Spectral changes in photosynthesis. In: McElroy, W.D., Glass, B. (eds.) Light and Life, pp. 397–423. Johns Hopkins Press, Baltimore (1961)

Kornberg, H.L.: John Rodney Quayle (1926–2006). Photosynth. Res. **89**, 59–62 (2006)

Kortschak, H.P., Hartt, C.E., Burr, G.O.: Carbon dioxide fixation in sugarcane leaves. Plant Physiol. **40**, 209–213 (1965)

Krebs, H.: The discovery of carbon dioxide fixation in mammalian tissues. Mol. Cell. Biochem. **5**, 79–94 (1974)

Krebs, H.: Otto Warburg: Zellphysiologe, Biochemiker, Mediziner 1883–1970, Grosse Naturforscher; 41, Wissenschaftliche Verlagsgesellschaft. Stuttgart (1979)

Krebs, H., Henseleit, K.: Untersuchungen über die Harnstoffbildung im Tierkörper. Hoppe-Seyler's Zeitschrift für physiologische Chemie **210**, 33–66 (1932)

Krebs, H.A., Johnson, W.A.: Metabolism of ketonic acids in animal tissues. Biochem. J. **31**, 645–660 (1937)

Kresge, N., Simoni, R.D., Hill, R.: The discovery of ferredoxin and its role in photosynthesis: the work of Anthony San Pietro. J. Biol. Chem. **280**, e48 (2005)

Krogmann, D.W.: The golden age of biochemical research in photosynthesis. Photosynth. Res. **63**, 109–121 (2000)

Kuhn, H.G.: James Franck (1882–1964). Biogr. Mem. Fellows R. Soc. **11**, 51–74 (1965)

Larkum, A.W.D.: Contributions of Henrik Lundegårdh. Photosynth. Res. **76**, 105–110 (2003)

Latour, B., Woolgar, S.: The cycle of credibility. In: Barnes, B., Edge, D. (eds.) Science in Context: Readings in the Sociology of Science, pp. 35–43. MIT Press, Cambridge (1982)

Latour, B., Woolgar, S.: Laboratory Life. The Construction of Scientific Facts. Princeton University Press, Princeton (1985)

Laudan, L.: Science and Values. The Aims of Science and Their Role in Scientific Debate. University of California Press, Berkeley (1984)

Laurence, W.L.: Atomic key to life is feasible now, New York Times (9 October 1945) p. 6 (1945)

Laurence, W.L.: Dawn Over Zero. The Story of the Atomic Bomb. Knopf, New York (1946a)

Laurence, W.L.: Is atomic energy the key to our dreams? Saturday Evening Post (13 April 1946) pp. 9–10, 36–37, 39, 41 (1946b)
Löb, W.: Zur Kenntnis der Assimilation der Kohlensäure. Landwirtschaftliche Jahrbücher **35**, 541–578 (1906)
Lehmann, J.: Kohlenstoff-14 und die Assimilation des Kohlendioxids. Chemie in unserer Zeit **2**, 67–73 (1968)
Lehninger, A.L.: Phosphorylation coupled to the oxidation of dihydrophosphopyridine nucleotide. J. Biol. Chem. **190**, 345–359 (1951)
Lemmerich, J.: Aufrecht im Sturm der Zeit: Der Physiker James Franck (1882–1964). Verlag für Geschichte der Naturwissenschaften und Technik, Diepholz (2007)
Levitt, L.S.: Photosynthesis as a photoelectric phenomenon. Science **118**, 696–697 (1953)
Levitt, L.S.: The role of magnesium in photosynthesis. Science **120**, 33–35 (1954)
Liebig, J.: Die organische Chemie in ihrer Anwendung auf Agricultur und Physiologie. Vieweg & Sohn, Brunswick (1840)
Liebig, J.: Chemistry in its Applications to Agriculture and Physiology, 2nd, enlarged edn. Taylor & Walton, London (1842)
Liebig, J.: Die Wechselwirthschaft. Annalen der Chemie und Pharmazie **46**, 58–97 (1843)
Lineweaver, H., Burk, D.: The determination of enzyme dissociation constants. J. Am. Chem. Soc. **56**, 658–666 (1934)
Lipman, T.O.: Vitalism and reductionism in Liebig's physiological thought. Isis **58**, 167–185 (1967)
Lipmann, F.: Metabolic generation and utilization of phosphate bond energy. Adv. Enzymol. **1**, 99–162 (1941)
Lipmann, F.: Wanderings of a Biochemist. Wiley-Interscience, New York (1971)
Loach, P.: A remembrance of Melvin Calvin. Photosynth. Res. **54**, 1–3 (1997)
Loeb, J.: Studies in General Physiology. The University of Chicago Press, Chicago (1905)
Loew, O.: Über Formaldehyd und dessen Condensation. Journal für praktische Chemie **33**, 321–351 (1886)
Loew, O.: Einige Bemerkungen über Formose. Berichte der deutschen chemischen Gesellschaft **20**, 141–144 (1887)
Loew, O.: Über die Condensation des Formaldehyds unter verschiedenen Bedingungen. Berichte der deutschen chemischen Gesellschaft **21**, 270–275 (1888)
Loew, O.: Über Bildung von Zuckerarten aus Formaldehyd. Berichte der deutschen chemischen Gesellschaft **22**, 470–478 (1889)
Lohmann, K.: Über die Pyrophosphatfraktion im Muskel. Naturwissenschaften **17**, 624–625 (1929)
Loomis, W.E.: Historical introduction. In: Pirson, A., Ruhland, W. (eds.) Handbuch der Pflanzenphysiologie, vol. 5, Teil 1, pp. 85–114. Springer, Berlin (1960)
Losada, M., Trebst, A.V., Arnon, D.I.: Photosynthesis by isolated chloroplasts XI: CO_2 assimilation in a reconstituted chloroplast system. J. Biol. Chem. **235**, 832–839 (1959)
Lundegårdh, H.: On the oxidation of cytochrome *f* by light. Physiologia Plantarum **7**, 375–382 (1954)
Machamer, P., Darden, L., Craver, C.F.: Thinking about mechanisms. Philos. Sci. **67**, 1–25 (2000)
Macrakis, K.: Surviving the Swastika. Scientific Research in Nazi Germany. Oxford University Press, New York (1993)
Magee, J.L., Witt, T.W., Smith, E.C., Daniels, F.: A photocalorimeter. The quantum efficiency of photosynthesis in algae. J. Am. Chem. Soc. **61**, 3529–3533 (1939)
Mann, T.: David Keilin, 1887–1963. Biogr. Mem. Fellows R. Soc. **10**, 183–205 (1964)
Manning, W.M.: Photosynthesis. In: Hugh, S.T. (ed.) Third Report of the Committee on Photochemistry: Reprint and Circular Series of the National Research Council, number 108, pp. 117–156. National Academies Press (1938)
Manning, W.M., Stauffer, J.F., Duggar, B.M., Daniels, F.: Quantum efficiency of photosynthesis in *Chlorella*. J. Am. Chem. Soc. **60**, 266–274 (1938)
Maquenne, L.: Action de l'ozone sure les carbures. Bulletin de la Société Chimique de Paris **37**, 298–300 (1882)

Martin, A.J.P., Synge, R.L.M.: A new form of chromatogram employing two liquid phases. Biochem. J. **35**, 1358–1368 (1941)

Maruyama, K.: The discovery of adenosine triphosphate and the establishment of its structure. J. Hist. Biol. **24**, 145–154 (1991)

Mauskopf, S.H., Schmaltz, T.M. (eds.): Integrating History and Philosophy of Science: Problems and Prospects. Springer, Dordrecht (2011)

May, M.: Kausales Schlieβen. Eine Untersuchung zur kausalen Erklärung und Theorienbildung, dissertation, University of Hamburg (1999)

Mayaudon, J.: Study of association between the main nucleoprotein of green leaves and carboxydismutase. Enzymologia **18**, 345–354 (1957)

Mayaudon, J., Benson, A.A., Calvin, M.: Ribulose-1,5-diphosphate from and CO_2 fixation by *Tetragonia expansa* leaves extract. Biochimica et Biophysica Acta **23**, 342–351 (1957)

Mayer, J.R.: Die organische Bewegung in ihrem Zusammenhange mit dem Stoffwechsel. Drechsler, Heilbronn (1845)

Mazé, P.: Recherches sur l'assimilation du gaz carbonique par les plantes vertes. Comptes Rendus de l'Académie des Sciences **171**, 1391–1393 (1920)

McElroy, W.D., Glass, B. (eds.): Light and Life. Johns Hopkins Press, Baltimore (1961)

McMullin, E.: Scientific controversy and its termination. In: Engelhardt, H.T., Caplan, A.L. (eds.) Scientific Controversies. Case Studies in the Resolution and Closure of Disputes in Science and Technology, pp. 49–91. Cambridge University Press (1987)

Melis, A., Buchanan, R.B.: A Tribute to Daniel I. Arnon, Photosynthesis Research 46 (1–2). (Special issues dedicated to Daniel Arnon) (1995)

Mendelsohn, E.: The political anatomy of controversy in the sciences. In: Engelhardt, H.T., Caplan, A.L. (eds.), Scientific Controversies. Case Studies in the Resolution and Closure of Disputes in Science and Technology, pp. 93–124. Cambridge University Press, Cambridge (1987)

Metz, C.B., Clapp, P.L. (eds.): The Naples Zoological Station and the Marine Biological Laboratory. Symposium Supplement to the Biological Bulletin, vol. 168, No. 3 (1985)

Mitchell, P.: Chemiosmotic coupling in oxidative and photosynthetic phosphorylation. Biochem. J. **79**, 23B–24B (1961a)

Mitchell, P.: Coupling of phosphorylation to electron and hydrogen transfer by a chemi-osmotic type of mechanism. Nature **191**, 144–148 (1961b)

Moore, B., Webster, T.A.: Synthesis of formaldehyde from carbon dioxide and water by inorganic colloids, acting as transformers of light energy. Proc. R. Acad. Lond., Series B **87**, 163–176 (1913)

Morrison, M.: Where have all the theories gone? Philos. Sci. **74**, 195–228 (2006)

Mortenson, L.E., Valentine, R.C., Carnahan, J.E.: An electron transport factor from *Clostridium pasteurianum*. Biochem. Biophys. Res. Commun. **7**, 448–452 (1962)

Morton, O.: Eating the Sun. How Plants Power the Planet. Fourth Estate, London (2007)

Moses, V., Calvin, M.: The path of carbon in photosynthesis XXI. The identification of carboxy-ketopentitol diphosphates as products of photosynthesis. Proc. Natl. Acad. Sci. U S A **44**, 260–277 (1958)

Moses, V., Moses, S.: The Calvin Lab: Bio-organic chemistry group at the University of California, Berkeley, 1945–1963. An oral history conducted 1995–1997. Regional Oral History Office, The Bancroft Library, University of California, Berkeley (2000)

Myers, J.: Evidences for a specific participation of chlorophyll *b* in photosynthesis, Recent advances in botany, Volume II: From lectures and symposia presented to the IXth International Botanical Congress, Montreal, August, 1959, University of Toronto Press, Toronto, Canada, pp. 1073–1076 (1961)

Myers, J.: Conceptual developments in photosynthesis. Plant Physiol. **54**, 420–426 (1974)

Myers, J.: Bessel Kok (1918–1979). Biogr. Mem. Natl. Acad. Sci. U S A **57**, 124–149 (1987)

Myers, J.: The 1932 experiments. Photosynth. Res. **40**, 303–310 (1994)

Myers, J.: In one era and out the other. Photosynth. Res. **73**, 21–28 (2002)

Myers, J., French, C.S.: (1960a) Evidences from action spectra for a specific participation of chlorophyll *b* in photosynthesis. J. Gen. Physiol. **43**, 723–736

Myers, J., French, C.S.: (1960b) Relationships between time course, chromatic transient, and enhancement phenomena of photosynthesis. Plant Physiol. **35**, 963–969

Nachmansohn, D.: Biochemistry as part of my life. Annu. Rev. Biochem. **41**, 1–28 (1972)

Needham, J., Dunn, W., Baldwin, E. (eds.): Hopkins and Biochemistry: 1861–1947. Hefner and Sons, Cambridge (1949)

Nef, J.U.: Dissoziationsvorgänge in der Zuckergruppe (Zweite Abhandlung). Über das Verhalten der Zuckerarten gegen Aetzalkalien. Annalen der Chemie **376**, 1–119 (1910)

Nef, J.U.: Dissoziationsvorgänge in der Zuckergruppe (Dritte Abhandlung). Annalen der Chemie **403**, 204–383 (1913)

Nickelsen, K.: Wissenschaftliche Pflanzenzeichnungen—Spiegelbilder der Natur? Botanische Abbildungen aus dem 18. und frühen 19. Jahrhundert. Bern Studies in the History and Philosophy of Science, Bern (2000)

Nickelsen, K.: Botanists, draughtsmen and nature. The construction of eighteenth-century botanical illustration. Stud. Hist. Philos. Biol. Biomed. Sci. **37**, 1–25 (2006a)

Nickelsen, K.: Draughtsmen, Botanists and Nature. The Construction of Eighteenth-Century Botanical Illustrations. Archimedes, Kluwer Academic, Dordrecht (2006b)

Nickelsen, K.: Otto Warburg's first approach to photosynthesis. Photosynth. Res. **92**, 109–120 (2007)

Nickelsen, K.: Ein bisher unbekanntes Zeitzeugnis—Otto Warburgs Tagebuchnotizen Feb.–April 1945. NTM. Zeitschrift für Geschichte der Naturwissenschaften, Technik und Medizin **16**, 103–115 (2008a)

Nickelsen, K.: From leaves to molecules: botany and the development of photosynthesis research. Ann. Hist. Philos. Biol. **12**, 1–40 (2008b)

Nickelsen, K.: The construction of a scientific model: Otto Warburg and the building block strategy. Stud. Hist. Philos. Biol. Biomed. Sci. **40**, 73–86 (2009)

Nickelsen, K.: "For Methods go to Naples": Die Stazione Zoologica Anton Dohrn. In: Schmundt, H., Véc, M., Westphal, H.(eds.), Mekkas der Moderne. Pilgerstätten der Wissensgesellschaft. Böhlau. Wien. Böhlau (2010)

Nickelsen, K.: From the Red Drop to the Z-scheme of photosynthesis. Annalen der Physik **524**, A157–A160 (2012a)

Nickelsen, K.: The path of carbon in photosynthesis: how to discover a biochemical pathway. Ambix **59**, 266–293 (2012b)

Nickelsen, K.: Kooperation und Konkurrenz in den Naturwissenschaften. In: Jessen, R. (ed.) Konkurrenz in der Geschichte. Praktiken—Werte—Institutionalisierungen, pp. 353–379. Campus, Frankfurt (2014)

Nickelsen, K., Govindjee: The Maximum Quantum Yield Controversy: Otto Warburg and the "Midwest-Gang". Bern Studies in the History and Philosophy of Science. Bern (2011)

Nickelsen, K., Graßhoff, G.: In pursuit of formaldehyde: causally explanatory models and falsification. Stud. Hist. Philos. Biol. Biomed. Sci. **42**, 297–305 (2011)

Nickelsen, K., Graßhoff, G.: Concepts from the bench: Krebs and the urea cycle. In: Hon, G., Schickore, J., Steinle, F. (eds.) Going Amiss in Experimental Research, pp. 91–117. Springer, Dordrecht et al. (2008)

Nishimura, M.S., Whittingham, C.P., Emerson, R.: The Maximum Efficiency of Photosynthesis. Carbon Dioxide Fixation and Photosynthesis, pp. 176–210. Cambridge University Press, Cambridge, (1951)

Ochoa, S.: Chemical transformations in photosynthesis. Fed. Proc. **9**, 551–553 (1950)

Ochoa, S., Vishniac, W.: Carboxylation reactions and photosynthesis. Science **115**, 297–301 (1952)

Ochoa, S., Veiga Salles, J.B., Ortiz, P.J.: Biosynthesis of dicarboxylic acids by carbon dioxide fixation. III. Enzymatic synthesis of igr-malic acid by reductive carboxylation of pyruvic acid. J. Biol. Chem. **187**, 863–874 (1950)

Ohmura, T.: Oxidative and photosynthetic phosphorylation. Nature **176**, 467–468 (1955)

Oppenheimer, J.R.: Internal conversion in photosynthesis (Abstract). Phys. Rev. **60**, 158 (1941)

Osterhout, W.J.V.: A method for studying respiration. J. Gen. Physiol. **1**, 17–22 (1918)

Osterhout, W.J.V.: Comparative studies on respiration. VII. Respiration and antagonism. J. Gen. Physiol. **2**, 1–3 (1919)

Osterhout, W.J.V.: Jacques Loeb. J. Gen. Physiol. **8**, ix–lix (1928)

Osterhout, W.J.V., Haas, A.R.C.: On the dynamics of photosynthesis. J. Gen. Physiol. **1**, 1–16 (1918)

Osterhout, W.J.V., Haas, A.R.C.: The temperature coefficient of photosynthesis. J. Gen. Physiol. **2**, 299–304 (1919)

Pauli, W.: Wissenschaftlicher Briefwechsel mit Bohr, Einstein, Heisenberg u. a., Vol. 3: 1940–1949. Springer, Berlin (1993)

Pauly, P.J.: Controlling Life: Jacques Loeb and the Engineering Ideal in Biology. Oxford University Press, New York (1987a)

Pauly, P.J.: General physiology and the discipline of physiology, 1890–1935. In: Geison, G.L. (ed.) Physiology in the American Context, 1850–1940, pp. 195–208. American Physiological Society, Bethesda (1987b)

Pearl, J.: Causality: Models, Reasoning, and Inference. Cambridge University Press, Cambridge (2000)

Pearlstein, R.M.: Photosynthetic exciton theory in the 1960s. Photosynth. Res. **73**, 119–126 (2002)

Pfeffer, W.: Pflanzenphysiologie. Ein Handbuch der Lehre vom Stoffwechsel und Kraftwechsel in der Pflanze, 2., revised and amended edn. Engelmann, Leipzig (1897)

Pietro, A.S.: Memories: from protein synthesis to photosynthesis. Photosynth. Res. **96**, 185–199 (2008)

Pirson, A.: Sixty years in algal physiology and photosynthesis. Photosynth. Res. **40**, 207–221 (1994)

Pirson, A., Krollpfeiffer, I., Schaefer, G.: Leistungsfähigkeit und Fehlerquellen manometrischer Stoffwechselmessungen. Marburger Sitzungsberichte **76**, 3–27 (1953)

Pollacci, G.: (1902a) Intorno all' assimilazione clorofilliana delle piante. Seconda memoria. Atti dell'Istituto Botanico dell'Universitá di Pavia **8**, 1–66

Pollacci, G.: (1902b) L'assimilation chlorophyllienne. Deuxiéme memoire. Archives Italiennes de Biologie **36**, 446–448

Pollacci, G.: Elettricitá e vegetazione. Parte I. Influenz dell' elettricitá sulla fotosintesi clorofilliana. Atti dell'Istituto Botanico dell'Universitá di Pavia **13**, 1–152 (1907)

Porter, G.: Some reflections on the work of Theodor Förster. Naturwissenschaften **63**, 207–211 (1976)

Portis, A.R., Parry, M.A.J.: Discoveries in Rubisco: a historical perspective. Photosynth. Res. **94**, 121–143 (2007)

Prebble, J.: The discovery of oxidative phosphorylation: a conceptual off-shoot from the study of glycolysis. Stud. Hist. Philos. Biol. Biomed. Sci. **41**, 253–262 (2010)

Prebble, J., Weber, B.: Wandering in the Gardens of the Mind: Peter Mitchell and the Making of Glynn. Oxford University Press, Oxford/New York (2003)

Pringsheim, E.G.: Julius Sachs: Der Begründer der neueren Pflanzenphysiologie, 1832–1897. Fischer, Jena (1932)

Pringsheim, E.G.: Ernst Georg Pringsheim: Eine autobiographische Skizze. Medizinhistorisches Journal **5**, 125–137 (1970)

Quayle, J.R., Fuller, R.C., Benson, A.A., Calvin, M.: Enzymatic carboxylation of ribulose diphosphate. J. Am. Chem. Soc. **76**, 3610–3612 (1954)

Rabinowitch, A.: Founder and father. Bull. Atomic Sci. **61**, 30–37 (2005)

Rabinowitch, E.I.: Photosynthesis and Related Processes. Vol I: Chemistry of Photosynthesis, Chemosynthesis and Related Processes In Vitro and In Vivo. Interscience Publishers, New York (1945)

Rabinowitch, E.I.: Photosynthesis and Related Processes. Vol II, 2: Kinetics of Photosynthesis (continued); Addenda. Interscience Publishers, New York (1956)

Rabinowitch, E.I.: Robert Emerson (1903–1959). Biogr. Mem. Natl. Acad. Sci. **25**, 112–131 (1961)
Rabinowitch, E.I.: The Mechanism of Photosynthesis, Studies on Microalgae and Photosynthetic Bacteria, vol. 373, pp. 112–121. University of Tokyo Press, Tokyo (1963)
Rabinowitch, E.I., Govindjee: Different forms of chlorophyll *a* in vivo and their photochemical function. In: McElroy, W.D., Glass, B. (eds.) Light and Life, pp. 378–386. Johns Hopkins Press, Baltimore (1961)
Racker, E., de la Haba, G., Leder, I.G.: Thiamine pyrophosphate, a coenzyme of transketolase. J. Am. Chem. Soc. **75**, 1010–1011 (1953)
Reinhardt, C.: Shifting and Rearranging. Physical Methods and the Transformation of Modern Chemistry. Science History Publications, Sagamore Beach (2006)
Reinke, J.: Über aldehydartige Substanzen in chlorophyllhaltigen Pflanzenzellen. Berichte der deutschen chemischen Gesellschaft **14**, 2144–2150 (1881a)
Reinke, J.: Studien über das Protoplasma. III. Der Process der Kohlenstoffassimilation im chlorophyllhaltigen Protoplasma. Untersuchungen aus dem botanischen Labor der Universität Göttingen **2**, 197–202 (1881b)
Reinke, J.: Theoretisches zum Assimilationsproblem. Botanische Zeitung **40**, 289–297, 305–314 (1882)
Renger, G.: Horst Tobias Witt (1922–2007). Photosynth. Res. **96**, 5–8 (2008)
Rheinberger, H.-J.: Toward a History of Epistemic Things: Synthesizing Proteins in the Test Tube. Stanford University Press, Stanford (1997)
Rhodes, R.: The Making of the Atomic Bomb. Simon and Schuster, New York (1986)
Rhodes, R.: Dark Sun. The Making of the Hydrogen Bomb. Simon and Schuster, New York (1995)
Rich, P.: Special issue dedicated to Robin Hill. Photosynth. Res. 34 (3) (1992)
Rieke, F.F.: On the quantum efficiency of photosynthesis. J. Chem. Phys. **7**, 238–244 (1939)
Robinson, G.W.: Excitation transfer and trapping in photosynthesis. Brookhaven Sympos. Biol. **19**, 16–48 (1967)
Roelofsen, P.: On the metabolism of the purple sulphur bacteria. Proc. R. Acad. Sci. Amst. **37**, 3–12 (1934)
Roll-Hansen, N.: Experimental method and spontaneous generation: the controversy between Pasteur and Pouchet, 1859–64. J. Hist. Med. Allied Sci. **34**, 273–292 (1979)
Rosenberg, J.L.: The contributions of James Franck to photosynthesis research: a tribute. Photosynth. Res. **80**, 71–76 (2004)
Rossiter, M.W.: The Emergence of Agricultural Science. Justus Liebig and the Americans, 1840–1880. Yale University Press, New Haven (1975)
Rouge, E.: Recherche des premiers produits de l'assimilation chlorophyllienne du carbone. Journal Suisse de Pharmacie **59**, 1–9 (1924)
Rürup, R.: Schicksale und Karrieren. Gedenkbuch für die von den Nationalsozialisten aus der Kaiser-Wilhelm-Gesellschaft vertriebenen Forscherinnen und Forscher, (in collaboration with Michael Schüring). Wallstein, Göttingen (2008)
Rüskamp, W. (ed.): Eine vollkommene Närrin durch meine ewigen Gefühle. Aus den Tagebüchern der Lotte Warburg 1925 bis 1947, Druckhaus Bayreuth (1989)
Ruben, S.: Photosynthesis and phosphorylation. J. Am. Chem. Soc. **65**, 279 (1943)
Ruben, S., Kamen, M.D.: (1940a) Photosynthesis with radioactive carbon IV. Molecular weight of the intermediates and a tentative theory of photosynthesis. J. Am. Chem. Soc. **62**, 3451–3455
Ruben, S., Kamen, M.D.: (1940b) Radioactive carbon in the study of respiration in heterotrophic systems. Proc. Natl. Acad. Sci. U S A **26**, 418–422
Ruben, S., Kamen, M.D.: Long-lived radioactive carbon: C14. Phys. Rev. **59**, 349–354 (1941)
Ruben, S., Hassid, W.Z., Kamen, M.D.: Radioactive carbon in the study of photosynthesis. J. Am. Chem. Soc. **61**, 661–663 (1939a)
Ruben, S., Kamen, M.D., Hassid, W.Z., DeVault, D.C.: Photosynthesis with radio-carbon. Science **90**, 570–571 (1939b)
Ruben, S., Kamen, M.D., Hassid, W.Z.: Photosynthesis with radioactive carbon II. Chemical properties of the intermediates. J. Am. Chem. Soc. **62**, 3443–3450 (1940a)

Ruben, S., Kamen, M.D., Perry, L.H.: Photosynthesis with radioactive carbon III. Ultracentrifugation of intermediate products. J. Am. Chem. Soc. **62**, 3450-34-51 (1940b)

Ruben, S., Randall, M., Kamen, M.D., Hyde, J.L.: Heavy oxygen as a tracer in the study of photosynthesis. J. Am. Chem. Soc. **63**, 877–879 (1941)

Sabalitschka, T., Riesenberg, H.: Über die Ernährung von Pflanzen mit Aldehyden. II. Biochemische Zeitschrift **144**, 545–550 (1924a)

Sabalitschka, T., Riesenberg, H.: Über die Ernährung von Pflanzen mit Aldehyden. III. Biochemische Zeitschrift **144**, 551–555 (1924b)

Sabalitschka, T., Riesenberg, H.: Über die Ernährung von Pflanzen mit Aldehyden. IV. Biochemische Zeitschrift **145**, 373–378 (1924c)

Sachs, J.: Über den Einfluss des Lichtes auf die Bildung des Amylums in den Chlorophyllkörnern. Botanische Zeitung **20**, 365–373 (1862)

Sachs, J.: Über die Auflösung und Wiederbildung des Amylums in den Chlorophyllkörnern bei wechselnder Beleuchtung. Botanische Zeitung **22**, 189–294 (1864)

San Pietro, A., Lang, H.M.: Accumulation of reduced pyridine nucleotides by illuminated grana. Science **124**, 118–119 (1956)

San Pietro, A., Lang, H.M.: Photosynthetic pyridine nucleotide reductase I. Partial purification and properties of the enzyme from spinach. J. Biol. Chem. **231**, 211–229 (1958)

Schickore, J., Steinle, F. (eds.): Revisiting Discovery and Justification: Historical and Philosophical Perspectives on the Context Disctinction. Springer, Dordrecht et al. (2006)

Schleifer, K.-H.: Professor Dr. Dr. h.c. mult. Otto Kandler: distinguished botanist and microbiologist. Bull. BISMiS **2**, 141–148 (2011)

Schmitt, F.F.: Socializing epistemology: an introduction through two sample issues. In: Schmitt, F.F. (ed.) Socializing Epistemology. The Social Dimensions of Knowledge, pp. 1–28. Rowman & Littlefield, Lanham, Maryld./London (1994)

Schoenheimer, R.L.: The Dynamic State of Body Constituents. Cambridge University Press, Cambridge (1940)

Schoenheimer, R.L., Rittenberg, D.: Deuterium as an indicator in the study of intermediary metabolism. I. J. Biol. Chem. **111**, 163–168 (1935)

Schoenheimer, R.L., Rittenberg, D.: Deuterium as an indicator in the study of intermediary metabolism. XI. Further studies on the biological uptake of deuterium into organic substances, with special reference to fat and cholesterol formation. J. Biol. Chem. **121**, 235–253 (1937)

Scholl, R.: Causal inference, mechanisms, and the Semmelweis case. Stud. Hist. Philos. Sci. Part A **44**, 66–76 (2013)

Scholl, R., Nickelsen, K.: Discovery from a causal point of view: Oxidative phosphorylation and the Calvin-Benson cycle. History and philosophy of the life sciences (2015). In print

Schüring, M.: Minervas verstoßene Kinder. Vertriebene Wissenschaftler und die Vergangenheitspolitik der Max-Planck-Gesellschaft. Wallstein, Göttingen (2006)

Schroeder, H.: Die Hypothesen über die chemischen Vorgänge bei der Kohlensäure-Assimilation und ihre Grundlagen, Fischer. Jena (1917)

Schroeder, H.: Methoden zur Bestimmung der Assimilation der Kohlensäure aus der Luft und aus dem Wasser. In: Abderhalden, E. (ed.) Handbuch der biologischen Arbeitsmethoden, vol. 3, pp. 653–714. Urben & Schwarzenberg, Berlin (1928)

Schulmann, R., Kox, A.J., Janssen, M., Illy, J.E.: The Berlin Years: Correspondence, 1914–1918, Vol. 5 der The Collected Papers of Albert Einstein. Princeton University Press, Princeton (1998)

Seaborg, G.T., Benson, A.A.: Melvin Calvin (1911–1997). Biogr. Mem. Natl. Acad. Sci. **75**, 2–21 (1997)

Seidel, R.W.: Accelerating science: the postwar transformation of the Lawrence Radiation Laboratory. Hist. Stud. Phys. Sci. **13**, 375–400 (1983)

Shen, Y.-K., Wei, J.-M.: The discovery of photophosphorylation. In: Kung, S.-D., Yang, S.-F. (eds.) Discoveries in Plant Biology, vol. 2, pp. 105–114. World Scientific Publishing, Singapore (1998)

Shevela, D., Eaton-Rye, J., Shen, J.-R., Govindjee: Photosystem II and the unique role of bicarbonate. A historical perspective. Biochimica et Biophysica Acta **1817**, 1134–1151 (2012)
Simoni, R.D., Hill, R.L., Vaughan, M.: The use of isotope tracers to study intermediary metabolism: Rudolf Schoenheimer. J. Biol. Chem. **43**, e1–e2 (2002)
Singleton, R.: Heterotrophic CO_2-fixation, mentors, and students: the Wood-Werkman reactions, J. Hist. Biol. **30**, 91–120 (1997)
Sloan, P.R.: Biophysics in Berlin. In: Sloan, P.R., Fogel, B. (eds.) Creating a Physical Biology: The Three-Man Paper and the Origins of Molecular Biology, pp. 77–124 . University of Chicago Press, Chicago (2009)
Smith, C.: Force, energy, and thermodynamics. In: Nye, M.J. (ed.) The Cambridge History of Science. Vol. 5: The Modern Physical and Mathematical Sciences, pp. 289–310. Cambridge University Press, Cambridge (2003)
Solomon, M.: A more social epistemology. In: Schmitt, F.F. (ed.) Socializing Epistemology. The Social Dimensions of Knowledge, pp. 111–134. Rowman & Littlefield, Lanham, Maryld./London (1994a)
Solomon, M.: Social empiricism. Noûos **28**, 325–343 (1994b)
Solomon, M.: Social Empiricism, MIT Press (2001)
Spath, S.B.: C. B. van Niel and the Culture of Microbiology, 1920–1965, Thesis (Ph. D.), University of California, Berkeley (1999)
Spoehr, H.A.: Photochemische Vorgänge bei der diurnalen Entsäuerung der Succulenten. Biochemische Zeitschrift **57**, 95–111 (1913)
Spoehr, H.A.: The Theories of photosynthesis in the light of some new facts. Plant World **19**, 1–16 (1916)
Spoehr, H.A., McGee, J.M.: Studies in Plant Respiration and Photosynthesis. Carnegie-Institution, Washington (1923)
Stachel, J., Cassidy, D.C., Renn, J., Schulmann, R. (eds.): The Collected Papers of Albert Einstein. Vol 2: The Swiss Years: Writings, 1900–1909. Princeton University Press (1989)
Stemler, A.J.: The bicarbonate effect, oxygen evolution, and the shadow of Otto Warburg. Photosynth. Res. **73**, 177–183 (2002)
Stepka, W., Benson, A.A., Calvin, M.: The path of carbon in photosynthesis: II Amino acids. Science **108**, 304 (1948)
Stiles, W.: Photosynthesis: The Assimilation of Carbon by Green Plants. Longmans/Green, London (1925)
Stoklasa, J., Zdobnicky, W.: Photochemische Synthese der Kohlenhydrate aus Kohlensäureanhydrid und Wasserstoff in Abwesenheit von Chlorophyll. Biochemische Zeitschrift **30**, 433–456 (1911)
Stoll, A.: Über den chemischen Verlauf der Photosynthese. Naturwissenschaften **20**, 955–958 (1932)
Stowe, B.B.: Kenneth V. Thimann (1904–1997). Proc. Am. Philos. Soc. **143**, 502–509 (1999)
Strehler, B.L.: The luminescence of isolated chloroplasts. Arch. Biochem. Biophys. **34**, 239–248 (1951)
Strehler, B.L.: Photosynthesis—energetics and phosphate metabolism. In: McElroy, W.D., Glass, H.B. (eds.) Phosphorus Metabolism II, pp. 491–502. Johns Hopkins Press, Baltimore (1952)
Strehler, B.L.: Firefly luminescence in the study of energy transfer mechanisms II: Adenosine triphosphate and photosynthesis. Arch. Biochem. Biophys. **43**, 67–79 (1953)
Strehler, B.L.: Halcyon days with Bill Arnold. Photosynth. Res. **48**, 11–18 (1996)
Strehler, B.L., Arnold, W.A.: Light production by green plants. J. Gen. Physiol. **34**, 809–820 (1951)
Strehler, B.L., Hendley, D.: Continuous measurement of photosynthetic phosphorylation with the firefly luminescence assay system. In: McElroy, W.D., Glass, B. (eds.) Light and Life, pp. 601–608. Johns Hopkins Press, Baltimore (1961)
Strehler, B.L., McElroy, W.D.: Purification of firefly luciferin. J. Cell Compd. **34**, 457–466 (1949)
Strehler, B.L., Totter, J.: Firefly luminescence in the study of energy transfer mechanisms I. Substrate and enzyme determination. Arch. Biochem. Biophys. **40**, 28–41 (1951)
Sucker, U.: Das Kaiser-Wilhelm-Institut für Biologie. Seine Gründungsgeschichte, seine problemgeschichtlichen und wissenschaftstheoretischen Voraussetzungen (1911–1916). Steiner, Stuttgart (2002)

Tabery, J.: Synthesizing activities and interactions in the concept of a mechanism. Philos. Sci. **71**, 1–15 (2004)
Tagawa, K., Arnon, D.I.: Ferredoxin as electron carrier in photosynthesis and in the bacterial production and consumption of hydrogen gas. Nature **195**, 537–543 (1962)
Tamiya, H. (ed.): Mechanism of Photosynthesis (Fifth International Congress of Biochemistry. Moscow, 10–16 August 1961. General Editor: Sissakian, N.M. Vol. VI). Pergamon Press, Oxford (Polish Scientific Publishers, Warsaw) (1963)
Tanada, T.: The photosynthetic efficiency of carotenoid pigments in *Navicula minima*. Am. J. Bot. **38**, 276–283 (1951)
Teich, M.: Ferment or enzyme: what's in a name? Hist. Philos. Life Sci. **2**, 193–215 (1981)
Teller, E.: Memoirs: A Twentieth-Century Journey in Science and Politics. Perseus Publishing, Cambridge (2001)
Thimann, K.V.: The absorption of carbon dioxide in photosynthesis. Science **88**, 506–507 (1938)
Thorhaug, A., Berlyn, G.: A tribute to Lawrence Rogers Blinks (1900–1989): light and algae. Photosynth. Res. **100**, 129–141 (2009)
Thunberg, T.: Über einen neuen Weg von der Kohlensäure zum Formaldehyd. Ein Beitrag zur Theorie der Kohlensäureassimilation. Zeitschrift für Physikalische Chemie **106**, 305–312 (1923)
Timofeeff-Ressovsky, N.W., Zimmer, K.G., Delbrück, M.: Über die Natur der Genmutation und der Genstruktur, Nachrichten der Gesellschaft der Wissenschaften zu Göttingen. Mathematisch-Physikalische Klasse **1**, 189–245 (1935)
Todhunter, I. (ed.): William Whewell, D.D., Master of Trinity College, Cambridge; An Account of his Writings with Selections from his Literary and Scientific Correspondence. Macmillan, London (1876)
Tolmach, L.J.: Effects of triphosphopyridine nucleotide upon oxygen evolution and carbon dioxide fixation by illuminated chloroplasts. Nature **167**, 946–948 (1951a)
Tolmach, L.J.: The influence of triphosphopyridine nucleotide and other physiological substances upon oxygen evolution from illuminated chloroplasts. Arch. Biochem. Biophys. **33**, 120–142 (1951b)
Toon, A.: Models as Make-Believe. Imagination, Fiction and Scientific Representation. Palgrave (2012)
Trebst, A.V., Manuel, L., Arnon, D.I.: Photosynthesis by isolated chloroplasts X. Dependence of carbon dioxide assimilation on the photochemical reactions of chloroplasts. J. Biol. Chem. **234**, 3055–3058 (1959)
Trebst, A.V., Losada, M., Arnon, D.I.: Photosynthesis by isolated chloroplasts XII: inhibitors of CO_2 assimilation in a reconstituted chloroplast system. J. Biol. Chem. **235**, 840–844 (1960)
Underwood, E.A.: Lavoisier and the history of respiration. Proc. R. Soc. Med. **37**, 247–262 (1943)
Usher, F.L., Priestley, J.H.: The mechanism of carbon assimilation. II. The photolytic decomposition of carbon dioxide in vitro. Proc. R. Soc. Lond., Series B **78**(518), 318–327 (1906a)
Usher, F.L., Priestley, J.H.: A study of the mechanism of carbon assimilation in green plants. Proc. R. Soc. Lond., Series B **77**(518), 369–376 (1906b)
van Niel, C.B.: Photosynthesis in Bacteria, Contributions to Marine Biology, vol. 3. Stanford University Press, pp. 161–169 (1930)
van Niel, C.B.: On the morphology and physiology of the purple and green sulphur bacteria. Archiv für Mikrobiologie **3**, 1–112 (1931)
van Niel, C.B.: Photosynthesis of Bacteria. In: Harris, R.G. (ed.) Cold Spring Harbor Symposia on Quantitative Biology, vol. 3, pp. 138–150. The Biological Laboratory, New York (1935)
van Niel, C.B.: The bacterial photosynthesis and their importance for the general problem of phtosynthesis. Adv. Enzymol. **1**, 263–328 (1941)
van Niel, C.B.: The education of a microbiologist: some reflections. Ann. Rev. Microbiol. **21**, 1–30 (1967)
Vernon, L.P., Zaugg, W.S.: Photoreductions by fresh and aged chloroplasts: requirement for ascorbate and 2, 6-dichlorophenolindophenol with aged chloroplasts. J. Biol. Chem. **235**, 2728–2733 (1960)

Vierhaus, R., Brocke, B. vom. (eds.): Forschung im Spannungsfeld von Politik und Gesellschaft. Geschichte und Struktur der Kaiser-Wilhelm-/Max-Planck-Gesellschaft. Deutsche Verlags-Anstalt, Stuttgart (1990)

Vinogradov, A.P., Teiss, R.V.: Isotope composition of oxygen from different sources. Doklady Akademii Nauk SSSR **33**, 497–501 (1941)

Vinogradov, A.P., Teiss, R.V.: Novel determination of the isotope composition of oxygen in photosynthesis. Doklady Akademii SSSR **56**, 57–58 (1947)

Vishniac, W., Ochoa, S.: Photochemical reduction of pyridine nucleotides by spinach grana and coupled carbon dioxide fixation. Nature **167**, 768–769 (1951)

Vishniac, W., Ochoa, S.: Fixation of carbon dioxide coupled to photochemical reduction of pyridine nucleotides by chloroplast preparations. J. Biol. Chem. **195**, 75–93 (1952a)

Vishniac, W., Ochoa, S.: Phosphorylation coupled to photochemical reaction of pyridine nucleotides by chloroplast preparations. J. Biol. Chem. **198**, 501–506 (1952b)

von Gerhard, F.: Hans Kautsky (1891–1966). Chemie in unserer Zeit **15**, 197–200 (2004)

von Mohl, H., Wilhelm, M.: Untersuchungen über die anatomischen Verhältnisse des Chlorophylls. Gustav Bähr, Tübingen (1837)

von Mohl, H.: Dr. Justus Liebigs Verhältniss zur Pflanzenphysiologie. Fues, Tübingen (1843)

Walker, D.: Energy, Plants and Man. Oxygraphics Limited, Sheffield (1992a)

Walker, D.A.: Robert Hill. Photosynth. Res. **34**, 337–338 (1992b)

Walker, D.A.: One thing leading to another. Photosynth. Res. **46**, 45–46 (1995)

Walker, D.: Tell me where all past years are. Photosynth. Res. **51**, 1–26 (1997)

Walker, D.A.: "And whose bright presence"—an appreciation of Robert Hill and his reaction. Photosynth. Res. **73**, 51–54 (2002)

Walker, D.A.: From Chlorella to chloroplasts: a personal note. Photosynth. Res. **92**, 181–185 (2007)

Warburg, E.: Über die Bildung des Ozons bei der Spitzenentladung in Sauerstoff, Sitzungsberichte der Königlich Preussischen Akademie der Wissenschaften (Berlin). Phys.-math. Klasse (**II**), 712–721 (1900)

Warburg, E.: Einige Bemerkungen über photochemische Wirkung II. Verhandlungen der Deutschen Physikalischen Gesellschaft **11**(21), 654–660 (1909)

Warburg, E.: Über den Energieumsatz bei photochemischen Vorgängen in Gasen. II, Sitzungsberichte der Königlich Preussischen Akademie der Wissenschaften (Berlin) pp. 216–225 (1912)

Warburg, E.: Über den Energieumsatz bei photochemischen Vorgängen in Gasen. III, Sitzungsberichte der Königlich Preussischen Akademie der Wissenschaften (Berlin) pp. 644–659 (1913)

Warburg, E.: Über die Anwendung der Quantenhypothese auf die Photochemie. Naturwissenschaften **5**(30), 489–494 (1917)

Warburg, E.: Quantentheoretische Grundlagen der Photochemie. Zeitschrift für Elektrochemie **26**, 54–59 (1920a)

Warburg, E.: Die Quantenregeln in der Photochemie. Naturwissenschaften **12**(47), 1058–1063 (1924)

Warburg, E., Leithäuser, G., Johansen, E.: Über das Vakuumbolometer. Annalen der Physik **24**, 25–42 (1907)

Warburg, O.: Beobachtungen über die Oxydationsprozesse im Seeigelei. Hoppe-Seyler's Zeitschrift für physiologische Chemie **57**, 1–16 (1908)

Warburg, O.: Beiträge zur Physiologie der Zelle, insbesondere über die Oxydationsgeschwindigkeit in Zellen. Ergebnisse der Physiologie **14**, 253–337 (1914)

Warburg, O.: Über die Geschwindigkeit der photochemischen Kohlensäurezersetzung in lebenden Zellen. I. Biochemische Zeitschrift **100**, 230–270 (1919)

Warburg, O.: Über die Geschwindigkeit der photochemischen Kohlensäurezersetzung in lebenden Zellen. II. Biochemische Zeitschrift **103**, 188–217 (1920b)

Warburg, O.: Theorie der Kohlensäureassimilation. Naturwissenschaften **9**, 354–358 (1921)

Warburg, O.: Versuche über die Assimilation der Kohlensäure. Biochemische Zeitschrift **166**, 386–406 (1925)
Warburg, O.: Methodische Vorbemerkungen. In: Warburg, O. (ed.) Über den Stoffwechsel der Tumoren. Arbeiten aus dem Kaiser-Wilhelm-Institut für Biologie, Berlin-Dahlem, pp. 1–11. Springer, Berlin (1926)
Warburg, O.: Über die katalytischen Wirkungen der lebendigen Substanz, Über die katalytischen Wirkungen der lebendigen Substanz. Arbeiten aus dem Kaiser-Wilhelm-Institut für Biologie, Berlin-Dahlem, pp. 1–13. Springer, Berlin (1927)
Warburg, O.: Über den Quantenbedarf der Kohlensäureassimilation. Naturwissenschaften **33**, 122–122 (1945)
Warburg, O.: Kohlensäureassimilation I. Der Quantenbedarf der Photochemischen Reduktion der Kohlensäure. In: Kuhn, R. (ed.) Naturforschung und Medizin in Deutschland 1939–1946. Vol. 39: Biochemie, pp. 201–210. Dietrich, Wiesbaden, (1947)
Warburg, O.: Assimilatory quotient and photochemical yield. Am. J. Bot. **35**, 194–204 (1948)
Warburg, O.: Ein-Quantenmechanismus der Photosynthese. Angewandte Chemie **63**, 282–283 (1951)
Warburg, O.: Energetik und Photosynthese. Naturwissenschaften **39**, 337–341 (1952)
Warburg, O.: Photosynthesis. Experiments at the Max Planck Institute for cell physiology, Berlin-Dahlem. Science **128**(3315), 68–73 (1958)
Warburg, O., Burk, D.: 1-Quanten-Mechanismus und Energie-Kreisprozess bei der Photosynthese, Naturwissenschaften **37**, 560 (1950a)
Warburg, O., Burk, D.: The maximum efficiency of photosynthesis. Arch. Biochem. **25**, 410–443 (1950b)
Warburg, O., Christian, W.: Isolierung und Kristallisation des Proteins des oxydierenden Gärungsferments. Biochemische Zeitschrift **303**, 40–68 (1939)
Warburg, O., Krippahl, G.: Hill-Reaktionen. Zeitschrift für Naturforschung B **13**, 509–514 (1958)
Warburg, O., Lüttgens, W.: Experiment zur Assimilation der Kohlensäure. Naturwissenschaften **32**, 161–161 (1944a)
Warburg, O., Lüttgens, W.: Weitere Experimente zur Kohlensäureassimilation. Naturwissenschaften **32**, 301–301 (1944b)
Warburg, O., Negelein, E.: Über den Energieumsatz bei der Kohlensäureassimilation. Zeitschrift für Physikalische Chemie **102**, 235–266 (1922)
Warburg, O., Negelein, E.: Über den Einfluss der Wellenlänge auf den Energieumsatz bei der Kohlensäureassimilation. Zeitschrift für Physikalische Chemie **106**, 191–218 (1923)
Warburg, O., Uyesugi, T.: Über die Blackmansche Reaktion. Biochemische Zeitschrift **146**, 486–492 (1924)
Warburg, O., Burk, D., Schocken, V., Hendricks, S.: The quantum efficiency of photosynthesis. Biochimica et Biophysica Acta **4**, 335–349 (1950)
Warburg, O., Geleick, H., Briese, K.: Über die Messung der Photosynthese in Carbonat-Bicarbonat-Gemischen. Zeitschrift für Naturforschung **7b**, 141 (1952)
Warburg, O., Krippahl, G., Schröder, W.: Katalytische Wirkung des blaugrünen Lichts auf den Energieumsatz bei der Photosynthese. Zeitschrift für Naturforschung **9b**, 667–675 (1954)
Warburg, O., Krippahl, G., Schröder, W.: Wirkungsspektrum eines Photosynthese-Fermentes. Zeitschrift für Naturforschung **10b**, 631–639 (1955)
Warburg, O., Krippahl, G., Lehman, A.: Chlorophyll catalysis and Einstein's law of photochemical equivalence in photosynthesis. Am. J. Bot. **56**, 961–971 (1969)
Warner, C.H.: Formaldehyde as oxidation product of chlorophyll extracts. Proc. R. Soc. Lond., Series B **87**, 378–385 (1914)
Wassink, E.C., Kersten, J.A.: The absorption spectrum and the role of carotenoids in the photosynthesis of diatoms. Enzymologia **12**, 3 (1946)
Weber, B.: Glynn and the conceptual development of the chemiosmotic theory: a retrospective and prospective view. Biosci. Rep. **11**, 577–617 (1991)
Weber, M.: Philosophy of Experimental Biology. Cambridge University Press, Cambridge (2005)

Weber, M.: Experiment in biology. In: Zalta, E.N. (ed.) The Stanford Encyclopedia of Philosophy. http://plato.stanford.edu/archives/spr2012/entries/biology-experiment/. Accessed 2 June 2013 (2012)

Weigert, F.: Die chemischen Wirkungen des Lichts. Enke, Stuttgart (1911)

Weigert, F.: Der Ausnutzungsfaktor der Lichtenergie, Zeitschrift für wissenschaftliche Photographie. Photophysik und Photochemie **11**, 381–394 (1912)

Weissbach, A., Horecker, B.L., Hurwitz, J.: The enzymatic formation of phosphoglyceric acid from ribulose diphosphate and carbon dioxide. J. Biol. Chem. **218**, 795–810 (1956)

Weller, S., Franck, J.: Photosynthesis in flashing light. J. Phys. Chem. **45**, 1359–1373 (1941)

Werner, P.: Otto Warburg. Von der Zellphysiologe zur Krebsforschung. Neues Leben, Berlin (1988)

Werner, P.: Ein Genie irrt seltener. Otto Heinrich Warburg, ein Lebensbild in Dokumenten. Akademie Verlag, Berlin (1991)

Werner, P.: Otto Warburgs Beitrag zur Atmungstheorie: Das Problem der Sauerstoffaktivierung. Basiliskenpresse, Marburg (1996)

Werner, P.: Learning from an adversary? Warburg against Wieland. Hist. Studies Phys. Biol. Sci. **28**(1), 173–196 (1997)

Werner, P.: Humboldt als Vermittler: Schleiden und Mohl contra Liebig. Hist. Philos. Life Sci. **23**, 213–257 (2001)

Werner, P., Holmes, F.L.: Justus Liebig and the plant physiologists. J. Hist. Biol. **35**, 421–441 (2002)

Werner, P., Irmscher, A. (eds.): Fritz Haber. Briefe an Richard Willstätter 1910–1934. VWR Engel, Berlin (1995)

Whatley, F.R.: Photosynthesis by isolated chloroplasts: the early work at Berkeley. Photosynth. Res. **46**, 17–26 (1995)

Whatley, F.R., Allen, M.B., Arnon, D.I.: Cofactors of photosynthetic phosphorylation. Plant Physiol. **32**, iii (1957) (Suppl.)

Whatley, F.R., Allen, M.B., Rosenberg, L.L., Capindale, J.B., Arnon, D.I.: Photosynthesis in isolated chloroplasts V. Phosphorylation and carbon dioxide fixation by broken chloroplasts. Biochimica et Biophysica Acta **20**, 462–468 (1956)

Wheaton, B.R.: The Tiger and the Shark. Empirical Roots of Wave-Particle Dualism. Cambridge, UP (1983)

Wiesen, J.: The Richard Willstätter controversy: the legacy of anti-semitism in the west-German chemical industry. In: Lesch, J.E. (ed.) The German Chemical Industry in the Twentieth Century, pp. 347–366. Kluwer, Dordrecht (2000)

Wildman, S.G.: Early events along the trail leading to identification of Rubisco. Protein Sci. **1**, 303–304 (1992)

Wildman, S.G.: Discovery of Rubisco. In: Kung S.-D., Yang, S.-F. (eds.) Discoveries in Plant Biology, vol. 1, pp. 163–173. World Scientific Publishing, Singapore (1998)

Wildman, S.G.: Along the trail from Fraction I protein to Rubisco. Photosynth. Res. **73**, 243–250 (2002)

Wildman, S.G., Bonner, J.: The proteins of green leaves. I. Isolation of enzymatic properties and auxin content of spinach cytoplasmic proteins. Arch. Biochem. **14**, 381–413 (1947)

Willstätter, R.: Zur Erklärung der Photoreduktion von Kohlensäure durch Chlorophyll. Naturwissenschaften **21**, 252–253 (1933)

Willstätter, R.: Aus meinem Leben. Von Arbeit, Musse und Freunden, 2 edn. Verlag Chemie, Weinheim (1973)

Willstätter, R., Stoll, A.: Untersuchungen über Chlorophyll. Methoden und Ergebnisse. Springer, Berlin (1913)

Willstätter, R., Stoll, A.: Untersuchungen über die Assimilation der Kohlensäure. Sieben Abhandlungen. Springer, Berlin (1918)

Wilson, A.T., Calvin, M.: The photosynthetic cycle. CO_2 dependent transients. J. Am. Chem. Soc. **77**, 5948–5957 (1955)

Wimsatt, W.: Complexity and organization. In: Schaffner, K.F., Cohen, R.S. (eds.) PSA 1972. Proceedings of the Philosophy of Science Association, Reidel, Dordrecht, pp. 67–86 (1972)

Witt, H.T.: Untersuchungen der Photosynthese bei Anregung mit Blitzlicht. In: Pirson, A. (ed.) Handbuch der Pflanzenphysiologie, pp. 634–674. Springer, Berlin (1960)
Witt, H.T.: Functional mechanism of water splitting photosynthesis. Photosynth. Res. **29**, 55–77 (1991)
Witt, H.T., Müller, A., Rumberg, B.: Experimental evidence for the mechanism of photosynthesis. Nature **191**, 194–195 (1961a)
Witt, H.T., Müller, A., Rumberg, B.: Oxidized cytochrome and chlorophyll in photosynthesis. Nature **192**, 967–969 (1961b)
Wohl, K.: The mechanism of photosynthesis in green plants. New Phytol. **39**, 33–64 (1940)
Wohl, K.: On the mechanism of photosynthesis in purple bacteria and green plants. New Phytol. **40**, 34–55 (1941)
Wood, H.G., Werkman, C.H.: The utilization of CO_2 by the propionic acid bacteria in the dissimilation of glycerol. J. Bacteriol. **30**, 332 (1935)
Wood, H.G., Werkman, C.H.: The utilization of CO_2 in the dissimilation of glycerol by the propionic acid bacteria. Biochem. J. **30**, 48–53 (1936)
Wood, H.G., Werkman, C.H.: The utilization of CO_2 by the propionic acid bacteria. Biochem. J. **32**, 1262–1271 (1938)
Woods, D.D.: Albert Jan Kluyver (1888–1956). Biogr. Mem. Fellows R. Soc. **3**, 109–128 (1957)
Woodward, J.: What is a mechanism? A counterfactual account. Philos. Sci. **69**, S366–S377 (2002)
Wurmser, R.: Recherches sur l'assimilation chlorophyllienne. Archives de Physique Biologique **1**, 33–141 (1921)
Wurmser, R.: La loi de l'équivalent photochimique dans la photosynthèse chlorophyllienne. Le Journal de Physique et le Radium **7**, 33–44 (1926)
Wurmser, R.: Oxydations et réductions. Presses Universitaires de France, Paris (1930)
Yodh, A.G., Tromberg, B.J.: Celebrating Britton chance. J. Biomed. Opt. **5**, 115–118 (2000)
Zallen, D.T.: The "light" organism for the job: Green algae and photosynthesis research. J. Hist. Biol. **26**, 269–279 (1993a)
Zallen, D.T.: Redrawing the boundaries of molecular biology: the case of photosynthesis. J. Hist. Biol. **26**, 65–87 (1993b)

Index

A

Allen, Mary B. (1922-1973), 264, 279, 280
Arnold, William (1904-2001), 51, 93, 110, 139, 151, 262, 309
Arnon, Daniel I. (1910-1994), 173, 254, 256, 263, 264-266, 274, 279, 280, 282, 297, 308
Aronoff, Samuel (1915-2010), 218, 260
Avron, Mordhay (1931-1991), 266

B

Bassham, J. Alan (1922-2012), 216-218, 228, 229, 232, 241, 247
Baur, Emil (1873-1944), 28, 32, 33, 45, 49, 52
Bayer, Adolf von (1835-1917), 300
Barcroft, Joseph (1872-1947), 59, 84
Benson, Andrew A. (born 1917), 10, 13, 139, 201, 215, 216, 218-220, 224, 226, 227, 232, 233, 241, 242, 244-247, 250
Blackman, Frederick F. (1866-1947), 67, 68
Bonner, James (1910-1996), 158, 245, 265
Borsook, Henry (1897-1984), 110
Brackett, Frederick S. (1896-1988), 182, 184
Bredig, Georg (1868-1944), 28, 32-34, 45, 49, 52, 117, 119
Brody, Seymour S. (1927-2010), 161, 289, 290
Brown, Horace (1848-1925), 64, 77, 111
Burk, Dean (1904-1988), 152, 165, 314
Butlerov, Alexander M. (1828-1886), 23

C

Calvin, Melvin (1911-1997), 10, 13, 139, 173, 181, 195, 201, 214, 251, 260, 315
Chaikoff, Israel L. (1902-1966), 204, 206, 207
Chance, Britton (1913-2010), 268, 269, 288
Cornell, Samuel D. (1915-2010), 189, 190

D

Daniels, Farrington (1889-1972), 151, 181, 182, 184, 187, 189, 190, 199
Davenport, Harold E., 271-276
Delbrück, Max (1906-1981), 122-124, 140, 144, 283
Driesch, Hans (1867-1941), 83
Duggar, Benjamin M. (1872-1956), 151
Dutton, Herbert J. (1914-2006), 284
Duysens, Louis N. M. (PhD 1952), 193, 251, 285-290, 293, 294, 299-301, 304, 308, 309

E

Emerson, Robert (1903-1959), 12, 62, 93, 106, 108, 120, 139, 149, 152, 161, 165, 193, 209, 256, 265, 314
Erlenmeyer, Emil (1825-1909), 12, 28, 29, 31, 33, 39, 44, 45, 49, 51, 52

F

Fischer, Emil (1852-1919), 26, 29, 52, 56, 58
Förster, Theodor (1910-1974), 286
Franck, James (1882-1964), 11, 12, 93, 95, 96, 121, 137, 139, 152, 206, 209, 255, 314
French, Charles Stacy (1907-1995), 12, 62, 93, 124, 137, 139, 264, 285
Frenkel, Albert (born 1919), 158, 165, 186, 226, 266
Fuller, R. Clinton (born 1925), 166, 239, 244, 246

G

Gay-Lussac, Joseph L. (1778-1815), 21
Gaffron, Hans (1902-1979), 12, 93, 95, 98, 105, 108, 113, 120, 123, 124, 139, 147, 151, 223, 282, 314
Gest, Howard (1921-2012), 266

K. Nickelsen, *Explaining Photosynthesis,* History, Philosophy and Theory of the Life Sciences 8, DOI 10.1007/978-94-017-9582-1

Goodman, Murray (1928-2004), 219
Govindjee (born 1932), 1, 77, 94, 108, 109, 124, 149, 150, 157-159, 161, 164-167, 177, 194, 195, 199, 218, 286, 293-296, 299, 302, 303, 306

H

Hassid, Zev (1899-1974), 206-209, 211
Hendricks, Sterling B. (1902-1981), 182, 184, 189
Hearon, John Z. (1920-2009), 165
Herzfeld, Karl F. (1892-1978), 98, 126-128, 130, 131, 140, 144, 293, 294
Hill, Robert (Robin) (1899-1991), 12, 93, 132, 139, 178, 185, 213, 251, 255, 267, 269, 270
Hoppe-Seyler, Felix (1825-1895), 25, 52
Horecker, Bernard L. (1914-2010), 246, 249

J

Jagendorf, André T. (born 1926), 254, 265, 266

K

Kalckar, Herman (1908-1991), 211, 257, 258
Kamen, Martin D. (1913-2002), 12, 130, 201-204, 208, 209, 211-214, 258, 313, 315
Kandler, Otto (born 1920), 262
Kautsky, Hans (1891-1966), 93-95, 99, 101, 104, 106-108, 140
Kay, Lorel L. (1926-2013), 234, 235, 239, 241, 242, 248, 250
Keilin, David (1887-1963), 132-134, 270
Kok, Bessel (1918-1979), 173, 185, 196, 251, 290-292, 302-305, 308, 309
Krebs, Hans (1900-1981), 56, 57, 80, 173, 212, 221
Krehl, Ludolf von (1861-1937), 57

L

Laurence, William L. (1888-1977), 170, 180, 214
Lawrence, Ernest O. (1901-1958), 202, 213, 214, 217, 299
Lawrence, John H. (1904-1991), 203
Lewis, Charlton M. (1905-1996), 152-157, 160, 162, 171, 176, 181, 191, 196, 209, 284, 316
Lipmann, Fritz (1899-1986), 211, 257, 258, 266
Liebig, Justus (1803-1873), 15, 16, 19
Loeb, Jacques (1859-1924), 56, 83, 109, 116
Löb, Walter (1872-1916), 26, 27, 30, 33
Lummer, Otto (1860-1925), 78

M

Manning, Winston M. (1910-2002), 29, 30, 61, 129, 151, 152, 284
Massini, Peter (born 1921), 228, 232, 237, 238, 248
Mayaudon, Jacques (born 1924), 244, 245
Mayer, Julius R. (1814-1878), 16
Moses, Vivian (PhD 1953), 216, 218

N

Negelein, Erwin (1897-1979), 75-77, 103, 104, 107, 112, 125, 140, 142, 144, 149, 150
Nishimura, M. Shimpe (1911-1989), 158, 165, 172

O

Ochoa, Severo (1905-1993), 253, 254, 256, 260, 266
Oppenheimer, J. Robert (1904-1960), 284, 285, 286
Osterhout, Winthrop J. V. (1871-1964), 108, 109

P

Pirson, André (1910-2004), 90, 91, 188
Pfeffer, Wilhelm (1845-1920), 26
Planck, Max (1858-1947), 78
Pringsheim, Ernst G. (1881-1917), 16, 61, 89, 175

Q

Quayle, Rodney (1926-2006), 219, 244

R

Rabinowitch, Eugene (1901-1973), 96, 124, 139, 140, 158, 161, 252, 303
Randall, Merle (1888-1950), 213
Ridenour, Louis N. (1911-1959), 173
Rieke, Foster F. (1905-1970), 152, 154
Ruben, Samuel (1913-1943), 130, 204, 258

S

Sachs, Julius (1832-1897), 16, 17, 30
Scarisbrick, Richard, 136, 270, 271, 274
Schleiden, Matthias J. (1804-1881), 20
Schoenheimer, Rudolph L. (1898-1941), 205
Spiegelman, Sol (1914-1983), 182
Stacy French, Charles (1907-1995), 12, 62, 121, 124, 137, 139, 264, 285
Stepka, William (1917-?), 226
Steward, Frederick C. (1904-1993), 166, 170, 199

Stoll, Arthur (1887-1971), 12, 15, 28, 34, 55, 99, 293
Strehler, Bernhard L. (1925-2001), 51, 262, 263

T
Thimann, Kenneth, V. (1904-1997), 110, 220, 221
Tippo, Oswald (1912-1999), 165, 166
Tolbert, Anne G. (née Tolbert) (born 1930), 235, 241, 242
Tolmach, Leonard J. (1923-1991), 253-256

U
Urey, Harold (1893-1981), 205

V
Vennesland, Birgit (1913-2001), 195
Vinogradov, Alexander P. (1895-1975), 214
Vishniac, Wolf V. (1922-1973), 232, 253, 254, 256, 260, 261
von Mohl, Hugo (1805-1872), 20, 22
van Niel, Cornelius B. (1897-1985), 12, 93, 113, 209, 259, 264, 278

W
Warburg, Emil (1846-1931), 56, 61, 64, 78, 79, 81, 82, 89, 139
Warburg, Otto (1883-1970), 11-13, 55, 56, 58, 60, 78, 80-83, 109, 116, 139, 149, 150, 175, 214, 251, 257
Weigert, Fritz (1876-1974), 52, 81, 82
Werkman, Chester H. (1893-1962), 212
Whittingham, Charles P., 172, 174, 182, 185, 188
Wildman, Samuel (1912-2004), 245
Willstätter, Richard (1872-1942), 12, 15, 28, 99
Wilson, Alexander T., 238-240, 241
Witt, Horst T. (1922-2007), 251, 292, 303, 304
Wohl, Kurt (1896-1962), 120, 122, 123, 282, 314
Wood, Harland G. (1907-1991), 173, 212

MIX
Papier aus verantwortungsvollen Quellen
Paper from responsible sources
FSC® C105338

If you have any concerns about our products,
you can contact us on
ProductSafety@springernature.com

In case Publisher is established outside the EU,
the EU authorized representative is:
Springer Nature Customer Service Center GmbH
Europaplatz 3, 69115 Heidelberg, Germany

Printed by Libri Plureos GmbH
in Hamburg, Germany